Systems Approaches for Improvement in Agriculture and Resource Management

Kathleen Wilson *University of Hawaii*

George E. B. Morren, Jr. *Rutgers University*

Macmillan Publishing Company
NEW YORK

Collier Macmillan Publishers
LONDON

Macmillan Publishing Company
866 Third Avenue, New York, New York 10022

Collier Macmillan Canada, Inc.

Library of Congress Cataloging-in-Publication Data

Wilson, Kathleen Karah.
  Systems approaches for improvement in agriculture and resource management /
Kathleen Wilson, George E. B. Morren, Jr.
     p. cm.
  Includes bibliographical references.
  ISBN 0-02-428411-4
  1. Agriculture — Research.   2. Agriculture — Study and teaching.
3. System analysis.   4. Natural resources — Management.   I. Morren,
George E. B.   II. Title.
S540.A2W54   1900
338.1′01′1 — dc20   89-28310
                                                                              CIP

Printing: 1 2 3 4 5 6 7 8   Year: 0 1 2 3 4 5 6 7 8 9

Preface

This book was written for students pursuing courses and majors in the agricultural, food, and natural resources sciences at U.S. agricultural colleges and universities. It is intended to introduce new perspectives to professionals-in-training who will, upon graduation, be called upon to deal with complex and ever-changing problems and environments. Given these intentions and aims, the book is also a product of its history and context as well as the work of specific contributors. Some of this is outlined subsequently.

In 1981, the National Higher Education Committee of the U.S. Department of Agriculture's Joint Council on Food and Agricultural Sciences set a national agenda for improving instruction in the nation's public land grant colleges and colleges of agriculture. It was widely perceived that this vital sector of U.S. higher education was in some difficulty due to declining enrollments, loss of mission, institutional and political pressure, and criticism from the agricultural and business communities it was supposed to serve.

We live in a complex and dynamic world. The students we educate and train must be prepared to deal with this complexity no matter what their specific vocational and personal goals. Although founded in the nineteenth century under radically different circumstances, our system of public higher education has, up till now, proven to be resilient and adaptable in the face of rapid social, political, economic, and technological change. We are currently undergoing yet another round of challenge and response.

Educational leaders now recognize that, in addition to farmers, agricultural colleges train business men and women, scientists, and other professionals for positions in the food, agricultural, and natural resources industries. Yet many colleges across the nation are plagued with declining enrollments in agricultural, food, and natural resource programs. Industry experiences difficulty recruiting graduates who possess, not only solid scientific training, but also the breadth and depth required of future executives. Managers of modern high-technology companies must understand, solve and improve a wide variety of problems, including those arising from the public's concern about food, agriculture and the environment.

In 1982, the U.S. Department of Agriculture, the educational community, and the food and agricultural industry launched a nationwide project to strengthen higher education to meet those needs. Since 1982, the mission of the National Agricultural and Natural Resources Curriculum Project has been to create new curricula, instructional approaches, undergraduate courses and course modules, and methods to evaluate curricula for colleges throughout the country. The project has received grants and other support from the U.S. Department of Agriculture and 23 industry groups and foundations such as the Exxon Educational Foundation; R.J. Reynolds Industries, Inc.; the Upjohn Co.; Pfizer; Curtice-Burns; Agway; Tennessee Eastman; and Hershey, as well as from the university community. Directed by Dr. Richard H. Merritt, a plant scientist, former dean, and professor at Rutgers, The State University of New Jersey, the project's first task was to survey educational institutions to assess the major concerns and deficiencies identified by faculty and administrators. On this basis, six long-term priority areas were identified for curriculum development, including systems analysis in food, agriculture, and natural resources; problem solving; ethics and public policy; cultural and social aspects of domestic and international agriculture; energy use in food and agriculture; and integrated reproduction management.

This book is the product of the flagship venture, the Systems Team of the National Agricultural and Natural Resources Curriculum Project. In addition, one other national team is currently being coordinated on ethics and public policy. Both teams are multidisciplinary, with food and agricultural scientists, resource specialists, economists, social scientists, philosophers, theologians and engineers from universities and the business community.

The Systems Team began its work with a formal meeting at the University of California at Davis in May 1984. We are grateful to Dean Charles Hess for hospitality and practical support at this meeting. Its membership included Richard J. Bawden, a parasitologist by training and dean of the faculty of agriculture of Hawkesbury Agricultural College, Australia; Lovel Jarvis, an economist from the University of California, Davis; George E. B. Morren, an anthropologist from Rutgers University; Mark Singley, an agricultural engineer from Rutgers; Robert Sowell, an agricultural engineer from North Carolina State University; Donald Vietor; an agronomist from Texas A & M; and Kathleen Wilson, a community resources development specialist from the college of tropical agriculture and human resources of the University of Hawaii.

The formal purpose of the Davis meeting was to discuss how to respond to the perceived need for a course on systems analysis. In a very real way, the meeting also began a process of debate and accommodation among strongly opinionated people with diverse backgrounds and interests. Although we were selected for the team because we had used and taught systems approaches in our respective fields in the past, there was no general agreement on basic methodological questions.

Bawden and Wilson emerged as leaders in their advocacy of Peter Checkland's soft systems approach and of the centrality of theories of learning. In the short run, this mystified some team members who were comfortable with the hard systems approaches that had emerged from systems engineering and economics. Eventually, however, it proved to be revelatory and has directly shaped the form of this book. Then, it initiated a learning process that advanced over subsequent years. The group emerged from that meeting lacking a clear consensus on details but agreeing that the succeeding year would be spent dealing with conceptual issues and, in particular, that this would be done through the practical vehicle of developing case materials.

Having spent the intervening year exchanging materials, the team reconvened at California State Polytechnic University, Pomona in May 1985. We are grateful to Dean Allen Christensen for assistance and generous hospitality. A subsequent meeting was held in conjunction with the annual meeting of the Agricultural Division of the National Association of State Universities and Land Grant Colleges in Honolulu, Hawaii, in 1986. After the formal meetings, the team was able to meet and tentatively assign responsibilities for further developmental work and the writing of this book.

We have also benefited from the opportunity to offer faculty training workshops around the country. The first of these was held at Colorado State University in Fort Collins in late spring of 1986. Attending were 42 faculty members from 29 institutions, representing 15 academic disciplines. In addition to beginning to communicate the team's ideas to the larger educational community, this and subsequent workshops are very important field trials for the methodology and the approach to teaching it that we advocate. We came away from the Colorado gathering with many criticisms and suggestions and also a growing network of cooperating colleagues, some of whom, such as Greg Clary, Vern Cardwell, Tom Ruehr, Leroy Barker, and Ernie Schusky, have gone on to organize additional workshops. Other colleagues are cited subsequently for their specific contributions. Since that 1986 workshop, others have been offered at North Carolina State University; Asilomar, California; Atlanta, Georgia; Saint Louis, Missouri; and Reston, Virginia; and more are scheduled for the immediate future. In addition, the approach has been presented to the international higher education community through various activities sponsored by the U.S. Agency for International Development.

Acknowledgments

Due to the collective nature of the curriculum development activities of the systems team, it is particularly necessary to sort out and credit all of the diverse contributions to the project and this book. Wilson and Morren are the lead writers and overall editors, and undoubtedly their voices prevail on many issues. Wilson drafted chapters 1, 2, 4 and 5, and Morren edited and contributed materials to them. Morren organized and drafted chapters 3, 7, and 8, and Wilson edited and contributed materials to them. In addition, Richard Bawden and Mark Singley drafted portions of the section of Chapter 3 on modeling. Chapter 6 was written by D. M. Vietor and H. T. Cralle, and Morren edited it and contributed additional materials.

The authors and compilers of the case studies and case materials (which play such an important role in the text and also are presented in the appendix) deserve special recognition. Chatham River, our first really "good" case, which also was extensively used in faculty training workshops conducted throughout the U.S., was assembled and written by Robert Sowell with the assistance of David Miller. They also developed the computer software to accompany this and other cases currently under development. Jarvis and Morren also contributed some of the materials for Chatham River. The Mucho Sacata Ranch case was assembled and written by Jerry Stuth, Wayne Hamilton, and D. S. Vietor. The Papua New Guinea co-op case was written by Morren. The Floral Growers' Associations case, used in Chapters 5, 7, and 8, was developed by Wilson with the assistance of Territa Davide.

A number of other colleagues and students made individual contributions of materials appearing in the text, including Richard Bawden for the "Farmer Takes a Loan" scenario, other examples, several figures, and some of the shape of Chapter 2; Matthew Hongo, Jerry Domingo, and Don Lunde-Jenkins for "mind maps" appearing as figures in Chapter 4; David Fish for contributions to the section on the mind map technique and for drafting the "Rancher Dennis" and "Zeke the Grass Farmer" scenarios, all in Chapter 4.

David Kolb's "Learning Styles Inventory"(1976), discussed extensively in Chapter 2 and referred to throughout the book, is available in quantity (together with useful documentation) to course instructors from McBer & Co., 137 Newberry Street, Boston, MA 02116.

Numerous other individuals have made specific intellectual and practical contributions to this book. In addition to members of the systems team, Michael Schulman, Ernest Schusky, Larry Busch, Dick Merritt, Bonnie McCay, Bill Seagraves, Pam Mills-Pacho, and David Fish read and criticized drafts of chapters or other materials. Peter Checkland, Colin Spedding, and George Nieswand shared their perspectives as invited guests of the team in Honolulu in July 1985, when the administration, faculty, and staff of the East-West Center and the College of Tropical Agriculture and Human Resources, University of Hawaii, also provided hospitality and shared insights and experience. We are particularly grateful to Dean Noel Kefford of the University of Hawaii for his enthusiasm and support for systems approaches in undergraduate education. Deborah Brighton, former planner for the Green Mountain National Forest, shared experience and documents on public participation. Dr. Frances Aaron Brooks of New Jersey Farm Bureau provided insights on farmer advocacy and the relations among farmers and the larger community.

Greg Clary, Vern Cardwell, S. G. Cornelius, Tom Ruehr, Lucas Calpouzos and Kathy Wilson's systems agriculture class were the final reviewers of this book in manuscript. All have provided invaluable comments and corrections. The authors are responsible for any remaining errors and excesses.

George Morren's research in Papua New Guinea (in the course of which information on the community cooperative referred to in the book was collected), has been sponsored by the National Science Foundation, the Wenner-Gren Foundation for Anthropological Research, and Rutgers University. Other research has been supported by the New Jersey Agricultural Experiment Station. Some of the research contributed to this book by Kathy Wilson and her graduate students has been supported by the U.S. Department of Agriculture under CSRS Special Grant No. 84-CRSR-2-2379, administered by the Pacific Basin Advisory Group (PBAG).

Kathleen Wilson
George E. B. Morren

Contents

3 An Introduction to Systems Thinking and Practice   67

4   Making Sense Out of Situations: Stages 1 and 2 of the Soft Systems Approach   116

8 Implementation: Stage 7 of the Soft Systems Approach

Appendix: Student Case Materials

Glossary

List of Figures

List of Tables _____

Managing Complexity and Change in Food, Agriculture, and Natural Resources

Making decisions about managing enterprises in changing environments or developing and using new technologies is a difficult task. In addition, the general public, private organizations, and government agencies are no longer willing to view food, agricultural, and natural resources enterprises as strictly private affairs. Realistically, they are seen as multifaceted — producers of essential commodities and services, users of common resources, and subject to social values and ethical standards such as those associated with safety and quality of life . . . as well as profit-making organizations.

Newly minted degree holders in food, agriculture, and natural resources increasingly find that their new jobs place them in messy and dynamic situations. "Messy" means that the parties involved in a situation do not agree on either the definition of problems or on what technological and management improvements should be developed, or both. Thus, the introduction of a technological package or the reallocation of resources to correct or improve a situation is not straightforward. "People factors" loom large. Similarly, "dynamic" means, quite simply, that things keep changing; that yesterday's problem definitions or solutions involve factors that have now changed or no longer exist. Accordingly, in order to deal with these kinds of situations, professionals entering the field need new competencies and inquiry methodologies.

These competencies and methodologies go beyond the problem-solving approaches and limits to inquiry that are traditionally taught in most of our schools and colleges. Nor are they usually provided in undergraduate general education core requirements. Graduates need an array or "tool bag" of inquiry approaches to help them understand and act on the interactions, impacts, and complexity of situations in ways that are conceptually valid and practically effective.

The Four Approaches to Inquiry

This book provides an overview of four approaches to inquiry that are currently used by managers of food, agricultural, and natural resources enterprises: basic science, applied science, hard systems, and soft systems. The soft systems approach to inquiry receives particular emphasis. The objective is to enhance the reader's ability to create and apply knowledge in real-world settings. The main thesis of the book is that food, agricultural, and natural resource professionals face many challenges and opportunities. Responding to them requires new competencies and a broad array of inquiry methodologies and knowledge. This book aims to help students — professionals-in-training — to develop such knowledge and competencies.

The Soft Systems Approach

Each of the four approaches to inquiry deserves its own text to adequately cover the detail needed to understand and practice it. There are four reasons for this book's emphasis on the soft systems approach.

First, the other three approaches are well represented by textbooks, particularly in fields and disciplines that predominately use one or two of them. Basic science inquiry is represented in chemistry and biology texts, for example. Applied science is exemplified by textbooks in the animal and agricultural sciences, medicine, and engineering. There are also numerous texts in economics, systems engineering, and management that present hard systems modeling approaches. This book is the first to bring the soft systems approach to undergraduate students in the food, agricultural, and natural resources sciences.

Second, quite a few of the complex situations that many graduates will be called upon to handle require the knowledge, sensitivities, skills, and specific inquiry techniques that are part of the soft systems approach. This is particularly relevant to such food, agricultural, and natural resources job categories as management, service, field and sales representation, merchandising, technician, information, education, communications, and agricultural production. According to the U.S. Department of Agriculture, approximately 70 percent of projected jobs in the industry will be found in these categories (Coulter, Stanton, and Goecker, 1986).

Third, it is a central thesis of this book that, in tackling a typical situation, the soft systems approach should be used first before proceeding to use one or more of the others. This is because the approach helps to develop agreement on such key issues as the nature of the problems, what will constitute an improvement, and the kinds of technologies and other techniques appropriate to achieving specific goals. Such agreement cannot be achieved without significant involvement and input from all the parties involved in a situation. The soft systems approach provides the procedures and tools to help reach such agreements. This contrasts with the other three approaches, which tend to assume that the objective of inquiry (the improved state being sought) is not in question. This statement is not intended as a criticism of the other approaches, but rather describes a major difference in the assumed starting point of inquiry. Soft systems analysis is carried out in order to reach a point where parties agree on the improved state of affairs they will seek and on the means that are appropriate to achieve it.

Fourth, the management of the biological and physical world for economic purposes, as well as to sustain or improve the quality of people's lives, requires effective and efficient organizational systems. Without such systems, people do not stay in business or are subject to constant and costly pushes and pulls from external forces beyond their control. The soft systems approach provides procedures and techniques for the development, assessment, and revision of the organizational systems of a given food, agricultural, or natural resource management situation. All graduates should gain employment that involves the operations of one or more organizations. Familiarity with the procedures of soft systems analysis will help graduates to contribute competently to the ongoing effectiveness, efficiency, responsiveness, and strategic planning of the food, agricultural, or natural resources enterprises employing them.

Systems Thinking: A Holistic Approach

The book provides an expanded perspective that may strike production-oriented readers as too public or outsider oriented. This is unavoidable if the objective is to develop the idea that *we live in a complex world that people in groups and organizations of various kinds view very differently.* This is part of the challenge faced by professionals. So the issue is not ignoring a production orientation but strengthening that perspective by understanding the others and, more importantly, *learning how to gain an understanding* of other perspectives. The next section of this chapter discusses a series of food, agricultural, resource, and management concerns that is currently the subject of debate and divergent opinion. Its presentation is intended to be illustrative and provocative rather than exhaustive and final. Those readers who are tackling the book as part of a course may wish to spend time, even a formal session, debating and futher widening the perspectives offered on particular issues presented here. The strategy of such debates is to be comprehensive and accommodative rather than narrow and exclusionary. The purpose is to practice approaching questions openly and to develop ways of better accounting for differing perspectives.

Another reason for introducing the book with a discussion of challenges and opportunities is to help students who have little or no work experience in food, agriculture, and natural resources to begin with a broader, more holistic view of the field and the activities associated with it. The following sections are presented in a way that suggests interactions among organizations of people and biological and physical properties. A holistic view implies paying attention to such interactions, to the controls that are present, to communications between parts, to properties that emerge as a result of real or projected interactions, to the kinds of operations that are present or absent, and to the impacts of interactions between the parts on the environment. Thus, a key first step toward systems thinking and practice is to develop one's ability to explore experience holistically. Systemic thinking is essential to your future ability to develop improvements or solutions to the complex problems faced by most agricultural and natural-resource enterprises.

This chapter offers illustrations of such interactions. Subsequent chapters will present techniques for developing abilities to see and record them. This chapter concludes with a summary of the key points found in the rest of the text, chapter by chapter. In addition, each chapter begins and ends with summaries of key points. Reading a summary first may help some readers to clarify the more detailed substance of a chapter. Others who find the summaries repetitive are free to read ahead.

Management and Systems Thinking

Managing enterprises in a changing natural and socioeconomic environment is far from simple. Management requires that professionals make wise decisions about things they can manage and accommodate and respond to the natural and socioeconomic forces in the environment that they cannot control. As challenges and opportunities are discussed subsequently, note the kinds of forces at work. The next section begins with an overview of the nature of food, agricultural, and natural resources industries and their connections to and impacts on the rest of society. Following this, selected challenges and opportunities are presented to give readers a sense of the importance and value of careers in the field. The examples presented illustrate some of the complexity and messiness of managing enterprises in the field and, hopefully, also highlight the need for training that will assist readers to tackle such situations effectively in the future.

The Primary Industries

Agriculture

The food, agricultural, and natural resources industries are often referred to collectively as *primary* industries because they provide the energy and raw materials upon which the rest of a modern industrial economy depends. Here the concern is with how modern societies are powered and provisioned with energy, food, fiber, timber, minerals, water—the most basic inputs that modern consumers and citizens tend to take for granted. Without these essential inputs, or even if one or two of them are only marginally interrupted, our life-style pauses until supplies are restored or other kinds of adjustments are laboriously pieced together; jobs, travel, necessities, as well as comforts, all hang in the balance. Accordingly, the enterprises involved in this vast provisioning system are not going to go away easily.

These industries contribute vitally to the U.S. economy and people's sense of well-being. In fact, American agriculture alone is the world's largest commercial industry, with assets exceeding $1 trillion. The industry employs approximately 20 percent of America's labor force—more than 20 million people. The agricultural industry includes the producing, as well as the transporting, processing, manufacturing, retailing, and research and development of food and fiber. It accounts for 20 percent of the nation's gross national product ($610 billion in 1985).

These formidable figures show that agriculture's contribution to the total economy depends on a tremendously complex and interdependent system of producers, suppliers, transporters, distributors, researchers, and information disseminators. And their employees, in turn, directly depend on the food and fiber system for a living.

Agriculture and Codependent Industries

The people of this vast network engaged in food and agricultural activities are also consumers. Their purchases for their agricultural enterprises benefit local businesses — stores, banks, repair services, fuel distributors, and so on. They are also private consumers looking out for home and family. A recent "NBC Reports" television program estimated that for every five farms that go out of business, one supporting service enterprise, such as a bank, clothing store, food store, or drug store goes out of business. Agriculture creates jobs, not only in local communities, but across the nation. These purchases are in addition to the basic $140 billion that farmers and agribusiness firms spend annually for the goods and services used to produce our nation's food and fiber products.

These purchases affect the viability and sustainability of many other industries. For example, purchases by the agricultural and food industries generate 37 percent of the jobs in the glass and glass products industry, 69 percent of metal container industry employment, and 26 percent of the jobs in the paper products industry. The agricultural and food industry consumes 45 percent of the output of the paperboard container and box industry, 26 percent of the chemical and chemical products industry, and 25 percent of the crude petroleum and natural gas industry. In addition, the petroleum and related industries, the rubber and plastics industries, the iron and steel manufacturers, coal mining, transportation, and warehousing businesses (rail, truck, barge), printing and publishing businesses, plastics and synthetic materials industries, machine shop products industry, and motor vehicle and equipment manufacturers are all strategically tied to the agricultural sector's use of their products (Phillips, 1985).

Thus, every region of the U.S. has some portion of its employment base linked to agriculture. No matter where a person lives in the United States, about 20 percent or more of employment is so dependent. On a national basis, that is one job in five.

Opportunities in Agriculture and Natural Resources

The security, sustainability, and viability of the United States economy and the well-being of its people rely on the continued stewardship and development of scientific and professional expertise in the food, agricultural, and natural resource sciences. A recent study conducted by the U.S. Department of Agriculture, however, reported that there will be a shortfall of graduates trained in these fields during the next decade in relation to job openings. Through 1990, scientists, engineers, managers, sales representatives, and marketing specialists will account for three-fourths of the total annual U.S. employment openings for new college graduates with expertise in agriculture, natural resources, and veterinary medicine (Coulter, Stanton, and Goecker, 1986). These figures are good news for readers who are contemplating careers in food, agriculture, or natural resource management. They are less than good news for those concerned about the future of these sectors of our economy.

America's food, agricultural, and natural resource system is one of its greatest success stories. In the past, national leaders have recognized that agriculture is the foundation on which overall development of the nation must rest. Investment by both the public and private sectors in the United States guaranteed the development of a scientific base and research and extension capability. Returns on the dollars thus allocated to agricultural development have been very high (Ruttan, 1982). Great strides in technology development have been made. Tremendous breakthroughs in the understanding and alteration of biological, chemical, and physical properties of water, plants, soils, and animals have led to many new agricultural, natural, and synthetic resource technologies and applications. Indeed, for the reader who is contemplating a career in one or another of the food, agricultural, or natural resource sciences, there are many opportunities and challenges. These fields of applied science have had an important, productive past and look to a promising future.

Current and Future Challenges

The future of these industries, however, presents many challenges. These challenges are complex and demand new competencies of those who choose a career in a related field. As noted previously, the food, agricultural, and natural resource sectors are connected in very complex ways with other major industries in the U.S. What happens in one industry affects what happens in another. Agricultural and natural resource production can no longer be thought of as isolated areas of endeavor. Food, agricultural, and natural resources enterprises constitute an interconnected complex of human activities involving transactions between people and their environments in the broadest sense. Bad or insular decisions and choices have effects far beyond the bounds of a single firm, organization, or agency.

Throughout this text, improvements in agricultural and natural resource management are conceptualized as *human activity systems*, with the people involved — the farmers, growers, ranchers, public employees, food process technicians, agribusiness men and women, employees of natural resource industries, researchers, or families — as the subjects of our concern. There is an important role for professionals to help clients such as these learn to manage change more effectively so that their life-styles and businesses are more sustainable no matter what happens to the environments in which they operate. Sustainability results from the quality of interaction among the family; the business enterprise; and the physical, biological, and socioeconomic environments. Effectiveness in dealing with change is a function of the learning of family members and employers and employees involved in the management of those environments. Peter Checkland, who developed the concept of the human activity system, notes that natural and designed physical and biological systems are "inevitably linked closely to the human activity" associated with these systems (Checkland, 1971).

The growing recognition of the complexity of the management of agricultural and natural resource enterprises locally, statewide, nationally, and internationally has led some to postulate that the role, value, and responsibilities of professionals within the agricultural and natural resource sectors have changed significantly. No longer do many see the role, value, and responsibilities as only producing as much food, fiber, and natural resource products as possible or economically viable. Rather, many see the greater challenge to be learning how to manage food, agriculture, and natural resources in a responsible and sustainable manner, so that the impacts on people, families, nations, and environments are as positive as possible. This is no small challenge!

The Boom–Bust Cycle: Policy and Politics

Farm Management

The management challenge is one that we have not yet dealt with very well. The television stories and newspaper headlines of the past decade all too often remind us that our attempts to improve or support agriculture in the U.S. and around the world have had less than desirable effects. The *Time* magazine cover for February 18, 1985, captures the beginning of a bitter debate over the government's role and ability to facilitate a sustainable and viable agricultural base. The cover headline, "Going Broke—Tangled Policies—Failing Farms," well summarized the situation.

The 1970s were a decade of high prosperity in the croplands. Worldwide demand for U.S. grain and fiber boomed. The government encouraged farmers to plant as much as they could. The farmers complied, were able to sell at well above the federal support levels, and borrowed a lot of money to buy more land to plant more crops and buy more machinery. The more people bought, the more the land prices rose. Then the world recession of 1981 struck. Worldwide demand for crops declined abruptly. Crop prices and land values tumbled. Deflation hit, yet growers were managing their businesses based on an inflation mentality (which is to maintain an asset–equity ratio in the 30–70 percent range). Farmers, largely those who grew between $40,000 and $500,000 worth of food or fiber a year, began losing a lot of money. A crisis of mounting debts, foreclosures, and a fast-changing debate and projection about the sustainability and viability of agriculture in the U.S. rages on even as we write in 1989.

Medium-sized farm enterprises got caught in the middle of changing policies and their own inability to sort out competing views of where the world was heading. The cautions of Schumacher's (1973) popular book *Small Is Beautiful* seemed not to reach the agribusiness manager during this time, as so many real estate, financial, and government advice-givers were pushing a "big and more is better" philosophy. So by 1986, farmers in America owed more money than the nations of Mexico and Brazil combined. Perhaps the impacts of government policies on the farmer are best summed up by the remarks of Don Paarlberg, former chief economist of the U.S. Department of Agriculture under presidents Nixon and Ford, "We were speeding at 80 miles an hour, then, instead of slowing down as prudent drivers should, the government slammed on the brakes. It plastered farmers against the windshield" (Robbins, 1985). While policy decisions were based on current and anticipated conditions, many people have questioned not only the approach selected but the level of debate that led up to it.

Growth Pressures

The 1970s also saw "boom and bust" in other primary industries, such as energy and minerals. Space does not permit even a cursory overview, but the general pattern resembles and is probably linked to the agricultural crisis discussed earlier. The important point to note is that national and regional economies, as well as communities and individual enterprises, that are dependent on such commodities as petroleum, coal, copper, and the like are highly vulnerable to swings of world markets. Boom times, when world market conditions are favorable, subject such communities to extraordinary growth pressures. These pressures result from (1) the rapid influx of migrants, including people from other countries, who are also new residents, neighbors, consumers, and employees; (2) the pressure to rapidly expand governmental services and infrastructure in order to help new businesses become established and to accommodate more residents; (3) the emergence of new environmental problems; and (4) the need to meet the concerns and complaints of existing residents, who may see themselves as not benefiting from growth.

All bubbles ultimately burst, and the effects on dependent nations, communities, firms, and individuals of price declines for their marketable commodities are devastating; impacts cited in the literature include unemployment, emigration, bankruptcy, individual stress, governmental insolvency, and the irreversible and uncompensatable environmental and cultural impacts associated with the past development-boom phase.

Risks

Financial Risks. The North American Arctic, embracing both Alaska and northern Canada, has been caught up in a boom–bust cycle for more than a century. The area has had anything but a stable economy (Dixon, 1978). The first boom, in the mid-nineteenth century, was the fur trade. The second was the gold rush, beginning in 1897 in the Canadian Klondike (London, 1900) and in 1903 in Alaska. Another major boom involved military construction, starting in the 1940s and extending into the 1950s, with the construction of air-defense facilities, including the DEW Line of radar stations along the Arctic Circle. The most recent boom has been associated with the development of north-slope (Alaska) and Beaufort Sea (Canada) oil and pipeline construction. Construction of the trans-Alaska pipeline officially commenced in 1974, in turn setting off another cycle. Each of these periods of frantic growth, even including the latest one, has been followed by hard times. In other words, this pattern of development seems not to foster sustainable development, but rather the reverse.

One can only begin to imagine what it is like to manage an enterprise and attempt to adjust to the ever-changing circumstances that have characterized the North American Arctic region. The consequences of the oil boom in Alaska have been documented as they have occurred (and continue to occur). One of the problems with periods of rapid economic growth is that they are difficult to anticipate and prepare for, especially in small-sized communities and family businesses. According to Dixon (1978), what happened in Fairbanks, Alaska, was that although everyone knew that oil development was going to happen *sometime,* state and local government and most of the business community either lacked capital or was reluctant to invest it and other resources to prepare for the boom. Yet, when pipeline construction seemed imminent in 1970, some forward-thinking business people did invest heavily in facilities and inventory. When the pipeline was delayed by lawsuits, the forward-looking businesses were burned. According to the mayor of Fairbanks, "Because of lack of commitment by the oil industry, it was difficult for people here to justify spending money, either their own or the public's, on preparation for the boom that might never come" (cited in Dixon, 1978:135).

This conservative, wait-and-see attitude adopted by many people in this example may be attributed to the long history of boom–bust cycles in the area and a wise strategy at the time. Such insights are important, since the various inquiry approaches described in this book are intended to help people plan the changes they want and need. However, real life does not always proceed in the ways we anticipate or for which we plan.

Social Risks. One possible confounding factor in this and many other situations is that some of the key parties in a situation may choose to act in ways that influence the success of the other parties' actions. For example, the specific source of delay in construction of the pipeline was the conflicting views of oil companies, environmentalists, Native Americans, state officials, and other development interests regarding specific impacts, ownership of needed basic resources, and the distribution of potential costs and benefits. Matters were tied up in the courts for several years and ultimately were resolved only through a seemingly incongruous alliance of oil companies and Alaskan natives to drive a unique legislative package through the Congress in Washington.

Other events occurred that might have been handled better if certain governmental and business organizations had implemented appropriate planning procedures and thought more holistically about the emerging situation. Three months after construction of the pipeline was begun, Fairbanks ran out of telephone numbers! And it would take two years to obtain the equipment required to meet existing, to say nothing of projected, demand. The next crisis was in housing, followed by electrical supply, schools, transportation, water supply, and sewage treatment. Crime, ranging from labor racketeering to murder to drug and alcohol abuse to prostitution, escalated. Finally, outside speculators, with better access to capital, were able to build housing and start businesses when the locals could not. This too generated much bitterness at the time.

Other Themes of Concern

The headlines and social critiques of agricultural and natural resource activity of the past decades also reveal major themes of concern and issues that confront current and future professionals in these fields. Each of these themes raises countless questions about how best to manage food, agricultural, and natural-resource enterprises. They cover the broad issues of the effects of agricultural management practices on the biological, physical, social, economic, and cultural environment. They also cover themes of concern regarding how these factors and environments affect food, agricultural, and natural resource activities.

Perhaps the most prominent themes of concern deal with the massive surpluses of some agricultural commodities, the continued increase in costs of production relative to food prices, and the environmental and health impacts of agricultural chemicals. Other major themes of concern deal with the consequences of current soil management practices on public health, water quality, and water availability and the long-term sustainability of the physical environment.

Sustainability

Practices that cause soil erosion or fertility loss through biological, physical, and chemical stresses cannot be sustained indefinitely into the future. The persistence of chemical biocides in the same soils and the salinization of the water that runs through it are further threats to its integrity. In 1972, the Rural Development Act was established in response to these concerns and in order to establish guidelines for development and conservation of rural land and water resources and to prevent and abate agriculturally related pollution in rural areas. Yet, in 1989, as reported in the news media, in the records of regulatory agencies and courts, and in legislative hearings, abuses persist.

Land Use. Major choices confront us. Much of the potential cropland is now used for grazing or is in forest. Conversion of range or forest to cropland would increase our nation's capability to grow crops, but this would reduce our potential to increase forage or timber production. It would also decrease wildlife habitats. Moreover, the conversion of relatively dry rangelands to crop agriculture increases a region's vulnerability to drought and soil erosion, as occurred in the dust-bowl years of the 1930s. Because of the rapid rise in foreign demand for agricultural commodities and farmers' growing dependence on the international market in order to remain economically viable, the choice about how to manage land resources is far from simple. Indeed, in some regions of the country, the choice is between devoting land to cropland or to housing, water conservation, waste disposal, industrial and commercial facilities, and the infrastructure to support them.

Water. Effective use of water has been at the center of crises and controversies in various regions of the United States throughout the 1970s and 1980s and exemplifies another kind of complexity faced by agricultural and natural resource managers. These situations have involved four kinds of changes: (1) drought, (2) overutilization, (3) declining water quality, and (4) proposals and programs to develop or reallocate water resources. More often than not, a given local or regional situation involved a combination of these. For example, drought has two meanings: (1) a prolonged period of lower-than-normal rainfall and (2) insufficient water to support normal human uses and activities. Thus, if a rainfall deficit occurs in a region that is experiencing population and economic growth, the water-supply system will be very quickly stressed. Moreover, alternative sources may be unusable due to pollution, and authorities will need to make political decisions regarding allocation priorities. Access, protection, conservation, and utilization of water are concerns that many agricultural and natural resource enterprises must tackle immediately.

Agriculture is now the nation's largest consumer of water. More than 250 billion gallons a day are withdrawn from surface waters (U.S. Department of Agriculture, 1982). Total groundwater withdrawals exceed 80 billion gallons a day, including 21 billion gallons that are not replaced through natural recharge. Four times more water is used for producing food and fiber than for all other purposes combined. Irrigation water is a major concern. Irrigated crops, grown on only about 12 percent of the nation's croplands, account for about 25 percent of the total value of crop production in the U.S. How and for what purposes agricultural and natural resource managers use water are now everybody's concern. As discussed earlier, urban, residential, and industrial users compete with farmers and natural resource enterprises for water in many parts of the country.

Inadequate and mismanaged water-supply systems are leading to degraded water quality, wasted water and energy, reduced instream flows, and increased production costs for farmers and other industries. Return flows to streams from irrigated fields are important to downstream water users. Even in average years, stream flows in the Colorado, Rio Grande, and Great Basin regions are inadequate to meet offstream uses and instream needs for aquatic life.

Soils. Soils, another natural resource, are also under pressure. Soil erosion remains a threat to the sustainability and productivity of 13 percent of all croplands in the United States. Annual soil loss and the average annual rate of sheet and rill erosion on cropland in the corn belt are double that in any other region. The average rate of erosion on cropland in Hawaii and the Caribbean area is triple that of any other region, but the average rate also exceeds five tons per acre in the Appalachians, the corn belt, the Mississippi Delta region, and the Southeast. Holding erosion at a level of under four or five tons of topsoil per acre annually on croplands, pasturelands, and forest lands so that sustained use of the soil is possible is a major national goal (U.S. Department of Agriculture, 1982; Brown, 1984).

Flood. A significant portion of the nation's agricultural and natural resource base rests on flood-prone lands. More than 29 percent of the forest lands, 28 percent of the rangelands, 20 percent of the croplands, and 11 percent of our pasturelands are flood-prone. New efforts are needed to protect these lands from flood damage.

Pollution. Agricultural activity may be the most widespread source of non-point source pollution in the nation. Bacteria, nutrients, dissolved solids (e.g., salinity); suspended solids (sediment); and toxic materials are the most common and most serious nonpoint source pollutants in the U.S. This type of pollution comes mainly from runoff or irrigation return flow. Water runoff often increases levels of bacteria, sediment, nutrients, and pesticides in surface and subsurface water supplies that serve homes and industry. Irrigation return flow generally increases the level of dissolved solids, nutrients, and pesticides in streams and rivers. Large-scale animal operations have impacts on water quality similar to the impacts of untreated human sewage. Consequently, more drinking-water sources have become polluted; coastal waters are degraded; and aquatic, marine, and water-dependent wildlife is less abundant in many areas. This issue has taken on international dimensions in the case of treaty-protected migratory waterfowl such as ducks and geese.

Wetland Loss. Because of the pressures to increase agricultural production, thousands of acres of wetlands annually have been drained. Wetlands provide a habitat for a variety of wildlife. They also serve as sediment and nutrient traps for the stream flows into and through them and thus enhance downstream water quality. It is becoming clear that one person's, or family's, or agribusiness's use of an area can have far-reaching effects on the land, water, and human resources in that area. No longer are people willing to see agricultural and natural resource enterprises strictly as private affairs. Hence, as noted earlier, in addition to being viewed as profit-making organizations, enterprises of various kinds are also seen as producers of essential goods and services, as users of common resources, and as subject to the social values and ethical standards associated with quality of life.

The Costs of Technology

Technological developments that have contributed to sustained increases in production per acre are also linked to increased costs. Major production costs for farmers are interest on debt, machinery, fuel, fertilizer, pesticides, and labor. While in the past most inputs needed to manage and operate a farm came from the farm family's human and natural resources, now most are purchased from outside their system. The cost of inputs such as fuel and energy, chemicals, and fertilizer have increased tremendously during the 1970 to 1989 period. In the same period, demand for and costs of farm machinery have also increased. As a result, farmers are at risk of receiving less for their products than their cost to produce them, as well as of being economically devastated by growing loan interest payments and tax increases.

The agribusiness manufacturers and supply firms, who have also been hard hit by the weak farm economy, stand ready to sell inputs to growers and producers. While such inputs sometimes increase efficiency, many managers have difficulty deciding whom to listen to and how to sort out the consequences. Those who earlier adopted a profit-expansion strategy seemingly were later the hardest hit. And for a while, the government continued to encourage the profit-expansion, consumption-oriented approach to agricultural development beyond the point where it no longer made sense to many observers. Indeed, there are many forces at work that make it difficult to decide how to manage agricultural and natural resource enterprises.

Chemical Impacts. Another concern is modern agriculture's reliance on nonrenewable resources that have the potential for destroying natural processes of decomposition and nutrient cycling. For example, use of synthetic biocides that unintentionally kill useful soil organisms has disrupted heterotrophic food chains needed to break down organic materials. It has also resulted in new strains of pests and pathogens resistant to established methods of control. The destruction of natural processes has also created interdependencies among nations, since the novel strains of pests and pathogens often require new inputs that must be imported.

Ethics. The contamination of dairy cattle feed, which resulted in the tainting of several states' milk supplies, is a costly reminder that we all depend on people in the food and feed industries not only to provide quality services and products, but also to act ethically and wisely. They must think not only of their own business profits, but of their moral and social responsibility. The different ways in which officials of the various states' departments of agriculture and health dealt with the knowledge of contamination are revealing and alarming. Some public managers do not appear to serve the public, but rather the industries involved. In the long run (and in some cases the short run too) such malfeasance may lower the quality of life in a state, region, or nation. Ultimately, the industry that officials were trying to protect may also be affected negatively. Once public trust is violated, it is hard to restore. When public officials fail in their duties, their right to manage is often removed or counterbalanced by the monitoring activities of other groups. For good or bad, increasing regulation does reduce management's control over the operation of its private enterprise and increases the complexity of its task.

Irradiation. Technological advances that seem promising to some people are sources of fear and concern to others. Irradiation of fruit and vegetables is a case in point. New irradiation procedures and technologies hold great promise for ridding produce of pests prior to packaging and shipping and for prolonging shelf life. They not only kill insects but also delay ripening and kill microorganisms. Many growers would prefer irradiation over the established double-dip or vapor heat method because, among other reasons, it would mean less time spent on harvesting.

Yet, irradiation technologies have many critics. The source of radioactivity used in the irradiator is Cobalt 60 or Cesium 137. Cesium 137 is a byproduct of plutonium production for bombs and other weapons systems. It can be obtained very inexpensively from the federal government, yet it is one of the deadliest of the high-level nuclear wastes. Some people are annoyed that the government would charge states and other potential users for waste products of which they cannot safely dispose. A "not in my backyard" (or NIMBY) attitude has developed, even among people who aren't opposed to irradiation technology on other grounds. Technical, governmental, and community groups are concerned about the possibility of leakage if a steel tank, in which the radioactive elements are submerged in water and stored, were to crack during an earthquake. There is concern about possible human error during transport or at the plant and the health effects of low-level radiation emissions. There are also concerns about the competence of plant managers and employees to oversee the use of this technology safely. All of the irradiation plants in one northeastern state were closed and their management prosecuted because radioactive materials were improperly handled and, in one instance, allowed to escape into the environment, and company records were falsified. The legislature of that state passed a law banning food irradiation, but the governor vetoed it. There are many viewpoints.

Fertilizers. Another example of technological development that has presented new dilemmas is the introduction of artificial fertilizer to agriculture. There were phenomenal per-acre increases in crop production after artificial fertilizers were introduced. While welcomed as a breakthrough, fertilizers significantly contributed to the accumulation of large surpluses of many commodities, which led to the need for more and bigger storage facilities, followed by a federal excess-disposal program. Surpluses contributed to the decline in prices of the commodities involved. Then legislative action, such as the Agricultural Acts of 1970–1973 and 1977, the Payment-in-Kind (PIK) program of 1983, and the set-aside program of 1987, was seen as necessary to deal with that problem.

Federal Policies

When excess disposal and acreage allotment programs were introduced in the U.S., world agricultural trade and market prices were affected. Accordingly, the active price support programs aimed at cushioning producers against shocks from market fluctuations ultimately did not have the effect designers intended. These supports are complex mixtures of government loans and cash payments to farmers who sometimes have to restrict planting to qualify for them. The supports keep U.S. prices of many major food commodities well above those prevailing in the rest of the world. They guarantee the farmer a minimum price per bushel, with the government making up the difference between that floor (or minimum price) and the price in the marketplace. However, some observers believe that the effect worldwide was to encourage farmers in Canada, Brazil, Argentina, the European Community, and elsewhere to expand production because they knew they could undersell the U.S. grower. The federal price-support program kept U.S. markets from cutting commodity and food prices to regain lost overseas sales. Why cut prices when they can get a higher price from their own government? As a result, export sales of U.S. food products have dramatically decreased. They went from $7 billion in 1970 to a peak of $43.8 billion in 1981, then fell more than 20 percent in the next two years, and then only slightly increased in 1985 to $36.5 billion (Church, 1985). Only in the past few years have policies been altered to encourage export sales.

Current agricultural policy challenges farmers to manage their enterprises in ways that will avoid entrapment in the cost/price squeeze that leads inevitably to surplus production. The latter effect occurs when farmers strive to maintain their incomes by increasing productivity and sometimes expanding production.

Trends in Consumption

As if the national policy and world market situations were not complex enough, the farmer is also beset by the fickleness of fashion and value shifts and their impacts on market prices for a range of food, fiber, and textile products. Eating patterns change. Concern about health can bring marked changes in the diet of consumers, which in turn aggregates into important shifts in demand for certain agricultural products. This at least partly underlies the drift of consumers toward chicken and away from beef and other red meats. Increasing concern about the environment by urban-based advocacy groups can lead to intersectoral pressure to change growers' agricultural practices. These range from worries about widespread pollution to the animal welfare debate.

The decline of rural communities and their agrarian values is a concern to many people. The decline is due, at least in part, to the emigration of members of farm families impoverished by their disadvantageous terms of trade. While arguments are raised about the need to preserve farmland, inevitable conflicts arise over pressures to use land in alternative ways.

Natural Forces

Not only do the actions of other nations and the efforts of agribusiness affect the well-being of Americans in communities depending on agriculture and natural resources, but nature also takes its toll. Thus, managers also must respond to natural changes. In 1985–1986, for example, there were major droughts in the Southeast and the Southwest. In June 1985, even as the sun belt dried up, heavy rainfall and floods washed away at least 80 million tons of Iowa's rich topsoil, probably the worst rate of water erosion in the state's history. Hailstorms destroy $700 million of crops each year. A massive grasshopper infestation hit the western states in 1985. What is the proper role of government in protecting agriculture against losses due to natural disasters? The Federal Crop Insurance Corporation and the Federal Disaster Assistance Administration offer some protection through the natural disasters payments program. But some business and government leaders do not believe that government should bail out producers.

Global Interdependence and Agriculture

Over the past decade, the complexity of the management problems of U.S. farmers has intensified due to changes in world trade. Many traditional food-importing countries are now self-sufficient in production, and some are producing sufficient surpluses to be net exporters of food, fiber, and other agricultural products. Sometimes this has been accomplished by substituting the production of food crops for inedible cash crops, although more commonly it results from the transfer of new crops and technologies, such as those of the "green revolution." Thus, new interactions and interdependencies have emerged, which U.S. managers must take into account as they plan improvements.

Farmers in China are producing record harvests of wheat, coarse grains, rice, oil seeds, and cotton. In Britain, new winter barley varieties have added a million tons a year to cereal production. Saudi Arabia is literally turning its desert green, using its immense supply of "waste" fossil fuels. In 1985, it set a world record by generating a wheat surplus. Potential new uses for agricultural products are being discovered almost daily. What were once weeds are now processed into sophisticated pharmaceuticals; waste products are now animal feed. Because farmers can produce in surplus, people have been freed from the quest for food.

What is a blessing is also a problem. Globally and nationally, we cannot agree on what to do about surplus production. Secretary of Agriculture Block said in a 1986 speech that if a farmer in North Dakota sneezes, a farmer in India catches a cold! Perhaps the correct simile now is if a farmer in India sneezes, a farmer in North Dakota catches a cold! The growers of the world are more interdependent, and we have less control of our food supply than in the past.

A Diversity of Perspectives on U.S. Agriculture

Because agricultural and natural resources enterprises in the United States face many challenges and dilemmas, several analyses have been made that focus on the nature of these industries and propose appropriate responses. Authorities swing between two extreme and seemingly conflicting views: that agricultural and natural resources systems have chronic, built-in overcapacities or that they are perennially subject to scarcity due to underproduction. *The Roots of the Farm Problem* (Heady, 1965) and *The Overproduction Trap in U.S. Agriculture*

(Johnson and Quance, 1972) are two key works that present convincing evidence in support of the chronic overproduction thesis. *Seeds of Change* (Brown, 1970) and *By Bread Alone* (Brown and Eckholm, 1974) support the underproduction stance.

Another perspective that has influenced governmental and industrial action is *An Inquiry into the Human Prospect* (Heilbroner, 1972). It is concerned that people's supposed inability to control rapid population growth will lead to catastrophic famine and disease in the developing world and unrestricted industrial growth and environmental collapse in the developed world.

The widely disseminated, discussed, and interpreted Club of Rome Report, also known as the Limits to Growth report (Forrester, 1971; Kahn, 1976; Meadows et al., 1972), does not paint as gloomy a picture. For example, it is projected that the challenge of the population explosion can be met by technological advances sufficient to feed twice the current world population by the year 2000.

In *The Human Interest*, Lester Brown (1974) advocates regulating population growth and conserving limited resources and food supplies. He purposely avoids the alternative of increasing conventional food production because of potential environmental impacts.

Still other works, such as several studies commissioned by the U.S. Department of Agriculture (National Academy of Sciences, 1974, 1975) propose a world of rigorous supply–demand management in which humanity can and must control itself and its environment, a world in which both technologies and human values change. According to this view, a balanced future is sought in which both quantity and quality of human existence are valued. Rather than rejecting the machine, having blind faith in science, or giving up in despair, they project a vision of a future in which science and people get on with the job of rational analysis and positive action.

New Professional Competencies and Approaches

These U.S. Department of Agriculture/National Academy of Sciences reports imply that we already have the competencies and tools necessary for the required inquiry and action. Yet people do not make decisions and behave entirely rationally or objectively. These reports paint a picture of people that is not borne out in research or everyday observation.

This book has adopted a different view of what is needed. According to this view, which is shared by others in the agricultural higher education community, there is a pressing need for professionals who can deal more effectively with complex, messy, value-laden situations of the sort that face producers, distributors, processors, and consumers of agricultural and natural resources products. Practitioners and professionals in the fields of food, agriculture, and natural resources need to use both systematic and systemic thought and approaches to inquiry. They need to be able to think rationally and intuitively themselves as well as to be able to deal with thinking that appears to be irrational or different from their own ways of viewing a situation. The point is that people bring diverse world views and experiences to any problematic situation, and therefore one person's rationality may be viewed by another as ignorance or worse. In fact, the various approaches to the situation are merely different rationalities, which need to be dealt with and accommodated in order to find satisfactory improvements.

Finally, there is a pressing need for food, agricultural, and natural resources professionals who can help the people they serve acquire a stronger orientation toward the future. This means that they must develop the ability to focus inquiry, not on changes that might be needed, but on defining the future *state* toward which specific changes are to be directed. "Don't put the car in gear until you know precisely where you want to go!" This involves learning how to examine an organization's or enterprise's mission and develop strategic plans that relate to particular concerns, opportunities, or challenges.

All natural resource, food, and agricultural organizations and enterprises need the capacity — essentially the personnel with the necessary competencies — to engage in meaningful, ongoing mission evaluation, priority setting, and strategic planning if they are to sustain and enhance their operations in constantly changing economic, social, biological, and physical environments. In introducing the soft systems approach, the objective of this book is to present an inquiry process that readers can use to fully explore their current problematic situations, to develop detailed pictures of improved ways of addressing them, and to produce the concrete strategic plans necessary to achieve the features of that improved state of affairs.

As the introduction to this chapter noted, these days, decisions concerning what technologies to develop and use and concerning how to manage enterprises in changing environments are far from simple. Predicting the impacts of a biological or physical management practice is equally difficult. Graduates in the food, agricultural, and natural resource fields are becoming increasingly involved in situations that are messy and difficult to define. People factors loom large. New competencies and involvements are needed to deal with these situations. These involvements transcend and call into question the adequacy of traditional problem-solving methodologies that are taught in our colleges. Graduates need an array of approaches to inquiry that will help them comprehend and take action on the interactions, impacts, and complexities of situations in ways that are conceptually valid and practically effective.

The Role of Professionals

The traditional role of professionals as interventionists is also changing. Until now, those heading for careers in food, agriculture, and natural resources have been trained to view themselves as intervening in the biological and physical systems of the world. Yet what about the agricultural and natural resource enterprises that respond to, develop, alter, and manage biological, physical, and human systems? Professionals, managers, and other practitioners need to broaden the range of factors to which they pay attention. In addition, in the past, a professional's stance on intervention has been that of expert or advice-giver, while a more appropriate role may well be that of facilitator of people's own inquiry into their concerns, opportunities, and challenges.

Professionals must also learn to assimilate varying viewpoints and to structure debate in ways that will lead to improvements in a melange of complex and dynamic factors in problematic situations. To do this, new professional competencies are required. It is not simply a matter of taking account of different attitudes because underlying the varied perspectives that are offered for the future development of agriculture and natural resources are profoundly different ways of viewing the same world and of tackling its problems.

Use of Technical Terms

Peter Checkland (1981:11) makes a useful distinction between jargon and technical terms. He says that "It is very noticeable that the word 'jargon' is hurled as a missile at any attempt to use technical terms in discussing human activity or social systems." The renowned sociologist C. Wright Mills (1959) has defined jargon as "a seemingly arbitrary and certainly endless elaboration of distinctions which neither enlarge our understanding nor make our experience more sensible." Following Mills' critique, Checkland argues for the need to use precisely defined terms so that the reader knows exactly what is being discussed. As much as possible, the number of technical terms used in this text will be limited. When a term is used, it will have a body of conceptual or applied research literature that justifies it. The underlying theories, concepts, and empirical findings will be discussed when the term is introduced. A glossary is also provided in the back of this book. Because the presentation of arguments in this book depends on an understanding of technical terms, it is suggested that you pause to commit each term to understanding as it is introduced and take care to check whenever you encounter a term or word that you don't understand. You will also find that a good collegiate dictionary is a useful tool.

The Aims of This Book

The challenge of writing this text is to provide sufficient guidance for you to understand and use the methodologies and techniques it introduces, while not conveying the message that "this is the only way to do it." In Chapters 4 to 8, however, the book details the whats, whys, and hows of soft systems inquiry. This is because it is a relatively new approach, and its applications to food, agricultural, and natural resources issues warrant particular attention. The hope is to communicate with sufficient detail and clarity to give you a base from which to begin your own *experience*. Experience is emphasized because this approach is meant to be used and to be learned by using it. If the authors were with you in person, they would facilitate your discovery of some things on your own before talking about them. Authors of a book are, however, limited by the written word. So, while the text may appear to assume a fairly direct and prescriptive style, the intention is to communicate approaches to inquiry so that you can begin your own practice. If you are learning the concepts and process introduced in this text with an instructor, you may actually learn through experience (simulated or real) and use the text as a summary of what was said and what you did in order to improve perception, thinking, and action.

Outline of the Book's Main Points

The discussion developed in Chapters 1 to 8 is summarized here:

Chapter 1

1. Chapter 1 illustrates the dynamism and complexity of the world in which we live and, in particular, of the situations in which we must perform as professionals in the food, agricultural, and natural resource sciences. The interdependence of enterprises, governmental bodies, other institutions such as science, and whole nations is by itself a source of difficulty.

2. Today's and tomorrow's food, agricultural, and natural resource professionals face new challenges and opportunities. New kinds of competencies and a broad range of inquiry methodologies and knowledge are needed in order to respond to these challenges. This book is devoted to helping graduates develop such knowledge and competencies.

3. The ability to focus, not on specific technical changes, but on a vision of an improved future state is particularly emphasized.

Chapter 2

1. Learning is the process by which experience is transformed into knowledge and knowledge is transformed into action. It is a dynamic process of adaptation and action in which we repeatedly interact with our social, biological, and physical environment. It follows that learning to tackle real-world problems is a key to being effective in our professional careers as well as in our daily lives.

2. Learning involves four major dimensions, describing types of activities: (1) *prehension*—how we explore experience and find information to determine the meaning of a given situation; (2) *transformation*—how we take action once we think we have understood a situation; (3) *methodology*—how much of the complexity in a given situation we choose to handle; and (4) *insight*—how rational or intuitive we are in deciding what the facts and meanings are and what courses of action should be taken.

3. Individuals develop and use markedly different styles of learning and problem solving in relation to the opportunities and challenges they face in food, agricultural, and natural resource situations. Our learning style is measured by how we deal with experience in relation to the two primary dimensions of inquiry:

prehension, involving the concrete experience ↔ abstract conceptualization axis and transformation, involving the reflective observation ↔ active experimentation axis.

4. When these axes are combined, four styles of learning are described: (1) diverging, (2) assimilating, (3) converging, and (4) accommodating. Each style represents a person's particular orientation toward finding out about and taking action in a given situation.

5. Not only do people have their own styles of learning, they also have unique ways of looking at the world. While these techniques are based partly on a person's preferred learning style, many other factors contribute to a person's world view. These factors include values, beliefs, morals, tastes, cultural tradition, past experience, language, attitudes, and personalities. A person's world view will markedly influence the types of observation that person makes, the kind of information he or she collects, the meaning that person attributes to experience, and the lines of action he or she selects. Thus the way people create knowledge, and the substance of that knowledge, are determined by their particular learning styles *and* world views.

6. A model of the learning cycle is applied to review methodologies used in pursuing inquiry into food, agricultural, and natural resource problems; basic and applied science methodologies are presented in Chapter 2, and hard and soft systems inquiry are reviewed in Chapter 3.

7. The distinction is drawn and discussed between *methodology,* the abstract logic justifying our inquiry process, and *technique,* the actual procedures employed to gather data and analyze them.

8. All the competencies associated with the four styles of learning are needed to carry out any given methodology or process of inquiry. While some parts of a given approach will be easier to handle because they seem to come naturally, others will require competencies found in styles of learning different from one's own. These inquiry competencies can be learned.

9. Professionals in the food, agricultural, and natural resource sciences will use at least four different kinds of inquiry methodologies in order to tackle the range of problematic situations typically encountered. Each methodology is better at addressing some types of questions than others and has its own distinct tasks, guided by a philosophical foundation.

Chapter 3

1. This chapter reviews key premises of systems thinking, including the principles of holism, transformation, control, communication, hierarchy, and emergent properties.

2. The stages and uses of hard and soft systems inquiry are discussed and compared.

3. The learning-cycle model and the various inquiry methodologies are now combined and conceptualized as a spiral of inquiry. A key professional competency is the ability to select methodologies appropriate to the questions people

are asking in a particular situation and to decide how definable that situation is. In the abstract, none of the four inquiry methodologies is superior. Rather, the well-rounded professional needs to be competent in the use of a range of methodologies in order to tackle the variety of themes of concern that will be present in most food, agricultural, and natural resource situations.

4. This chapter also introduces the basics of modeling in systems inquiry, including techniques useful in both hard and soft systems approaches.

5. The chapter concludes with an overview of the soft systems approach and compares it to the procedures of hard systems inquiry.

Chapter 4

1. This chapter presents stages 1 and 2 of the soft systems approach in detail.

2. Divergent learning-style competencies are employed in order to enter into situations and to understand them fully and without bias. Assimilative learning-style competencies are employed to synthesize and analyze the information collected.

3. The inquiry process starts by looking at problematic situations rather than at a specific problem. A situation is comprised of people as individuals and in groups, a historical context that bears on the present, key human activities, themes of concern, unease and opportunity, decision-making structures, environmental factors, relational climates, senses of improvement and impact, and differences among viewpoints.

4. A key task is collecting and synthesizing certain kinds of oral and written information. What information to collect and how to collect it are discussed. A beginning set of techniques that facilitates the collection, synthesis, and analysis of information and that is congruent with the philosophy of the work of stages 1 and 2 is suggested, and practice exercises are provided.

5. Striving to view a situation from many angles and from the perspective of the people involved leads to insights that can be used to develop ideas of meaningful change. This process is the first step in trying to develop world views that incorporate and transcend those of individuals and groups; thus, the process provides a basis for collaboration and cooperation.

6. The task of describing and analyzing the situation is carried out by the people involved. Inquiry thus becomes a mutual learning process in which useful knowledge is generated by all. People's capacity to analyze and design improvements is increased by the mutual learning relationship, contrasting with the traditional expert–client relationship in which ownership of the inquiry process remains with the expert.

7. Several techniques are presented for analyzing themes of concern and opportunity in a situation (mind mapping) and for synthesizing information regarding the nature of the groups, human activities, and relational climates present (cartooning). The end product is a situation summary, a record of the conclusions of stage 2, which will be useful throughout the remainder of the inquiry process.

Chapter 5

1. This chapter is about stages 3 and 4 of the soft systems approach.

2. Often people who are interested in change fail to develop an understanding of potential future improved conditions. This chapter discusses how, using systems concepts, future conditions and states can be conceptualized as human activity systems.

3. When members of a group begin to think systemically, they envision an improved situation as if its elements were functioning entities. Using rules and properties defined in formal systems literature, they then conceptualize these entities as constituting an integrated, functioning whole.

4. Assimilative and convergent learning competencies are used to carry out the inquiry tasks associated with stages 3 and 4. Assimilative abilities are used to form ideas about the kinds of transformations that may occur in the future in order to improve problematic features of the present situation. Convergent abilities are used to develop some of these transformations into human activity system models.

5. Participants design potential human activity systems that have specified inputs, transformations, subsystems, outputs, boundaries, measures of performance, actors, owners, environmental constraints, beneficiaries, and victims, as well as a world view that makes all of these features meaningful.

6. These human activity systems are developed as models. Modeling is the mental process we use to build a manageable replica of something. In the case of a human activity system model, the designers essentially are attempting to answer such questions as "What primary human activities will be occurring in an improved state?" "By and for whom?" "With what resources and anticipated outcomes?"

7. In addition to the mandatory human activity systems model of an improved state of affairs, other kinds of modeling may be done in order to communicate complex interrelationships, to determine how present features might behave if changed in certain ways, or to evaluate alternative strategies of change.

Chapter 6

1. This chapter presents stage 5 of the soft systems approach, which is concerned with the need to pause after completing one or more models to compare them with reality, as represented by the situation summary produced in stage 2.

2. Competencies associated with the convergent learning style are called for here. The ability to choose from among alternative solutions is particularly emphasized.

3. Accordingly, there is a strong temptation to slip back into the expert and technology development roles. This can be avoided if the mutual learning approach is maintained in discussions between the people involved in the situation and the professional facilitator.

4. Four specific techniques for comparing a human activity system model with a picture of reality are outlined: general discussion, question generation,

overlay, and historical reconstruction. The choice of which technique to use depends on the nature of the situation.

5. The overall objective of stage 5 is to verify the viability of the models before they are used to generate specific proposals for change and detailed implementation plans.

Chapter 7

1. This chapter presents stage 6 of the soft systems approach.

2. The central task is for the people involved in the situation to debate the desirability and the feasibility of the specific changes generated by the models of an improved future state. The two basic questions are "Is this what we want [it] to be?" and "Can we really do this?"

3. A debate is a discussion or examination of several sides of a question and an opportunity to consider positive and negative aspects of the changes being proposed.

4. Models of future human activity are seen as desirable when participants judge them to agree with their own world views.

5. Models of future human activity are feasible when participants agree that they can be carried out (1) with available or accessible resources, staff, technology, and individual and organizational capabilities and (2) without incurring unacceptable group and environmental risks and costs.

6. The debate phase has much in common with the comparison phase, and they are often done together. While the comparison phase emphasizes convergent learning, the debate stage favors accommodative competencies. The result of stage 6 is a set of concrete changes to be implemented.

7. Because the contexts in which professionals may operate vary so widely, this chapter also provides a detailed discussion of the arenas in which the debate phase, as well as other soft systems inquiry activities, might be carried out: small face-to-face communities, business firms, multiparty disputes, the courts, bureaucracies, and government. Debate techniques that seem appropriate to particular kinds of situations and contexts are also suggested.

Chapter 8

1. This chapter is about stage 7, formally the last phase of the soft systems approach.

2. The competencies associated with the accommodative learning style are particularly emphasized here, since this is the action phase of the approach.

3. The central tasks of this stage involve developing an implementation plan and then moving to implement specific changes.

4. Alternative scenarios, in which the soft systems approach may also be used to monitor or evaluate initiatives that are already underway, are also discussed.

Appendix

1. The book also includes case materials on three situations. While two of them have been fictionalized, all are based on actual cases.

2. These materials are used in discussions throughout the book, but none of them are brought to a hard and definite conclusion. This holds open the possibility that you, the active learner, can use them to work through specific questions or aspects of the approach.

Glossary

1. A glossary of technical terms introduced in the text, with selected cross-references, is provided.

REFERENCES

Brown, Lester R. *Seeds of Change: The Green Revolution and Development in the 1970s*. New York: Praeger, 1970.

Brown, Lester R. *In the Human Interest*. New York: Norton, 1984.

Brown, Lester R., and E. P. Eckholm. *By Bread Alone*. New York: Praeger, 1974.

Checkland, Peter. *Systems Thinking, Systems Practice*. New York: John Wiley, 1981.

Church, George. "Real Trouble on the Farm." *Time*. February 18,1985:24–31.

Coulter, Jane, M. Stanton, and A. Boecker. *Employment Opportunities for College Graduates in the Food and Agricultural Sciences*. Washington, D.C.: U.S. Department of Agriculture, 1986.

Dixon, Mim. *What Happened in Fairbanks? The Effects of the Trans-Alaska Pipeline on the Community of Fairbanks, Alaska*. Social Impact Assessment Series, No. 1. Boulder: Westview, 1978.

Forrester, Jay W. *World Dynamics*. Cambridge, MA: Wright-Allen, 1971.

Heady, Earl D. *Roots of the Farm Problem*. Ames, Iowa: Iowa State University Press, 1965.

Heilbroner, Robert. *An Inquiry into the Human Prospect*. New York: Norton, 1972.

Johnson, Glenn L., and Leroy Quance (Eds.). *The Overproduction Trap in U.S. Agriculture: Resource Allocation from World War II to the Late 1960s*. Baltimore: Johns Hopkins University Press, 1972.

Kahn, Herman *The Next 200 Years: A Scenario for America and the World*. New York: William Morrow, 1976.

London, Jack. "The Economics of the Klondike." *The American Monthly Review of Reviews*. January 1900:70–74; reprinted Seattle, Shorey Bookstore.

Magnuson, Ed. "Clinging to the Land." *Time*. February 18, 1985:32–39.

Meadows, Donella H., Dennis L. Meadows, Jorgen Randers, and William Behrens. *The Limits to Growth*. New York: Universe, 1972.

Mills, C. Wright. *The Sociological Imagination*. New York: Oxford University Press, 1959.

Mumford, Lewis. *The Pentagon of Power—The Myth of the Machine*. New York: Harcourt Brace, 1970.

National Academy of Sciences. *Agricultural Production Efficiency*. Washington, D.C.: National Research Council, 1974.

National Academy of Sciences. *World Food and Nutrition Study: Enhancement of Food Production for the United States*. Washington, D.C.: National Research Council, 1975.

Phillips, Vince. "Backgrounder." Office of the Secretary, Public Liaison, April. Washington, D.C.: U.S. Department of Agriculture, 1985.

Robbins, William. "The Debt-Ridden Farmers." New York Times News Service/ *Honolulu Star Bulletin*, 1985.

Ruttan, Vernon W. *Agricultural Research Policy*. Minneapolis: University of Minnesota Press, 1982.

Schumacher, E. F. *Small Is Beautiful: Economics As If People Mattered*. New York: Harper & Row, 1973.

U.S. Department of Agriculture. *A National Program for Soil and Water Conservation*. Summary 2–14. Washington, D.C., 1982.

Winteringham, F. P. W. *Environment and Chemicals in Agriculture*. New York: Elsevier Applied Science, 1984.

The Learning Dimensions of Professional Inquiry

Chapter 2 develops much of the theoretical framework for the remainder of this book and is built on the following point: Professionals spend most of their careers solving problems and improving situations. Problem solving is essentially a learning process. Research has shown that while every person's way of learning is unique to some degree, there are nevertheless patterns that provide bases for models describing how people carry out the learning process.

While it may appear novel to begin a text intended for the food, agricultural, and natural resources fields with concepts of learning, the subject is central to the task professionals face of improving typical situations. One must engage in a learning process in order to improve a situation. Learning can be equated with professional inquiry when it is carried out in order to transform ideas into actions or to alter people's actions and environment.

In addition, learning different inquiry approaches, along with their benefits and limitations, is central to professional development. This involves acquiring new competencies and methodologies. It will also be important to understand the premises, as well as the procedures, associated with a given methodology so that it can be put into practice.

This chapter presents four key learning dimensions. It provides a model that is useful for thinking about learning patterns and processes. It also integrates learning concepts with inquiry approaches useful to professionals in food, agricultural, and natural resources situations. The chapter concludes with an overview of two of these inquiry approaches, basic science and applied science inquiry. Others are introduced in Chapter 3.

What Is Learning?

If you want to deal effectively with food, agricultural, and natural resource problems, then you must understand how you and other people learn about and act on situations in the real world. The mark of educated people is that they command not only information, but also productive ways to find out about and take action on a wide array of problems. Learning is the process people use to make sense out of and gain some control over the ever-changing world. This chapter introduces a process of reorientation toward education, learning, experience, inquiry, and assisting others to improve their situations, as well as the interconnections among these subjects. It also begins the task of showing how science and technology fit into this picture.

Many people equate learning with what goes on in schools and classrooms. In their minds, schools are where you go to learn. The role of teachers is to teach, to serve not only as repositories of fact and expertise, but to actively fill the empty heads of students. Weighty and authoritative — and frequently very expensive — textbooks help too. The complementary and predominant role of students is to sit passively, take notes, wait for assignments, read books, take exams, and get grades. This integrated pattern of "studenting" and "teachering" has had negative consequences for many graduates, for the institution of education, and for society. In particular, it has not prepared students adequately for effective, let alone creative and adventurous, work in the real world. Hence, students and teachers who have followed this model of learning in the past should benefit from the reorientation proposed in this book.

The Active Learner

It can be said that we "learn our way through" new experiences both by adapting ourselves to changes and by using our new understandings to change the situation we are experiencing. Our environments shape us and we, in turn, shape our environments. Accordingly, learning is a dynamic process of adaptation and action in which we have recurring experiences involving other people and the physical environment. It follows that in learning, your own concrete experience has greater impact on your understanding than does reading about, listening to, or watching someone else's concrete experience, learning, or knowledge generation.

Learning involves much more than memorizing facts and acquiring intellectual understanding. Because it is an adaptive process, learning includes our ability to act as well as to understand and attribute meanings. In addition, learning must engage our feelings as well as our thoughts and actions. We learn by doing, and we "do" in relation to changes in our environment. We change our environment by our acts; we observe their consequences, absorb their implications, plan further acts, and thereby produce knowledge *for ourselves*. Thus, the basis of learning is experience, and the best kind of experience for learning is that which evokes feelings. Of course, feelings may be positive or negative. "Opportunities" are experiences associated with positive feelings. Experiences associated with negative feelings are called "problems."

If you have developed appropriate competencies, learning becomes a process through which experience is transformed into knowledge. Learning to tackle real-world problems is a powerful way for us to discover how to be effective learners in our professional careers as well as in our daily lives.

There are several reasons for emphasizing learning in this text. First, it focuses on a process rather than a set of concepts or topics to be "covered." The kinds of situations that learners might want to explore in the context of food, agricultural, and natural resources management are unlimited. The first objective, then, is understanding and mastery of a set of procedures for discovering what situations are about and what actions to take.

Second, the learning process is made explicit to help students to become self-conscious about it. Ignorance is not bliss in any field. Every student's objective should be autonomy. An autonomous learner is self-motivated and self-directed. This means that he or she sets personal standards and goals and asseses his or her own strengths, weaknesses, and progress.

Third, recent research suggests that individuals develop and use markedly different styles of learning and problem solving in relation to the opportunities and challenges they face. Because of their preferred styles of learning, people possess different strengths, weaknesses, and excesses. In groups made up of diverse learners, these differences may sometimes be complementary, and at other times they may give rise to conflict. Some major conflicts concerning the current status and future developments in food, agriculture, and natural resources may have arisen because the individuals involved have approached a similar issue in differing ways based on their learning styles (Miller, 1985). The features of these styles are patterned and measurable, and they reflect much more than just differences of background, attitudes, expertise, or viewpoint. They affect the very ways people apprehend new experiences, how they grasp the issues involved, and how these issues are eventually transformed into knowledge.

Succeeding sections of this chapter present and discuss a model in which learning or problem solving is shown as a cycle of activity involving many dimensions. The most important of these are called "concrete experience, abstract conceptualization, reflective observation and active experimentation" (Kolb, 1986). This model is particularly useful, even powerful, not only because it facilitates self-consciousness about one's own learning, but because it matches — even underpins — the inquiry approaches presented in the present and the next chapters. A way to think about this is that the learning cycle is the general model, and each inquiry methodology is a subcase of or a deduction from it. The learning model is also fundamental to the presentation of systems thinking and serves to distinguish the approach presented in this book from others.

The learning cycle model and a technique deriving from it called the learning styles inventory are useful in revealing our preferred styles of learning, inquiry, and problem solving. Later you will see that everybody is more or less strongly oriented toward one or more of the four dimensions referred to, and the strength of this orientation describes our own characteristic learning style. Implied is a marked bias in favor of some kinds of inquiry activities over others. Our biases not only inhibit learning in general and the development of particular approaches to inquiry, but they also lead people, as individuals and in groups, into conflict over the interpretation of the meaning of new situations and into disputes over what to do.

People can and do change. This change can be self-directed. Individuals can be helped to improve their learning styles over time by finding out about their own strengths and weaknesses and by seeking exposure to new kinds of problems, opportunities, and challenges. You can learn investigative, analytical, planning, and action techniques that can serve as effective guides to improving situations regardless of your original preferred approach to problem solving.

Perhaps the most important lesson of this chapter so far is that the more one knows about learning and problem solving, the more evident is the need to use all of the competencies found in all the learning styles! As you begin to apply some of the inquiry approaches presented subsequently, you soon will find that, while you may be good at one stage of a given inquiry process, you will be not as good at several other stages until you educate yourself to deal with those inquiry techniques that don't come naturally!

Four *dimensions* of learning, describing aspects of people's behavior as they learn, are discussed in the next section of this chapter. The first is the prehension dimension of learning; the second, the transformation dimension. These two dimensions are combined in a grid to form a cyclical model of learning that can also be used to describe people's preferred learning styles. The notion of learning styles is based on the observation that when individuals approach inquiry, they tend to emphasize one sector of the learning cycle in preference to the others. There are specific competencies associated with each sector of that cycle.

Learning Dimension 1: Prehension

This dimension is an axis describing how people convert real-world experience into ideas. Two learning *modes* are involved, learning through *concrete experience* and learning through *abstract conceptualization*. Later this will show up in Kolb's learning-cycle model as

Concrete Experience (CE) $\longleftrightarrow$ Abstract Conceptualization (AC)

When we newly confront the concrete experience of a situation, we engage in a process of inquiry activities that seem to follow a regular path. First, we watch, listen, and obtain information from a variety of sources and people available to us. We may look at it from many angles as we try to understand the situation. The pioneering educator John Dewey (1910; see also Kolb, 1986) called this process *apprehending*.

The root of the word *apprehension* is *prehension*. Prehension means to arrest something by getting it in hand or grasping it. In this context, it is gaining understanding by gathering facts and impressions and then making them your own in incorporating them into your organized memory, or mental map. As we apprehend a situation, we try to discriminate between familiar elements and strange ones. In particular, as we sort out the elements of the situation, we puzzle over the unfamiliar. This reflection occurs internally, although different people puzzle over novelty in different ways. Some talk their way through the giving-it-meaning phase with themselves or with other people. Others think it through silently. This move from apprehended facts and impressions to some kind of understanding is called *comprehension* (Kolb, 1986). It is a bit like constructing and using a crude map to explore new territory. Often we can use old, previously developed maps to find our way around new ground successfully. From time to time, however, we need to revise our maps when we get lost or discover aspects of the territory that we have never before encountered. On other occasions, the features of a situation are so novel, unprecedented, or widespread that the old maps are quite inadequate and need to be redrawn entirely. It is also true that this replacement process seems too great a challenge to some people, and so they stick with the old guides, no matter how restricted or misleading they are.

Inquiry into any kind of situation can thus be seen to involve alternating between apprehending (i.e., perceiving or grasping) basic facts and impressions from our senses or external sources of information and internally comprehending (i.e., conceiving or conceptually mapping) the situation as we care to understand it.

Most readers can recall situations in which the apprehending–comprehending process was going on. Something is seen or a dilemma arises, and you use all your senses to hear, see, feel, smell, and taste it. If other people are present, lots of talking occurs. This is the apprehending process at work. At the same time, the mind is using everything it knows based on past experience and formal or informal learning to give the situation meaning.

The apprehending and comprehending process is a constantly recurring flux between exploring reality and mapping it. Exploration is an apt metaphor for how the process of learning begins, in that learning can be seen as a flux between experiences in the real world and abstract interpretation of them through the use of maps and meanings provided to us by society or through the use of new maps and meanings that we develop ourselves.

This view of learning is supported by almost a century of research. Over eighty years ago, John Dewey (1910) proposed that "our intellectual processes consist . . . in a rhythm of direct understanding — technically called apprehension, with an indirect mediated understanding technically called comprehension." Experiences are grasped through a continuous cyclic process of perceiving meaning through direct experience and designing or modifying conceptual maps that we carry around internally (Figure 2.1). By the late 1970s, brain research confirmed Dewey's insights (Edwards, 1979). The right brain and the left brain specialize in the two different modes of knowing about the world that Dewey and his successors, such as David Kolb (1986), call apprehension and comprehension.

In summarizing Edwards' work, Kolb (1986:48) notes the following characteristics of left- and right-brain functions:

> Left-mode function corresponds to the comprehension process. It is abstract, symbolic, analytical, and verbal. It functions in a linear sequential manner much like a digital computer. The right-mode function, corresponding to the apprehension process, is concrete, holistic, and spatial. Its functioning is anologic and synthetic, drawing together likenesses among things to recognize patterns.

Recent brain research has also improved understanding of that mysterious process called intuition. Intuition is the immediate knowing of something without the use of conscious reasoning. Some brain research suggests that intuitive be-

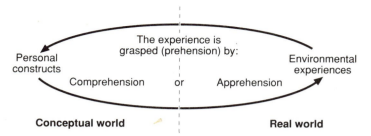

FIGURE 2.1 Dewey's conceptualization of the intellectual process: the comprehension–apprehension axis.

havior is guided by affective judgment (the apprehension process) rather than cognitive judgment (the comprehension process), as previously thought (Zajonc, 1980).

Preferences precede inferences in intuition. When intuitive behaviors are at work, feelings and values seem to dominate our thinking. Thinking it through and feeling it through seem to be fundamentally different (even opposite) modes of grasping the world around us. Neither is good or bad; they are just different processes yielding different results.

The apprehension function involves our valuing behaviors, our subjective sense of what is important and good. It is our way of appreciating a situation. It is how we attend to and become interested in a variety of aspects of a situation in which we are involved. Elements of a situation must literally capture our attention. We judge the value of things and people in a situation as well as the facts associated with it. This valuing process fuels our selection of facts and influences whatever actions we take. Apprehension also affirms people and events in a situation, signaling that they belong. Ultimately, it is based on beliefs, convictions, and trust.

Contrastingly, comprehension — Kolb's abstract conceptualization mode of learning — is based on being able to objectify a situation by controlling or selecting that to which we will pay attention. Whereas apprehension of a situation rests on our feelings, values, and impressions, comprehension rests on dispassionate analysis of a situation. While apprehension utilizes our willingness to believe, trust, and have strong convictions about a situation, comprehension rests on our criticism, doubt, and skepticism about the reliability and verifiability of critical factors in a situation.

Knowledge and truth thus result from the flux between these two modes of learning. Most people have difficulty self-consciously distinguishing between their apprehension behavior and their comprehension behavior.

Apprehension of a situation is a very personal process, only known to others when we are able or choose to communicate it. What you would be attempting to communicate are your very personal and idiosyncratic assumptions, beliefs, values, and appreciations about people and events. Communicating comprehension of a situation, in contrast, rests on your ability to use socially and culturally meaningful words, symbols, and images to convey understanding.

The conceptual maps or models that are the instruments of, or that result from, the comprehension process are merely representations of a situation. They may be highly personal and unique to the persons who constructed them, even when they are sufficiently general as to be understood by, and useful to, others. The point to remember, however, is that no two people's conceptual maps of a situation are the same. That is, how you interpret a situation and how another person interprets a situation will be different to some degree or another, even if you share an organized approach to knowledge, such as a particular scientific discipline.

It is also entirely possible that the two functions, apprehending and comprehending, can become socially devisive. This is because some people excel at the abstract map-making function, while others revel in collecting information from many sources and talking their way through a situation and a course of action. Such preferences inevitably lead to biased behavior, viewpoints, and attitudes about the respective importance of people's contributions. Those who are action-oriented are skeptical of abstract thinkers and their perceived inactivity in the face of change. The conceptualists, in their turn, are troubled when changes are made in situations without sufficient forethought. Some individuals develop viewpoints that are so extreme that they completely reject other ways of looking at the world. And if this is true for one extreme position, then it must be equally true of the other extreme position that represents an absolutely opposite stance.

So, with respect to the apprehension–comprehension dimension, it is possible to imagine two people who differ vastly in the manner by which they learn to adapt to their changing worlds. In practice, of course, it is rare to find such a simple distinction between extreme ways of doing things. Yet each of us does seem to adopt a position somewhere in between that reflects a bias toward one pole or the other. As we shall see later, this has profound implications for learning how to tackle problematic situations with respect to both our personal and professional behavior.

The prehension dimension, involving the flux between concrete experience (apprehension) and abstract conceptualization (comprehension), is but one of several dimensions that have been associated by psychologists with the process of learning. Each of these dimensions is described in a similar way, as a spectrum of positions adopted by individuals between two extreme poles representing diametrically opposite characteristics.

Learning Dimension 2: Transformation

This dimension is an axis describing how people extend intentions to actions. Two learning *modes* are involved, learning through *reflective observation* and learning through *active experimentation*. This will appear in Kolb's learning cycle model as

Reflective observation (RO) <—> Active experimentation (AE)

There is another dimension of learning that seems at some stage to intervene in the flux between apprehending and comprehending activities. If we only confined our explorations of new experiences to building up perceptions and abstract representations of them, we would not get anything done! And ultimately, action is the issue. At some point in this process, we need to do something with our understandings. This is to change focus from the *what* and the *so what* as perceived and conceptualized to the *now what* of intention.

During the now-what process, the appreciations, perceptions, and conceptual maps previously developed are transformed into practical knowledge to be extended to others and used to change situations (Figure 2.2). In other words, knowledge grasped from experience is transformed into action (Kolb, 1986:41). On this new axis, knowledge and meaning first arise through active experimentation (AE) and the grounding of our ideas and experiences in the external world through reflection (RO) about their possible meaning and attributes (Kolb, 1986:52).

There are at least four reasons why we extend our understanding of a situation to others (transforming our ideas by extending them):

1. To test our interpretations of a situation against those of others.
2. To present our views as a basis for decisions about acting on a situation.
3. To use our concepts as vehicles for debate about change with others who are involved in a situation.
4. To use our conceptualizations in generating new ideas.

Likewise, while reflecting on the values, beliefs, appreciations, and concep-

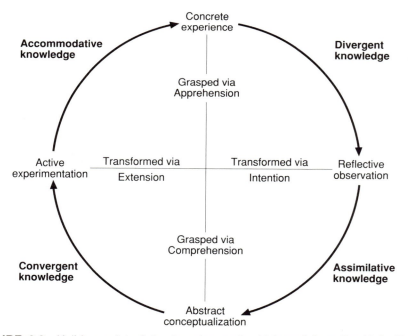

FIGURE 2.2 Kolb's model of the learning cycle. (Adapted from David A. Kolb, *Experiential Learning*, © 1984, p. 42. Reprinted by permission of Prentice-Hall, Inc., Englewood Cliffs, New Jersey.)

tualizations that we have formed of the situation, we may watch people and the flow of events in a situation a little longer in order to test whether our notions hold up. Issues to reflect upon include

- Are beliefs, values, and concepts worth sharing?
- Are they accurate?
- Do they contribute to a fuller understanding of the present situation or an improved view of the future?

How do you take action on a new situation? Do you tend to act immediately and work out the situation through trial and error? Or do you tend to be cautious, move slowly, and choose a course of action only after much thought or consultation? How anxious are you about the possibility of making a mistake or about sharing your beliefs? Do you wish to avoid error at all costs, or do you plunge in, expecting to make mistakes and valuing getting on with it? How much certainty must you have before you begin to take action? Does acting quickly take priority over questing after the ideal, or vice versa? Your answers to these questions begin to show your personal orientation toward transforming ideas, values, and beliefs into action.

As discussed previously, apprehending–comprehending capabilities seem to be correlated with right- and left-brain function. There is also evidence that intention–extension capacities are connected with the functions of the front and back of the brain (Edwards, 1979).

The psychologist David Kolb has integrated what have been labeled dimensions 1 and 2 into a single model of the learning process. Drawing particularly on the works of Dewey (1910), Piaget (1970), and Lewin (1951), Kolb displays the learning process as a cycle that involves grasping meaning and transforming it into socially shared knowledge. Thus, as illustrated in Figure 2.2, the learner alternates between the two modes of prehending new situations through apprehension and comprehension, and the two modes of transforming the resultant perceptions and notions of a situation via intention (watching the world to see if our ideas hold) and extension (actively testing our ideas). The result of this prehension and transformation process is the creation of knowledge. This conception is called experiential learning because the process is grounded in our responses to transactions with the people and situations we encounter.

The learning cycle is not a model of what goes on inside people's heads as they learn, but rather it assimilates the kinds of human activities people carry out as they learn. Figure 2.3 presents the learning cycle as patterned human activities. It assigns types of activities to each of the four major cells representing styles of learning.

Learning Styles

Research by Kolb and associates indicates that each of us forms a unique approach to learning. Using the two learning dimensions (which incorporate four major modes of learning), Kolb has developed a Learning Styles Inventory (LSI)

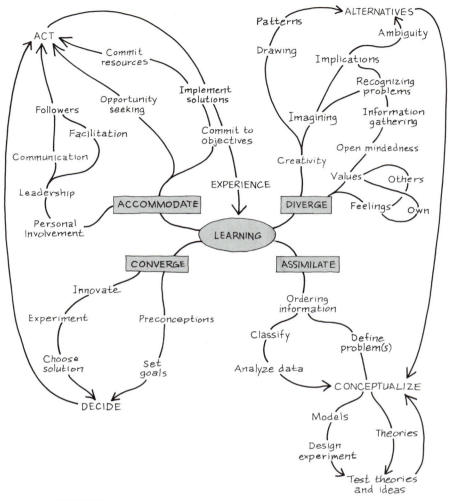

FIGURE 2.3 The learning cycle as patterned human activity.

consisting of a twelve-item self-description questionnaire. Respondents rank-order four words or phrases in each of the twelve item-sets in a way that best describes their approach to learning.

> One word in each item corresponds to one of the four learning modes — concrete experience (sample word, feeling), reflective observation (watching), abstract conceptualization (thinking), and active experimentation (doing). The LSI measures a person's relative emphasis on each of the four modes of the learning process — concrete experience (CE), reflective observation (RO), abstract conceptualization (AC), and active experimentation (AE) — plus two combination scores that indicate the extent to which the person emphasizes abstractness over concreteness (AC–CE) and the extent to which the person emphasizes action over reflection (AE–RO). [Kolb, 1986:68]

Kolb's team tested men and women of varied background and age. Several patterns developed. First, Kolb found that people tend to have a preferred way of grasping the meaning of situations (learning dimension 1), which is combined with a preferred way of acting on that meaning (learning dimension 2). People's learning styles are calculated by plotting their scores for the two dimensions on a graph. This graph has four quadrants, representing the four learning styles. The quadrant where a person's plot falls identifies the pertinent learning style.

Kolb's second finding was that the four modes of learning are related to the possession of definite learning competencies. These findings are further explained here.

Prehension Abilities

How do people who pursue the meaning of a situation through direct experience operate? They learn by involving themselves fully in new experiences — learning by encounter. They are able to listen empathetically to others (i.e., understand what others believe the situation to be). They don't allow their own beliefs, values, judgments, and appreciations to stand in the way of understanding the viewpoints of others. They seek diverse information, some conflicting and some that doesn't seem to make sense. They tolerate the complexity and ambiguity of situations as they unfold.

Those with a predominant orientation toward abstract conceptualization create concepts (mental maps, models, frameworks) that integrate their observations of a situation into logically sound theories. Theory means an abstract conceptual or symbolic framework that integrates information and observations about a situation in order to explain it.

Transformation Abilities

People oriented toward reflective observation as a preferred way of acting are able to compare their ideas about actual situations or events with those of other people. In essence, they can view a situation from many different perspectives. People who are oriented toward active experimentation tend to be intolerant of ambiguity, to choose a more active form of creating knowledge, and to test their understanding of a situation against reality.

Our Preferred Learning Orientations

A person's learning style is evident, not only when he or she takes the LSI questionnaire, but when that person tries to find out about and take action on a situation. Research also indicates that peoples' career choices are associated with where they locate themselves in the LSI along the prehension and transformation dimensions. The intersection of these two dimensions generates four primary learning styles, as illustrated in Table 2.1. While the following descriptions of learning styles are presented as if they were fairly clear-cut, the reader should be warned that *there is room for significant variation within each type.* For example, distance from the origin along one or both dimensions of the graph is the measure of the strength of an individual's bias or preference. Note also that the LSI should not be used for stereotyping people.

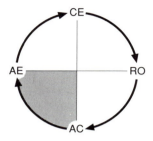

The Convergent Learning Style

Convergers prefer the abstract conceptualization and active experimentation modes of learning. They have particular strengths in problem solving, decision making, and the practical application of abstract principles and ideas. They are most comfortable with problems that have a single correct answer or solution. They reason deductively and are at ease formulating and using hypotheses about the meaning of situations. The process of deduction involves going from the general to the specific and forming propositions. Thus, they extract from the complexity of a real situation only those features they feel (possibly on the basis of the conceptual framework they are using) are most important or valuable in understanding the situation. Convergers tend to exhibit tight emotional control and prefer dealing with technical tasks and problems rather than with social and interpersonal ones. The converger's greatest excess is a tendency to identify *the* problem or *the* solution prematurely and, hence, occasionally to be guilty of solving the *wrong problem* with immense discipline and energy.

TABLE 2.1 Four Learning Styles Based on the Primary Dimensions

		Modes of Prehension	
		Apprehension	Comprehension
Modes of Transformation	Intention	DIVERGENT	ASSIMILATIVE
	Extension	ACCOMMODATIVE	CONVERGENT

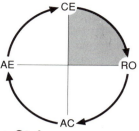

The Divergent Learning Style

Divergers prefer the concrete experience and reflective observation modes of learning. They have strengths that are the opposite of those attributed to convergers. Divergers excel in imagination and feelings and in their concern for values. The diverger is able to develop many alternative views of a situation and to weigh the associated values and meanings. These varying meanings, values, and views of the situation are combined into an impression or "gestalt." Divergers adapt to situations through carefully watching and observing rather than through taking direct action. Divergers are great idea generators and can see the range of implications that emerges as a situation unfolds. Moreover, divergers are strongly people-oriented and sensitive to the views and feelings of others. Divergers may experience difficulty with abstractions and appear to others as indecisive.

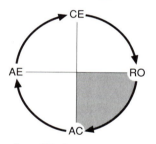

The Assimilative Learning Style

Assimilators prefer the reflective observation and abstract conceptualization modes of learning. They have particular strengths in inductive reasoning, formulating theory, and related activities such as modeling and concept building. As masters of induction, assimilators begin with various observations and logically arrive at a general statement explaining those observations. Like convergers, assimilators are less focused on people as sources and conveyers of knowledge and are more concerned with the development of (their own) ideas and concepts. Assimilators have little interest in action except possibly as an object of study. If the conceptualization of a situation doesn't seem to fit reality, the assimilator will expend great effort to modify or reexamine how that conceptualization was developed in the first place. In other words, assimilators feel more pressure to explain things than to solve problems and to act. Hence, the excess of assimilators is the tendency to build castles in the sky, ornately logical edifices that may have little to do with reality.

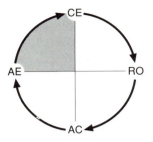

The Accommodative Learning Style

Accommodators prefer the active experimentation and concrete experience modes of learning. The strengths of accommodators are the opposite of the assimilators. They are oriented toward doing things, carrying out plans and tasks, and getting involved in new situations. They seek opportunity, take risks, and act! In situations where one must act quickly and adapt to fast-changing circumstances, the accommodator shines. If a theory or plan doesn't fit the situation, the accommodator is apt to throw it out. Accommodators move through life day by day through trial and error. They rely heavily on other people for information, ideas, and explanations, rather than on their own observational and analytic abilities. They feel comfortable being around people but may seem impatient or pushy to others.

The special idiosyncratic style of learning that each of us possesses is the result of many interacting factors. Evidence is now accruing that genetic inheritance has a role, along with early experience and education and even the relative development and use of different parts of the brain. As a result, each of us develops a style of learning that we prefer to use over all others. And we often hold this preferred style so strongly that it can dictate the types of problem situations that we select to tackle (and the ones we avoid) even to the extent of influencing the career paths we select. It seems also to affect our interpersonal relations, influencing our choice of friends and also our ability to cooperate with people who share our places of employment and other problematical situations.

Thus on both counts, the individual strengths and weaknesses issue and the interpersonal one, you need to be concerned that individual learning styles resist change. Your long-term ability to operate effectively in the real world as agricultural and natural resources managers can be strengthened through deliberate exposure to the features of learning styles, objective measurement of your own strengths and weaknesses, willingness to confront your weaknesses and excesses, *and* awareness of the preferred learning styles of others. There are techniques for dealing with all of these issues. It is possible to *improve* your learning style by seeking out situations and experiences that you might otherwise avoid. It is also possible to learn to work in groups consisting of people of diverse learning styles and not only to reduce sources of conflict, but also to take advantage of the diversity of strengths that these diverse styles offer.

Closely related to learning style is the notion of world view. Before turning to learning dimensions 3 and 4, the next section introduces this important concept, which will figure prominently in the rest of the book. The case is made that throughout the inquiry process to be presented, the world views of the people involved, including your own, must be identified and accommodated in any plans that finally emerge. One cannot state a problem or formulate an improvement adequately without taking into account the ways in which participants in the situation and process view the world.

Learning Style and World View

You may have noticed that in various arenas (for example, politics, environmental management, or foreign affairs) different people construe the same events in markedly different ways. It is profoundly important for you, both as a citizen and a practitioner-in-training, to be able to assess the status of such arguments. To what extent do disputes revolve around matters of substance? Do they involve conflicting ideological or theoretical positions regarding a generic issue? Is it conceivable that sometimes fights are rooted in *conflicting learning styles*?

This possibility may be attractive, or at least worth considering, if you have followed the argument in this chapter so far. It is suggested that because people approach inquiry into situations differently, the way in which knowledge is created and the substance of that knowledge are different.

The general point is that when people encounter concrete situations, they view, filter, sort out, and give meaning to their experience using characteristic mental frameworks. A particular mental framework is referred to as *Weltanschauung,* a German word only partially translatable into English as "world view." By some accounts, *Weltanschauungen* (the plural) consist of the experiences, feelings, emotions, attitudes, values, morals, beliefs, tastes, and personalities of individuals, as well as their patterns of reasoning and intelligence and their store of knowledge. The importance of the concept is signaled by the slogan "How we view the world [partly] determines how we act in the world." The capital letter **W** will be used throughout this text as a shorthand label for *Weltanschauung.*

A situation viewed by one person as a problem may be seen by another as an opportunity or challenge. One person's terrorist is another's freedom fighter. The **W** we hold will markedly influence the types of observations we make of a new experience and thus will also affect the information we collect. Our **W** will be particularly important as we use this information to attribute meaning to our experience, and as we create or recreate abstract models in our minds. Our **W** will certainly come out when we present ideas about or models of an experience. Thus our acts will also express our **W**s.

As suggested at the beginning of this section, the components of **W**s include (1) the experience of the individual carrying it and (2) the individual's preferred learning style. A person's *experience* consists of a mix of uniquely individual events and interactions felt in the past as well as social ones shared with many others. Social scientists talk about socialization or enculturation (near synonyms, for all practical purposes), the process whereby individual members of a given society acquire the basics of behavior that permit them to function as normal members. Language is the most obvious and basic cultural pattern we assimilate as infants. It is relevent to the discussion of **W**s precisely because language is a system of patterned sound *with meaning* and is simultaneously a device for communicating with others and a means for individuals *to classify their ongoing experience*. With regard to the last feature, in other words, language is a more or less ready-made framework for sorting out experience. We play Name That Tune every time something changes in our surroundings. If we cannot put a name on something and are thus unable to link it to our existing store of experience, we are *uneasy*. If, by the same token, we too readily identify a novel event as familiar and apply a convenient noun to it, we may have to live with the consequences of an inappropriate or maladaptive act.

Language is strongly linked to other social elements of **W**s, including our acquired sense of what is "good, true, and beautiful," or "moral" or "natural." "Moral" refers to beliefs leading to the imposition of meanings about the rightness of acts that affect other people. "Natural" concerns our ways of interpreting, or attributing meaning and value to, the interconnections between our selves and things, acts, and events in surrounding space and time.

While individuals are the carriers of **W**s, some of the experiential elements may be characteristic of groups, large and small. This makes sense in relation to the foregoing discussion of language; people who share a language thereby also share important elements of their **W**s. Moreover, other elements of national cultures and also the social systems of subgroups in society contribute to the **W**s carried by their members. These elements express themselves in every sphere of activity, including politics, communications, and interpersonal relations.

The **W**s of scientists are particularly relevant to agricultural and natural resources managers. These **W**s are often contained in or largely consist of what are called scientific paradigms. Paradigms are the meaningful body of knowledge currently used by a particular group of scientists to explain their observations. In other words, they are part of a group's framework for viewing natural, and sometimes also moral, order. We tend to vest science with a mantle of objectivity, as if somehow it were separate from other aspects of our culture and the way we learn. Scientists do too. Moreover, morality enters the picture in scientists' use of phrases such as "good science" (versus "bad science"). This is all part of our **W** in a technological society.

In fact, there are cases in which shifts of scientific paradigm have been resisted by some people precisely because the new knowledge was seen to challenge the moral order of society. The widely publicized debate on the teaching of evolution in the schools is a manifestation of such a controversial paradigm shift that has been going on for more than a century. A much broader and more profound issue than the theory of evolution is involved: people's view of humanity's *place in nature* and the premises for manipulating, as well as understanding, it.

Close to home are ongoing debates about our management or manipulation of nature, represented by ecosystems or recombinant DNA.

Individuals come to be members of nations, communities, ethnic groups, families, and professions — groups having characteristic **W**s — by chance (e.g., birth), assignment, or choice. By joining a group they also gather some elements of the group's **W**s. You should keep in mind, however, that in practice no two people on the face of the earth have identical **W**s. One important source of variability is the learning styles people come to prefer. Learning style is an element of an individual's overall **W** and influences how people develop the experiential elements as well.

Given that each of us develops a characteristic world view partly based on our preferred style of inquiry, it follows that we will tend to act in ways that reinforce that style rather than challenge it. As we age, this selective reinforcement encourages the conviction that our learning style, our own point of view, and the way in which they were generated are "right" and that all other learning styles, viewpoints, and associated processes of generation are "wrong," seriously flawed, or at least difficult to tolerate!

According to the philospher Hegel, this conviction becomes so strongly held that people may view those who take the opposing stance as their "deadliest enemy." Thus, the mere act of two people with contrasting learning styles joining together to tackle a problem situation may be loaded with tension and conflict. From the moment they start to identify the issues in the problem situation, their opposing world views will become apparent to each other and almost inevitably will result in conflict over the way the problem is even approached, unless one or both has learned how to work with people having differing styles. Note that the degree of incipient conflict depends on the strength of the contrast in learning style. A rule of thumb is that the bearers of opposing learning styles on the graph, specifically the converger–diverger and the assimilator–accommodator sets, will harbor the greatest potential discord because, according to the two dimensions of the model, they have little in common and tend to tackle problems in strikingly different ways.

Does this mean that people with differing learning styles and **W**s should not try to work with each other? Will they not, as Hegel suggested, be at each other's throats?

The answer to both questions is a qualified no. It is suggested that each party will have to learn how to approach team or group inquiry in light of the individual strengths each can bring to it. We all must learn how to appreciate other people's **W**s. Futhermore, whenever a number of people attempt to tackle the same problematic situation in a team or in other arenas, the various, and often conflicting, *Weltanschauungen* must be exposed so that they can be accommodated. At some point, improvements in complex problem situations will also involve confrontation of conflicting **W**s and their attempted resolution through the joint adoption of viewpoints that transcend the original differences. This is much easier said than done, of course. Attempts to resolve problems of the "us"-versus-"them" type by trying to change either or both protagonists' **W**s will rarely be successful. At the original level of debate, the differences in world view represent distinctly opposing positions. A transcendant **W**, by definition, is a proposal that breaks through to a viewpoint both parties can accept without abandoning or seriously compromising their original tenets. The procedures of the soft systems approach help develop transcendent **W**s. Other related issues are discussed in Chapter 8.

The Cycle of an Inquiry Process

An effective inquiry process requires using the competencies associated with all four learning styles. Theoretically, when confronting a situation, we need to cycle through from mode to mode. In the first phase, we involve ourselves fully, without bias, in a new experience and observe that experience from as many viewpoints as possible (diverging competencies). We then (re)create mental maps or models that give some meaning to our observations (assimilation competencies). Next, based on those conceptualizations, we develop plans and make decisions on possible courses of action (convergence competencies). Finally, we test this constructed meaning against the reality of the situation by implementing our plan and taking action (accommodative competencies). This final step, to take action, is intended to change the situation or to "improve" it. Whether or not it is successful, this results in a new or altered situation, which again demands investigation. Even if we don't change the situation, however, a new one will arise. So the process recurs as we are continually driven by an intrinsic need to explore our ever-changing environment by means of appreciating, observing, conceptually mapping, and doing.

Like all abstractions, the learning model simplifies complex reality. Learning is fraught with strong feelings, at once exciting and gratifying; yet often it is also filled with frustration, fear, and even anger. It involves us in a cycle of activities between finding out and thinking out, and between reflecting and taking action, and each of these activities seems to demand that we play confusingly different roles in the same situation.

At one moment, we are actors heavily immersed in the situation, and at another moment we are remote, analytical observers. At times we crave the facts of the matter, and at other times, we indulge our emotions, feelings, and values. You should be reassured to know that very few people are able to play all of these roles with equal ease.

At the heart of this learning model is the notion that effective learning involves alternating between conflicting ways of dealing with things. Like the armature in an electric motor chasing alternating magnetic poles, we seem to "flash" around between different modes of dealing with reality, driven by an internal tension generated between opposing modes. Unlike the armature, however, we are able to deny the drive to continue the entire cycle to completion and may choose instead to stick with one mode or another. In other words, we often resolve the tension associated with opposing ways of dealing with our world by allowing our preferred mode to dominate while suppressing or avoiding the others.

Other Dimensions of Inquiry

Before pursuing other learning style issues, it is pertinent to explore two additional dimensions of the learning process. These do not describe learning processes in the same way as the prehending and transforming dimensions. Rather, learning dimension 3 describes methodological choices involving a continuum of approaches to inquiry. Both seem to be important in affecting the ways by which we try to make sense out of and act to improve problematic situations as complicated and dynamic as those we encounter in agriculture and other areas of environmental management (or life in general). They also will bring us closer to understanding the place and role of science outside of the laboratory.

Learning Dimension 3: Methodology

As with the two learning dimensions discussed earlier in this chapter, this dimension, which is concerned with the *logic* of inquiry, is defined by two poles, reductionist separation and holistic integration. This can be thought of as producing the axis

$$\text{Reductionism} \longleftrightarrow \text{Holism}$$

As defining an additional dimension of learning, the two extremes or poles can be described in terms of the associated approaches to tackling a problematic situation. First we can break reality down into pieces and study each piece as a representation of the whole (reductionism). Alternatively, we can examine an issue as if it were an irreducible whole (holism). As with the prehending and transforming dimensions, we tend to adopt a preferred position on the reductionism–holism dimension, and this influences everything we undertake.

Assuming a reductionist stance affects how we approach concrete experience, reflective observation, abstract conceptualization, and active experimentation. How much of a situation we choose to explore, what we separate out to observe, the concepts we choose to develop, and the kind and extent of the experimentation we engage in will all refer to separate parts of the issue. The assumption that underlies the preference for reductionist logic is that if each small issue can be explained, or each small component problem solved, then the original situation will be understandable in its complexity and will be susceptible to overall improvement.

The holist methodological preference is the opposite of the reductionist approach. The assumption here is that any complex situation has certain aspects or properties that are immediately lost when it is broken up into its component parts. The holist stance assumes that no matter how much the individual parts are studied, the emergent properties possessed by the whole cannot be understood, nor can the original complex situation be improved, unless it is studied in its entirety.

Like the two primary dimensions of the learning cycle, the methodology dimension is a source of conflict and difference within the individual learner and publicly between people with disparate orientations. The often vituperous debates that have occurred between "reductionist scientists" and "systems people" reflect adherence to the two respective poles of the dimension, along with an attempt to resolve the conflict by suppressing the opposing mode. Take the debate between the organismic biologists and the ecologists. The "tree people" just say that the ecologists don't know anything about trees. The "forest people" say that the organismic types don't know the important things about trees. Just as this chapter argued earlier for more effective learning through the development of all four primary modes, it now proposes that developing both reductionist *and* holistic abilities will enhance your scientific exploration of new situations. Figure 2.4 is a simple illustration of the way the interrelationship among these three dimensions — prehension, transformation, and methodology — might be conceived.

The methodology dimension to inquiry is shown in Figure 2.5 as a spiral of

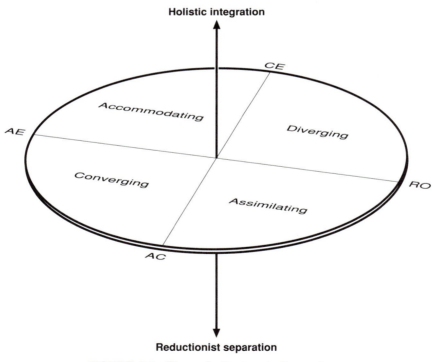

FIGURE 2.4 The reductionism–holism axis.

Holism

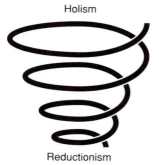

Reductionism

FIGURE 2.5 The learning spiral, describing shifting methodological orientations.

different kinds of learning or knowledge production, each able to handle varying amounts of complexity in different ways. As we work through complex situations, we may find it useful to move in either direction on the spiral. Sometimes it will be useful to reduce complexity for particular purposes, and at other times we must be able to deal with a more realistic slice of the whole of a situation.

At any level of complexity in the spiral, we cycle through the phases of learning by using techniques appropriate to the chosen methodology. As we cycle, we add to the relevance of and need for knowledge at other levels of complexity. Sometimes we will feel the need for more understanding of the context in which a particular problem rests, and at other times we will need an in-depth understanding of a few key factors.

Some will say that this conception of the way to handle complexity is not new. "It's common sense, isn't it?" Yet the basic and applied scientific disciplines tend to either overtly or covertly prescribe the amount of complexity they believe should be handled at one time. Thus the dominant inquiry methodologies and techniques learned by professionals-in-training, together with the corresponding **W**s, build in a sense of how much complexity should be tackled. Furthermore, disciplines and specialities are fairly narrow spans of knowledge that do not correspond very well to what is needed in the way of information and action in real-world situations. "I can't do that. It's not my speciality!" is a frequent refrain.

The practical consequences of this are recognized from time to time, especially in the proliferation of applied-science disciplines and specializations and also in pleas for the desirability of multidisciplinary approaches. The best approach to these difficulties is to train scientists and other professionals who are comfortable moving along the spiral as well as through complete cycles of prehension and transformation. In doing so, the artificial boundaries between disciplines are crossed readily.

It is also important to emphasize that this dimension does not involve merely differences in viewpoint or attitude. On the contrary, as with the other two dimensions, our effectiveness at integration and separation reflects the combination of our beliefs, skills, behaviors, attitudes and intellect: When we show preference toward one pole or the other on any of these dimensions, we are expressing our uniqueness, the result of our social learning, psychological makeup, and life experience, as well as our formal education.

Learning Dimension 4: Insight

Learning dimension 4 involves the insight process and describes yet another continuum:

$$\text{Rational} \longleftrightarrow \text{Intuitive}$$

This additional dimension included in the text's model of learning differs significantly from the other three. As noted earlier, recent brain research has shed light on the nature of intuition. Intuitive behavior seems to be guided by affective judgment (i.e., the apprehension process). In the learning cycle, our conceptual maps or comprehensions are a source of rational guidance regarding what we choose to apprehend—our perceptions. Our apprehensions, in turn, validate our comprehensions. And so on. While investigators are far from understanding how the intuitive process works, at least part of what happens involves our ability to size up the whole of a situation. Based on that *gestalt* (personal view of the whole), we feel that we should move in a certain direction or that the situation has a definite meaning for us. Hence, an intuition appears to be more of an emotion-driven, visual flash (the light bulb in the comic strip) than a rationally calculated thought. Accordingly, it is now widely recognized that knowledge is generated through both notional and intuitive processes.

Fritjof Capra (1975), a physicist and philosopher, represents a group of scientists who have recently recognized the two distinct ways of "knowing"—the rational and the intuitive. According to his argument, rational or intellectual knowledge can only reflect a representation of reality in a greatly reduced form. Intuitive knowledge, on the other hand, is associated with a complete view of the world. Capra emphasizes the complementarity of the two and contends that, just as it is possible to learn how to be a better synthetic, analytical, and rational learner, so too, in quite a different manner, is it possible to greatly develop intuitive ability. Most scientific researchers do recognize the importance of those intuitive insights that seem to enter consciousness in a flash and provide especially novel and creative ideas and meanings.

The Farmer Takes a Loan: A Scenario

This chapter now turns to a fictional scenario in order to illustrate the potential for improved understanding of a complex situation provided by the foregoing framework for learning and inquiry.

A dairy farmer feels that his way of life is threatened by the concrete reality of the declining income of his farm. As an *accommodator*, he approaches the situation by seeking advice from his neighbors. Taking their views into consideration, he has decided to increase the size of his milking herd. With the decision made, he asks his dairy extension agent to drop in to help him prepare an application for a bank loan.

As a technical specialist, the county agent has a marked *convergent* learning style. While they have a friendly mutual relationship, there is tension whenever the farmer and his agent discuss anything. On the one hand, the agent believes that the farmer is a fairly poor manager who could raise his income by reducing inefficiencies associated with obsolete equipment and improving questionable husbandry practices.

The farmer, on the other hand, thinks that the agent is really a puppet of the "theoretical" research scientists over at the College of Agriculture and is out of touch with the realities of farming. Sure enough, as soon as they meet to discuss the proposal, the agent and the farmer start to disagree. The agent believes that farm income could be significantly increased with the present herd if only the farmer would improve efficiency. The farmer believes that the agent's figures on per-head performance are based on unrealistic data from the experiment station. Thus he argues that increasing the scale of his operation is the only way to be successful. Needless to say, he does not accept the charge that he is an inefficient manager.

Here is a situation that is characterized by (1) different, though not necessarily conflicting, learning styles, (2) antagonistic **W**s, and (3) the clash of holistic and reductionist approaches to a situation.

The farmer views his farming and management practices as highly appropriate to his relatively relaxed and gratifying way of life. He is suspicious of adding more electronic gadgetry and depending on scientific models. The truth is that he enjoys pampering his cows! He dislikes measuring things objectively and certainly abhors keeping physical and financial records. He trusts his own judgment that his cows are all performing optimally, as well as his sense of how much he should feed each cow (and of course he has his favorites)!

The agent believes that his view of the situation is objective and knows that the farm could produce more and be more efficient if the standard of management were improved. He knows other farmers who do better than this client with the same resources. He resents the way the farmer refuses to accept recommendations to keep better records as a basis for more effective decision making. Yet he also knows that ultimately his job depends on favorable reports from his clients!

The farmer's **W** has farming as an all-encompassing way of life. His farm is simultaneously a household and a business that provides the satisfaction of independence and fosters family values. Contrastingly, the agent's **W** is that the farm is a production enterprise that should be operated efficiently so that its performance can be optimized. Both men, however, share the **W** that farming dairy cows is a really worthwhile thing to do.

Eventually, the farmer goes ahead with the loan application to purchase additional cows. At the same time, the farmer also accepts the agent's advice that he needs to improve record keeping and maybe change some farming practices. He realizes that he has to gain the support of his agent, who is friends with the banker.

Note that the accommodator's (farmer's) learning style does not actually oppose the converger's (extension agent's), as both share a preference for active experimentation over reflection. However, the farmer trusts his subjective experience over the objective scientific data preferred by the agent.

In the next act of this small drama, the farmer makes an appointment to see his banker to discuss prospects for a loan based on the case prepared with the assistance of the extension agent. Here lies the potential for yet another clash of **W**s, for the bank manager has a *divergent* learning style. To her, the farmer is a potential source of revenue to be gained by investing bank funds deposited by other customers. Her major goal is to insure that the loans are secured and will return a profit to the stockholders of the bank. To do this and also justify her decision to her bank directors, she requires that loan applicants provide a mountain of data regarding the overall health of the enterprise, their creditworthiness, and their competence.

In addition, she has known this family for years, as they have banked with her through good times and bad; yet her professional judgment in this case, as always, must reflect her view of the loan as an investment. Her **W** of the situation does not depend on her feelings for dairying or even for farming per se, but it is very much influenced by her affection for members of this particular family and by her overall commitment to the spirit of free enterprise.

Aware of her own ambivalence and knowing that the bank's economist is on vacation, she calls an agricultural economist friend at the university for his views on the future of dairying. Enter the fourth actor, who is an *assimilating* academic! As it happens, this professor's research lies in the field of stochastic simulation models. His professional world is one of mathematics, probability, and computers.

After 30 minutes or so on the telephone, the banker is not sure that she is any better off for having made the call. Her economist friend has told her that, on the one hand, with a given set of assumptions, his model predicts a fairly rosy future for dairying. Yet, if the values for a couple of unpredictable yet key parameters are changed, the model suggests that dairying is a threatened agricultural enterprise and that pig production would be a much better option. But then again, the economist had to admit that he was quite unhappy about the model itself and the nature of some of his equations. Under these circumstances, he hesitated to speculate on which way the industry would go or which enterprise would be more attractive.

Privately, the professor believes that most farmers have only a rudimentary understanding of the economic forces to which they are subjected. His **W** reflects his belief that any farmer who is not aware of the basic economic principles under which he operates should certainly not be encouraged in any way to stay in farming. As an assimilator, his major concern remains the construction of a satisfying model of agriculture. The real world is only important to him as a source of information to help that construction and as a test bed to check its output. The economist cannot comprehend the farmer's apparent determination to stay in dairy farming and especially to go into considerable debt to buy more unprofitable cows.

From the data provided by the farmer at the appointed meeting, it was clear to the bank manager that the projected new income generated by the additional cows would not even service the loan, let alone provide the extra farm revenue sought. Under the conditions and terms prevailing at her bank, the banker was not in a position to lend the money. From her objective posture, the problem was solved, or at least the decision not to approve a loan to support the purchase of more cows could clearly be made. Yet the strengths of the divergent learner are in identifying a broad range of alternatives through activities equivalent to brainstorming.

As a result of such activity, the farmer left the appointment with a whole range of options to examine. With his almost overwhelming love of dairy farming, however, he secretly doubted that he would be sufficiently motivated to examine objectively any of the options that excluded his involvement with cows. But he was also responsible enough to see the poor financial position into which he was slipping. Already he had asked his wife, a *diverger* like the banker, to gather information on the various alternative strategies that the banker had drawn up during their conversation. His wife had seemed to enjoy doing that kind of thing, and she had developed an intuitive grasp of the best way to go. Anyway, with the milking to be done twice a day, he doubted that he would have the time to do it himself! Procrastination won the day; the problem was left unresolved, at least temporarily.

In this story, conflict was minimized by avoiding interactions between people with opposing learning styles. Yet it clearly shows how various actors in the same situation view it from quite distinct perspectives, with each proceeding in quite different ways to address the issues, reflecting fundamental dissimilarities in their *Weltanschauungen*, learning styles, and other capabilities.

It is not difficult to imagine what might have happened if all four actors had been asked to work together directly on the farmer's problem. Evidence of the tensions between opposing learning styles and **W**s would have been there for all to see and attempt to accommodate.

Even though this was only a hypothetical example, it illustrates the difficulties of resolving even a seemingly simple problem when more than one actor is involved. Each of the viewpoints presented was a perfectly legitimate **W**, yet the actors saw the problem in ways that reflected their disparate stakes in the situation and their preferred approaches to learning about the situation and making decisions.

Although this situation might be perceived as private, the farmer's actions and some of their implications were public, in that they involved other actors and could spread beyond the farm gate. If the farmer had to change any key resource factors significantly, there could also have been implications for the physical state of the farm. Thus, more cows on the land might cause overgrazing, soil compaction, effluent disposal problems, or some other negative effect.

Moreover, in the real world, many other "actors" are directly or indirectly involved the moment a farmer considers changing the way he or she does things.

In this sense, seemingly simple problems very rarely are! Rather, they are exceedingly complex, and this complexity is a major reason why farmers (and others) encounter difficulty in our society.

Where the boundary of a problem is perceived to be, and thus at what level it should be tackled, depend on the **W** of the observer and the method chosen to study the problem. In one sense, the farmer in the example had the relatively simple problem of borrowing money to enable him to increase his dairy herd. Yet, from another viewpoint, his "real" problem was whether he had the ability to survive as a farmer. Thus, a seemingly simple issue about cow numbers becomes a more complex issue of the sustainability of his life-style.

At yet another level of complexity, the problem was merely another expression of the national family farmer problem. Similarly, the original issue of increasing cow numbers could also be examined in terms of its impact on the physical state of the entire farm. The adoption of more intensive methods of production might also attract the attention of animal welfare advocates which, in turn, escalates a simple cow number problem into the complex arena of ethics, environmental quality, animal rights, and the overall future of intensive livestock production.

Raising these sorts of questions emphasizes the importance of the third dimension (holism–reduction) in understanding differences in learning styles and **W**s. Almost since the birth of the scientific method, intellectuals have debated the relative merits of studying complicated phenomena by pulling them apart versus grappling with them "whole" in all their complexity. Judging by the passion of even current arguments and the scornful descriptions that each camp applies to the other, it can be concluded that holistic and reductionist methods are serious competitors. As with the modes on the two primary learning dimensions, adherents of any extreme position are convinced of their own superiority and the depth of error of their antagonists. The stance adopted in this book is that both methodological stances and associated viewpoints are equally valid and useful. Separation and conflict are the errors. As suggested earlier, both can be correct if it is understood that the two logical modes share a dialectical relationship. The well-rounded learner and effective practitioner alternates between the two, between the needs to separate and to integrate. Remember that the same argument prevailed in the construction of the model of learning.

The Spruce Budworm Case

The problem of what to do about the budworm pest in the spruce forests in eastern Canada provides another illustration of how different **W**s, learning styles, and methodological preferences give rise to different technical approaches and conflict (Miller, 1985). One party to the dispute saw the problem as an uncontrolled insect population, which was killing the spruce trees. Hence, the solution would be to bring the budworm epidemic under control. This group focused on the pest and pursued the development of a technical intervention, the design of chemical treatments, to kill it.

A second group in the controversy saw the situation quite differently. They focused, not on the pest, but on the unstable ecosystem. Their strategy for improvement was to design a more resilient ecosystem based on diversifying the "brittle" forest monoculture of spruce fir. They developed computer simulation models to predict what would happen if changes were made in any part of the ecosystem. Using these models, they designed and tested various integrated management strategies, including restructuring the entire forest monoculture, as well as integrated biological, physical, and chemical control techniques. Their ultimate goal was to manage the tree/pest interaction in ways that would avoid explosive outbreaks of the budworm pest of the sort that tended to occur when chemical controls alone were used.

A third group was unconcerned about how best to kill the pest or even how to develop more stable relationships between the insect and the softwood forest. For them, the issue was how to balance the needs of society for forest resources with the need for a sustainable exploitation regime. In their view, the budworm was only a problem to a tiny minority of society, the overspecialized pulp and paper industry. They emphasized the need to "get fresh ideas about how to resuscitate the forest economy given that it is shockingly clear that the spray program has failed as an effective management tool."

Here is yet another illustration of how people with different **W**s and learning preferences not only disagree with each other about solutions to problems, but also about what constitutes the problem in the first place. Under the circumstances of such vastly different *Weltanschauungen*, it becomes understandable why problematic situations as complex as the spruce budworm have taken a long time to resolve.

It should be emphasized that the illustrations presented so far, the fictional dairy farmer and the actual spruce budworm pest, are not isolated or unique. Contrasting learning styles, **W**s, and inquiry methods are common in real-world situations.

The Sustainable Agriculture Debate

Agriculturists, professionals from related fields, and humanists have increasingly been drawn into conflict over current agricultural practices and policies and the future sustainability of agriculture. Douglas (1984) has identified four alternative perspectives that are characteristic of participants in this debate about sustainable agriculture in a world of change. These perspectives also illustrate differences of **W** and methodology.

The first perspective equates sustainability with the ability of agriculture to continue to supply enough food to go around. Feeding the hungry around the world is merely a matter of supplying sufficient resources and assuring efficiency in production and distribution. What impacts such functions have on the physical and socioeconomic environments can then be assessed in terms of benefits, costs, tradeoffs, and so on.

A second group rejects this view as irresponsible in the face of the many detrimental effects of unsound farming practices that have already been observed. Instead, they advocate a more ecologically balanced agriculture with a focus on both permanence and productivity.

Yet a third group rejects both the emphasis on the biophysical aspects of agricultural ecosystems and the economic bias. The real issue for them is the viability and security of the rural way of life and the mores and morality of the agrarian ethic.

According to a fourth perspective, the problems of contemporary agriculture transcend issues of technology, economics, ecological integrity, and even the survival of rural communities. For them, the real issue is the way humanity lives, what foods are eaten, what natural resources are used, and the ethic underlying all of this. They call for a radical reappraisal of our entire attitude toward the human condition and humanity's relationships with the rest of nature. A particular concern is the rights of all living organisms to achieve their biological purposes.

Once again, this account of four extremely complex viewpoints in the debate on agricultural sustainability is necessarily compressed and simplified. While those who hold the most extreme views on the future of agriculture and resources use are few, many real conflicts have arisen over related issues, and people with somewhat less extreme but equally adamant postures have participated and prevailed. There is every reason to believe that professionals engaged in agriculture and natural resource management will increasingly face conflicting and complex situations such as the foregoing and will benefit from the use of the concepts of learning and the associated competencies discussed in this chapter.

It has been implied throughout this chapter that professionals in the food, agriculture, and natural resources field spend a good deal of their time solving problems and improving situations. Because of the organization of the industry and the nature of most of the available jobs, professionals carry out these activities for and in cooperation with others. Typically, this professional activity takes the form of *intervention* into ongoing organizational processes and even into people's lives. Therefore, the next section attempts to highlight some of the features of and choices involved in the interventionist role.

Intervention Styles

Acquiring new learning competencies also implies a new outlook on how professionals intervene in problematic situations in food, agriculture, and natural resources management. The established style of intervention, in which "experts" tell others what they ought to do, does not serve our society very well. There are alternative styles of intervention that follow directly from the principles of experiential learning presented to you already.

The challenge for the practitioner is to control the quality of his or her intervention in the lives of those to be served. *Intervention means interference.* According to John Heron (1975), interference may be *invited* or *uninvited.* Heron reserves the word *intervention* for the invited kind of interference. In turn, intervention may be of two broad types, *authoritative* and *facilitative.*

Simply stated, authoritative means "I'm the boss. I know it all. Forget about that stuff. Listen to me. Do what I say." As this book argues subsequently, authoritative interventions are not all bad. They are, however, abused in various ways, especially when they are uninvited.

A deduction from the principles of experiential learning is that successful directed learning promoted by a facilitator strives to simulate natural learning, what people do with their experience of the real world. The typical style of intervention by rural development workers overseas is *uninvited* and *authoritative.* This style is also common in education.

According to Heron, there are three basic types of authoritative interventions:

1. *Prescriptive* interventions explicitly direct the behavior of the client. An extreme and possibly abusive form would be to direct behavior that is beyond the bounds of the practitioner–client interaction, such as a doctor directing aspects of a patient's life that have no health implications.

Specific prescriptive activities include giving directions and advice, rendering judgment and criticism, and rating performance. For example, the U.S. Soil Conservation Service offers a soil conservation plan to any farmer who requests it. Although the intervention is invited, it is nevertheless prescriptive, perhaps of necessity.

2. *Informative* interventions seek to impart new knowledge to the client. An abuse would be to do this in a manipulative way for purposes unrelated to the well-being of the client.

Specific activities include lecturing and instructing, conveying information, and interpreting unfamiliar material. The traditional role of cooperative extension includes conducting demonstrations of new technology through short courses and demonstration plots, providing recommendations on crop varieties and agricultural chemicals, distributing pamphlets, and otherwise putting research findings in usable forms for farmers and other clientele.

3. *Confronting* interventions directly challenge the behavior, beliefs, and attitudes of the client. Typical abuses involve attempts to abrogate the client's autonomy.

Specific activities include arguing and questioning to provide direct and immediate feedback. For example, not all proposed changes appear to be realistic at first. Confronting interventions may provide the challenge clients need to consider "radical" changes seriously. Increasingly, modern large-scale growers must retain legal counsel for advice on problems involving farm labor due to very significant changes in the social and regulatory environment regarding such matters as the labor market, housing standards, workman's compensation, sanitation, and unionization. Good lawyers sometimes argue with their clients, especially in rapidly developing and potentially inflammatory situations.

"So what's the matter with authoritative interventions?" you ask. "It happens all the time." Nothing, if you take examples and instances in isolation. At the cost of sounding authoritative, however, it should be noted that this isolation is part of the problem, that alternative facilitative interventions are not part of the toolkit of many practitioners. So the point is not to avoid authoritative interventions, but to avoid abuses and balance authoritative interventions with facilitative ones, even to allow facilitation to predominate.

Heron goes on to describe three types of facilitative interventions:

4. *Cathartic* interventions seek to enable the client to relieve a painful repressed emotion.

Specific activities include encouraging strong feelings, for example, laughter, crying, rage. It will be recalled that some of our most profound learning takes place in the context of strong feelings. Local and regional farmer protests of loan foreclosures undoubtedly serve as learning experiences when participants join to support one another or otherwise search for ways out of their shared dilemma. In these instances, the interveners are agricultural leaders.

5. *Catalytic* interventions aim to help the client to learn and develop by self-direction and self-discovery in the context of the practitioner–client situation and beyond.

Specific activities include fostering reflection by example, encouraging purposeful problem solving, and eliciting information. Farmer organizations such as the National Grange, Farm Bureau, and numerous commodity groups, sometimes in concert with cooperative extension, sponsor regular meetings and other activities that serve as arenas for problem sharing and discussions of remedies, including political ones. Common functions are to share information, raise awareness of problems and actions, and build a consensus on important initiatives.

6. *Supportive* interventions affirm the worth and value of the client.

Specific activities include giving approval, confirming others' judgments, and validating the importance of others. Competitions and awards for technical accomplishment, as well as for leadership, community spirit, and staying power, have been common features of American rural communities.

In summary, instead of telling clients what to do, the intervener might do better to help clients figure out what they should do themselves, how to find out on their own, and how they must determine their own standards of performance. If authoritative interventions are necessary, then they should be invited (rather than imposed), and their scope should be narrowly defined. Soil conservation plans, which were referred to previously in connection with prescriptive interventions, seem to conform to this pattern because they are invited and of narrow scope. Likewise, in the educational arena, it seems perfectly legitimate, and possibly necessary, for a course instructor to advise students of the need to come to class if they are to participate in certain facilitative learning experiences. Similarly, there is no intention to deny the importance of gaining the scientific and technical competencies that are associated with the expert's role. Indeed, the remainder of this chapter and Chapter 3 discuss how science and technology can be incorporated into a much wider view of the world, involving both the principles of learning and systems thinking. Here the issue has been how expertise is applied in the process of intervening in the lives of people. The discussion of intervention styles will resume in Chapter 7.

The final section of this chapter discusses two inquiry approaches, basic science and applied science. These methodologies share similar premises, but the objectives and procedures are different. The section begins by discussing the meaning of such terms as *methodology* and *technique* in science. It then turns to the premises, objectives, and procedures of basic and applied science.

Methodology and Technique

It is important to distinguish between the terms *methodology* and *technique* (Checkland, 1981:161–2; Rudner, 1966:4–5; Morren, 1986:13–14). Methodology is the "logic of justification" (Rudner, 1966:5), a set of concepts, objectives, rules, and principles that guides the direction of an inquiry process from start to finish. In this book, methodologies are described in terms of phases of inquiry. In contrast, techniques are specific nuts-and-bolts procedures to use during a given phase of an investigation in pursuit of particular methodological objectives.

At least four major methodologies are used in the agricultural, food, and natural resource sciences: basic science, applied science, hard systems, and soft systems. The choice of which methodology to use is influenced by the objectives to be achieved, the context of the problem, and the abilities (related surely to the education received) of the problem solver. Consequently, the nature of a problem should prescribe which methodology to use, rather than the preferred methodology or competencies of the problem solver picking the problem!

Due to the demands of various situations, a number of different kinds of methodologies for inquiry come to be applied. Each tends to entail rather specific procedures. The path used to pursue inquiry using a particular methodology is fundamentally different from the path for any other. Each methodology has its own assumptions, procedures, and techniques. The kind and applicability of the results produced also vary. As shall be shown subsequently, however, some methodologies have received more explicit treatment than others.

The Premises of the Scientific Method

Our prevailing approaches to problem solving arose in the scientific revolution of the seventeenth century. Our educational system, as well as popular culture, are heavily influenced by scientific paradigms. The accepted standard for creating knowledge in modern science still relies on what is known as the "scientific method."

In the course of formal education, the premises or underlying assumptions of the scientific method are often more implicitly conveyed than explicitly defined. In fact, much would be gained if higher education did a better job of detailing the process of inquiry that is claimed to adhere rigorously to the premises of the scientific method. The following is a brief description of those premises, along with a presentation of two inquiry methodologies based on them: basic science inquiry and applied science inquiry.

Due to the influence of Greek thought and the writings of scientists of the age of Newton, many of today's disciplines view science inquiry as a process by which publicly testable knowledge of the world is acquired (Checkland, 1981:50). This process is characterized by applying rational thinking to experience, such as is derived from observation of deliberately designed experiments. The ultimate aim of basic science inquiry is to express concisely the laws that govern (or describe) the regularities of the universe. Where possible, these laws are expressed mathematically.

The phases of the science inquiry process commonly used are built on the principles of reductionism, replicability, and refutation. At the core of science inquiry is the ability to describe and conduct experimental happenings representing tests of hypotheses that other members of the scientific community can replicate (or repeat). "We reduce the complexity of the variety of the world in experiments whose results are validated by their repeatability, and we may build knowledge by the refutation of hypotheses" (Checkland, 1981:51). Yet in this attempt to simplify the real world, it is recognized that there are levels of complexity. At any given level there are characteristic properties that are irreducible. The trick is to avoid reducing something that should not be reduced without losing a vital aspect of the property under investigation. Checkland reminds us that the crucial problem that those using the scientific method face is its inability to cope with complexity (1981:59).

Because a major assumption of the scientific method is that knowledge is built by refuting hypotheses and thereby eliminating alternative explanations, there is general understanding that all knowledge is provisional.

Depending on the purpose of the inquiry, one of two basic approaches to problem solving is used. If the purpose is to answer the question "Why is this so?" then *basic* science inquiry is used. The focus of the problem is to understand a phenomenon or phenomena better. The objective of inquiry is not to apply what is found to an actual situation so that it is improved, but rather to discover new knowledge for its own sake. Inquiry for the purpose of application is left to others.

The second major inquiry approach is called *applied* research, applied science inquiry, or the technology development approach (OECD, 1976). As discussed subsequently, the process of applied science inquiry requires the investigator to go beyond the question "Why is it so?" to ask "How can it be done?" or "How can it be used in a particular context?" In some cases, those doing applied research discover new knowledge, or create new technology, or use existing technology and modify it to meet the conditions of a new context. In any case, a major objective of inquiry is to apply what is known to an actual situation.

Basic Science Inquiry

Figure 2.6 outlines the process of basic science inquiry. This methodology proceeds by analyzing the *problem* that initiates inquiry, conceptually reducing it to a systematic *collection of facts* that are initially thought to be relevant, and *creating hypotheses* suggested by the relevant facts. From there, the hypotheses are *tested experimentally* several times, and the competing hypotheses refuted (Northrup, 1983). This inquiry methodology relies on reducing the problem to testable proportions, experimentally testing the resulting hypothesis quantitatively, and reducing as many variables as possible. The ultimate output is deduced to statements in the form of "This is why," "These are the cause and effect," or "This is how things interact or are associated."

How hypotheses come into being is the subject of an interesting debate, which illuminates the overall inquiry process. The philosophers of science Morris Cohen and Ernest Nagel (1934) suggest that after initially understanding the problem, the real emphasis is to be placed on the proposal of hypotheses. Similarly, Fischer and Ottoson (1966) place "disciplined and directed thought" at the center of scientific inquiry, making hypotheses instruments for controlling the inquiry process. Data to be collected, therefore, are suggested by the hypotheses made.

Larabee suggests that someone who is in the process of generating hypotheses should "prepare — strive — wait" for the psychological leap made possible by his or her illuminating flash of insight (cited in Northrop, 1983). Thomas Kuhn (1962) also advances *insight* (learning dimension 4) as the factor that separates

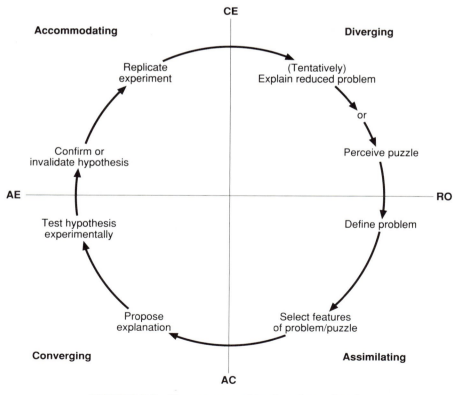

FIGURE 2.6 The process of basic science inquiry.

"problem solving science" from a revolutionary hypothesis. As indicated earlier, scholars recognize that the intuitive functions as well as systematic thought processes are at work in the use of the scientific method.

Northrop (1983) reminds us that the "most difficult portion of any inquiry is its initiation." Scientific inquiry can only begin when some observation is not adequately explained by traditional beliefs or when the observation is found to run counter to prevailing theories; that is, a *scientific* problem is defined. Northrop provides a taxonomy of scientific problems: (1) those of logical consistency, (2) those of issues of fact, and (3) problems of value. The first has to do with the relationship between theory and methodology, where it is understood that theory refers to the general laws of science. Here, resolution of the problem consists of deductively working out the consequences of a given set of assumptions. The second kind of problem has to do with the relations between methods and techniques, the validity of empirical evidence, and its relevance to the problem at hand. Finally, Northrop suggests that problems of value reflect questions of *what ought to be, compared with what is* — the quintessential problem of applied research. (Note that the scientific problems discussed in the foregoing should not be confused with the problems and problematic situations perceived by people in the learning and adaptational sense.)

Some authorities feel that the scientific method is at its weakest in the problem-definition stage of inquiry. Only when the problem is perceived and analyzed are the facts that are thought to be relevant defined. Yet a widely accepted standard set of rules and techniques for this phase seems to be missing. Northrop (1983) provides a thorough discussion of these issues. He suggests that the analysis is accomplished by inductively tracing the problem that began the inquiry to a "reduced factual situation." With the characteristics of the problem and its context revealed, the relevant facts are then used to postulate hypotheses.

While this procedure might be possible if the nature of the problem is one of either logical consistency or issues of fact, it is not adequate to deal with problems of value, although some people have tried. This is *the* problem of applied research. Consequently, problems of value tend to be converted to problems of fact and then reduced. As a result, the value element falls by the wayside, for it is assumed that true statements are answers to questions of fact or that truth is the ultimate value. This assumption, and the resulting process of inquiry, is fine under some circumstances, but remembering this assumption also helps to explain what frequently happens when scientific experts appear before the public and tie themselves in knots — and the ripe tomatoes fly!

Once the problem has been reduced to some factual basis, and in order for the analyst to describe what is observed, the analyst must draw on some previously derived knowledge that has been accepted as explanatory and meaningful. Kuhn uses the term *paradigm* to refer to the existing explanatory and meaningful body of knowledge scientists carry to explain observed phenomena. Therefore, rather than deriving the explanation from the thing observed, the person derives the explanation by calling up existing conceptual frameworks and fitting what is observed into that existing framework. Note that in this context, Kuhn's notion of paradigm closely corresponds to the notion of *Weltanschauung* previously discussed. Indeed, he explicitly refers to scientific revolutions as "changes of worldview" (1962).

Hypotheses are suggested possible solutions to or explanations of the problem at hand. Whether these hypotheses are true or not is the purpose of the inquiry. The marks of good hypotheses are that they must be plausible, that is, deductively consistent (i.e., with theory); they must be independently testable; and they must be simple and concise. By suggesting possible solutions, they are thought to press on the boundaries of knowledge. Their explanatory power must be relevant to the original purpose, that of solving a particular problem with adequate scope and depth. They should account for all variables that have come to light through the reduction of the problem and exclude those that are felt to be trivial or irrelevant. This is a creative process, reflecting the perceptual power of the analyst and requiring a full understanding of the problem. Hypothesis formulation is also understood to require persistence, for all hypotheses are considered tentative and subject to challenge. Cohen and Nagel (1934) remind us that this methodology differs radically from others by encouraging and developing the utmost possible doubt, so that what remains is always supported by the best available evidence.

Interrelationships between factors, especially if they cross disciplinary boundaries, are hard to deal with when using this kind of inquiry process. The fundamental assumptions and the techniques employed at each stage will tend to move people toward problem areas of which they have prior knowledge and away from trying to deal with complexity. It is not accidental that basic science flourishes in ivory-tower environments where leads can be pursued unconstrained by demands for immediate and practical results, and researchers feel free to find puzzles to solve. This kind of environment is nevertheless viewed as a necessity by many serious applied research institutions in academia, government, and industry because it is recognized that basic science provides the feedstock for applied science and technology.

The Job Market for Basic Science Researchers

The job market for basic research scientists remains so strong that the United States is not able to fill all openings from the ranks of native-born Americans. According to the U.S.D.A. statistics (Coulter et al., 1986), there are expanding career opportunities in research, engineering, and technical positions where the use of the basic science inquiry will be bedrock to the profession and to the nature of the problems that the scientists are asked to tackle. The strongest employment opportunities will be for persons having doctoral degrees or postdoctoral experience in molecular genetics, biochemistry, food science, food engineering, nutrition, biochemistry, and soil science. In addition, more than 800 technical job openings are projected annually for foresters and natural resource conservationists, an increasing proportion of which are likely to be in the private sector. For the next decade, there will be more jobs than trained graduates. It can only be hoped that higher education will train those aspiring to such careers well and rigorously in the premises and processes involved in the use of basic science inquiry. Such a rigorous education is not available in many institutions.

Applied Science Inquiry

Figure 2.7 presents the phases of investigation involved in applied science or technology development (Bawden and Macadam, 1983). This approach to problem solving is similar to basic science inquiry, in that it shares the same premises of reductionism, repeatability, and refutation. The differences lie in the nature of the problems it tackles and in some phases of its inquiry procedure.

Whereas the basic science approach tackles "Why is it so?" questions, the technology development approach tackles questions like "What is to be done?" and "What new thing and/or process can be created or redesigned to meet the requirements of this particular situation?" The research objective is to apply what is learned to altering critical components of a real situation.

This is the approach of most technical advisory services. The actual inquiry process used may vary slightly, depending on whether the technical solution being sought is an economic one; a restructuring of a co-op unit; the development of new machinery; or the design of an improved farm, soil, crop, or animal management practice. The typical objective is to optimize the factors under investigation.

There are numerous examples of the results of such applied science inquiry. Animal production benefits from technology development approaches, particularly in the ones of monitoring and controlled management and of improved production through breeding and hormonal manipulation. For example, water-flow meters, feed-weighing devices, variable-speed ventilating fans, and nutritional intake monitors are all vital to optimizing egg production. The development of bovine somatotropin (BST), a genetically engineered growth hormone, promises to increase milk production of individual cows by as much as 13.5 percent.

In plant agriculture, mechanical transplanters rapidly transfer seedlings from a flat to the field. New computer-controlled X-ray scanning systems count oranges, sort them by size, and check them for disease and injury. Automated harvesting and other "smart" food-processing and farm equipment may one day take over such chores as killing weeds and harvesting and inspecting crops.

In some instances, a new tool or process was developed because people expressed concerns, which led to an inquiry process. In other cases, inquiries were driven by technological developments in other fields, disciplinary concerns, or just the feeling on an investigator's part that "there must be some use for this thing out there."

How will new technologies affect employment opportunities? Will mechanization reduce the number of workers in any industry over time? What are the real costs and benefits of these technologies, and how are they to be measured? Is high-tech mechanization necessary to keep U.S. agriculture competitive in

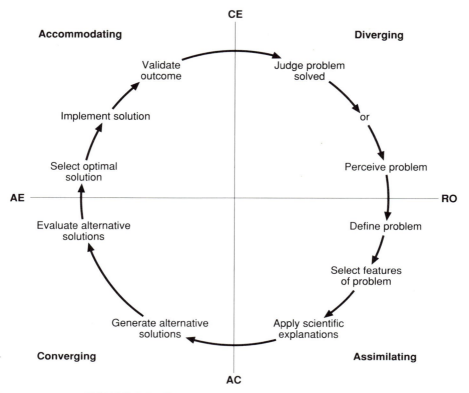

FIGURE 2.7 The process of applied science inquiry.

world markets, even to prevent whole commodity sectors from moving to other countries?

A number of studies have attempted to address these issues. Indeed, according to Vandermeer (1986), the tomato harvester is one of the best-documented cases in the history of agricultural mechanization. These questions actually began to be raised in the early 1970s, specifically about the mechanical tomato harvester (Schmitz and Seckler, 1970), and were widely publicized in Jim Hightower's *Hard Tomatoes, Hard Times* (1973). Hightower claimed that the tomato harvester eliminated farm workers and put Midwestern tomato growers out of business. Patrick Madden and Irene Johnston (1984), economists at Pennsylvania State University, found that the harvester actually saved jobs by revitalizing California's tomato industry. Without the harvester, they say, that state's tomatoes would have been unable to compete with imports (see also Raeburn, 1985). In another study, Martin and Olmstead (1985) of the University of California at Davis found that fewer jobs have been lost to mechanization than have been created by the resulting expansion of the industry. They estimated that 192,000 farm workers were employed in California in 1960 and that, despite increasing mechanization in the state, the number of farm workers reached 224,000 in

1980. Nevertheless, tomato "factory farms" exemplify the economic difficulties that large-scale corporate agriculture has faced (Kramer, 1987). This sector of agriculture might not exist at all without a whole new class of technology development.

In January 1979, the University of California at Davis was sued by the California Rural Legal Assistance on behalf of a consortium of farm workers' rights organizations in Alameda County Superior Court (Kendrick, 1984). The lawsuit was subsequently joined by small farmers and consumer groups. They argued that publicly funded institutions should not conduct research that would displace workers or favor large corporate farms. In November 1987, Judge Raymond L. Marsh's ruling in effect identified a generic weakness in experiment-station-style domestic and international applied research — the failure to recognize that the "application" itself should be the subject of empirical inquiry. Major dilemmas include the impacts of these changes on wider systems, whether innovations fill a real need, and whether they give rise to more socioeconomic or environmental problems than they solve.

The court's ruling, which did not apply to farm workers, said that the university had failed to take into account the impacts of its technology development research on small-scale farmers. The judge directed that the University of California set up a review process to insure that agricultural research paid for by federal funds would primarily benefit small-scale farmers (Sun, 1987).

While the applied science approach certainly has increased agricultural productivity, it still is inadequate for some situations because of compartmentalization and limitations of the scope of inquiry. Due to the process of reduction that occurs during the problem identification stage, not enough of the system is considered. It is only recently that technologists of different disciplinary backgrounds have started to work together in multidisciplinary teams to provide less reduced results. However, because the inquiry procedures used, particularly during the problem-identification stage, are based on the premises of the scientific method, teams have a hard time working together. Each specialist reduces the problem to focus on familiar components of the situation. Pulling all these bits together has worked only on rare occasions. As this chapter indicated earlier, multidisciplinary enterprises require at least one team member who has a broad, holistic perspective to pull everything together. Even this is no guarantee that the problems described will be avoided and that appropriate solutions or improvements will result.

The Job Market for Applied Science Professionals

According to the U.S. Department of Agriculture (Coulter et al., 1986), in the next decade there will be more job openings in the applied sciences than there are projected trained graduates, with approximately two-thirds requiring graduate degrees. Applied science professionals and technicians of many kinds will use applied science inquiry as one of their major methodological approaches, but they will also be involved in situations demanding systems thinking and the hard and soft systems approaches introduced in the next chapter.

REFERENCES

Bawden, Richard, and R. D. Macadam. "Problem Solving in Agricultural Systems — Innovations at Hawkesbury." Paper presented at the Biennial Conference of Principals/Directors of Agricultural and Horticultural Colleges of the Southwest Pacific. Richmond, New South Wales, Australia: Hawkesbury Agricultural College, 1983.

Capra, F. *The Tao of Physics*. New York: Random House, Fontana Collins, 1975.

Checkland, Peter. *Systems Thinking, Systems Practice*. New York: Wiley, 1981.

Cohen, M., and E. Nagel. *An Introduction to Logic and Scientific Method*. New York: Harcourt Brace, 1934.

Coulter, K. J., M. Stanton, and A. Goecker. *Employment Opportunities for College Graduates in the Food and Agricultural Sciences: Agriculture, Natural Resources and Veterinary Medicine*. Washington, D.C.: U.S. Department of Agriculture, Higher Education Programs, 1986.

Dewey, John. *How We Think*. New York: Heath, 1910.

Douglas, Gordon K., ed. *Agricultural Sustainability in a Changing World Order*. Boulder: Westview, 1984.

Edwards, Betty. *Drawing on the Right Side of the Brain*. Los Angeles: J. P. Tarcher, 1979.

Fischer, L., and H. Ottoson. "Hypotheses — Guides for Inquiry." In *Methods for Land Economics Research*, Gibson, Hildreth and Wunderlick, eds. Lincoln: University of Nebraska Press, 1966.

Heron, John. *Six Categories of Intervention Analysis*. Guildford, Surry, U.K.: Human potential research project, Centre for Adult Education, University of Surrey (mimeographed), 1975. .

Hightower, J. *Hard Tomatoes, Hard Times*. Cambridge, MA.: Schenkman, 1973.

Kendricks, J. B. "Agricultural Research on Trial." *California Agriculture* 38(5,6), 1984.

Kolb, D. A. *Experiential Learning: Experience as the Source of Learning and Development*. Englewood Cliffs, NJ: Prentice-Hall, 1986.

Kramer, Mark. *Three Farms: Making Milk, Meat and Money from the American Soil*. 2nd ed. Cambridge, MA: Harvard University Press, 1987.

Kuhn, Thomas. *The Structure of Scientific Revolutions*. Chicago: University of Chicago Press, 1962.

Lewin, Kurt. *Field Theory in Social Sciences*. New York: Harper & Row, 1951.

Madden, J. Patrick, and Irene Johnston. "A Case Study of the Mechanical Tomato Harvester." In *Agricultural Technology Delivery Systems*, I. Feller, ed. 4:4.1–4.69. University Park: Institute of Policy Research and Evaluation, Pennsylvania State University, 1984.

Martin, P. L., and A. L. Olmstead. "The Agricultural Mechanization Controversy." *Science* 227:601–6, 1983.

Miller, A. "Psycho-social Origins of Conflict Over Pest Control Strategies." *Agricultural Ecosystems and Environment* 12:235–251, 1985.

Morren, George E. B., Jr. *The Miyanmin: Human Ecology of a Papua New Guinea Society*. Ann Arbor: UMI Research Press, 1986.

Northrop, F. S. C. *The Logic of the Sciences and the Humanities*. Woodbridge, CN: Ox Bow Press, 1983.

Organization for Economic Cooperation and Development (OECD). *The Measurement of Scientific and Technical Activities* ("Frascati Manual"). Paris: Organization for Economic Cooperation and Development, 1976.

Piaget, Jean. *Genetic Epistemology*. New York: Columbia University Press, 1970.

Raeburn, Paul. "Automating America's Heartland." *High Technology* (December 1985): 48–55.

Rudner, R. S. *Philosophy of Social Science*. Englewood Cliffs, NJ: Prentice-Hall, 1966.

Schmitz, A., and D. Seckler. "Mechanical Agriculture and Social Welfare: The Case of the Tomato Harvester." *American Journal of Agricultural Economics* 52:569–78, 1970.

Shor, I., and P. Freire. *A Pedagogy for Liberation: Dialogues on the Transformation of Education*. South Hadley, MA: Bergin & Garvey, 1987.

Sun, Marjorie. "UC Told to Review Impact of Research." *Science* 238(4831) (November 27, 1987):1221.

Vandermeer, John H. "Mechanized Agriculture and Social Welfare: The Tomato Harvester in Ohio." *Agriculture and Human Values* 3(3):21–25, 1986.

Zajonc, R. B. "Feeling and Thinking: Preferences Need No Inferences." *American Psychologist* 35 (February 1980):151–175.

CHAPTER 3

An Introduction to Systems Thinking and Practice

Moving from disciplinary approaches to systems approaches and more expansive styles of problem solving requires a great leap. Scientists and other professionals have been searching for new approaches to meet the challenges presented by complex problems in which there is no opportunity to reduce the number of factors that needs to be handled. This need has led them to develop more holistic approaches based on systems thinking. These approaches to inquiry are based on different premises and processes from those of the scientific method, which were discussed in the previous chapter. This chapter presents several major systems approaches that have applications in agricultural and natural resources management and introduces the idea of modeling, which many people take to be synonymous with systems approaches. The chapter concludes with an overview of the soft systems approach, to which the remainder of this book is devoted.

Systems Thinking

The following extended passage from the foundation work of modern biology should be taken as a mini case study of a natural resources situation, as well as an exemplar of the highest level of original and creative thought. Even if you never before had a clue about systems thinking, after reading this piece from Darwin, you should begin to assimilate some basic premises.

> I am tempted to give one more instance showing how plants and animals remote on the scale of nature, are bound together by a web of complex relations. . . . I find from experiments that humble-bees are almost indispensable to the fertilization of the heartsease *Viola tricolor,* for other bees do not visit this flower. I have also found that the visits of bees are necessary for the fertilization of some kinds of clover; for instance, 20 heads of Dutch clover *Trifolium repens* yielded 2,290 seeds, but 20 other heads protected from bees produced not one. Again, 100 heads of red clover *T. pratense* produced 2,700 seeds, but the same number of protected heads produced not a single seed. Humble-bees alone visit red clover, as other bees cannot reach the nectar. It has been suggested that moths may fertilize the clovers; but I doubt whether they could do so in the case of the red clover, from their weight not being sufficient to depress the wing petals. Hence we may infer as

highly probable that, if the whole genus of humble-bees became extinct or very rare in England, the heartsease and red clover would become very rare, or wholly disappear. The number of humble-bees in any district depends in a great measure on the number of field-mice, which destroy their combs and nests; and Col. Newman, who has long attended to the habits of humble-bees, believes that "more than two-thirds of them are thus destroyed all over England." Now the number of mice are largely dependent, as everyone knows, on the number of cats; and Col. Newman says, "Near villages and small towns I have found the nests of humble-bees more numerous than elsewhere, which I attribute to the number of cats that destroy the mice." Hence it is quite credible that the presence of a feline animal in large numbers in a district might determine, through the intervention first of mice and then of bees, the frequency of certain flowers in the district. [Charles Darwin, *On the Origin of Species* (1859)]

Let's take a moment to "deconstruct" this piece to get at some of the basic ideas it expresses. In other words, let us read it in order to find out about the premises of Darwin's thinking.

The very first line announces Darwin's discovery that parts of a situation are interrelated, and he goes on to describe in some detail how a change in one part can bring about changes in other parts so related. This is the *principle of holism,* that it is good and *useful* to look at the world as if it were made up of the complex wholes we now call systems. Darwin goes on to show how, if you want to understand what's really going on, you can't just carry out an inventory of the life forms living in an area, naming, counting, weighing, and totaling them up. Even if you did this repeatedly over time, you might not have any idea why, for example, the abundance of clover was declining. The larger system is not just a summary of its component parts; because of the interrelationships between parts, the whole emerges as something quite distinct. This is called the *emergent properties premise.* Darwin also is concerned with observing changes of various kinds, how component species change in abundance, how inputs from outside the system get transformed as they move from part to part, and how whole systems may be transformed over time. Darwin expresses a clear understanding of how one part of a system can exert a controlling influence over another part, even controlling the reproduction of another species. Although he did not know it at the time because genetics had not yet been discovered, he also observes vitally important communications within the system: how certain insects carry genetic messages between individual members of flowering plant species. Finally, in his discussion of the village–cat–mouse–bee–clover connection, we see how people are a vital (if not dominant) component of many seemingly natural systems. This chapter will now turn to these major points.

System is an umbrella term covering a variety of ways of either viewing complex reality or designing approaches to deal with it. All systems approaches have in common the assumption, commonly referred to as the *holistic perspective,* that everything is or can be connected to everything else. The systems perspective encourages us to examine how things interact, interconnect, interrelate, or, in some sense, control each other. Systems approaches also share the idea that causality in nature, particularly in the living world, is circular (or recursive) rather than linear. The latter, linear causal thinking, is a feature of the scientific method.

By definition, a system is a set of parts that behave in a way that an observer *has chosen* to view as *coordinated* to accomplish one or more goals. Note that the concern is an observer's *choice* of parts to study. It is best not to think that systems are real. As is the case with the scientific method, using systems thinking is a way of imposing meaning on and shaping inquiry about experience.

The scientific method was developed to deal with the physical world, to make sense out of nonliving galaxies and billiard balls. It *was useful* to view these things involving linear patterns of forces and impacts. This paradigm has been carried over to the study of life, but it has approached several paradigmatic limits. A practical one is that, as already indicated, the scientific method cannot make sense out of complexity and dynamism. The second limit is conceptual; it seems that the world of living things, of crabs, redwood forests, and human societies, might be better understood in terms of differences and connections (Bateson, 1979:7–8). This is the principal justification for using systems approaches in the areas of food, agriculture, and natural resources.

Although it has existed, in some form at least, since Darwin's time, during the past forty years systems thinking has exerted a tremendous influence on all fields of science worldwide. Today, few would question the usefulness of systems approaches for dealing with many problematic situations in agriculture and natural resources. There are no biological, physical, or social sciences disciplines that do not have some scholars who identify themselves as systems specialists. What is meant by *systems approach,* however, is still quite diverse, depending on disciplinary background and other aspects of one's training.

The major premises introduced at the beginning of this chapter are reviewed here.

Holism

The first major premise of systems-based methodologies is that the world can be viewed as consisting of structured wholes, or systems, that maintain their identity or integrity under a range of conditions and that exhibit certain general properties emerging from their "wholeness." The proposition of the scientific method that the whole equals the sum of its parts is superseded in systems thinking by the proposition that the whole is *different* from the sum of its parts (Bateson, 1979). Every organism, every plant or animal, is a whole with a certain internal organization and a measure of self-regulation. In other words, organisms are organized complexities. Hence, saying we are approaching inquiry holistically means focusing our attention on understanding present and potential organized complexity, the effects of the activities and interactions of parts within the whole that is being investigated, and the impacts of factors lying outside the system that has been defined.

It is easy to accept organisms as whole things, but what of conceptualized things such as populations, communities, ecosystems, and human activity systems, which are less concrete? Hence the injunction that *systems are a way of viewing the world, and the entities cited are analytic constructs*. The systems perspective says only that it is useful to view the world *as if* it were composed of systems.

Transformations

The second major premise is that systems transform themselves continuously. If they fail to transform themselves, or if they are seriously disrupted by external forces, then systems may cease to exist. Thinking in terms of transformation processes that are at work in a situation is fundamental to systems practice. Inputs to a system are transformed through major functions that can be described or developed: as a result of such a transformation, an output from the system is produced. How particular transformations occur in specific parts of a system, such as energy flow, may be reducible to basic science principles.

Control

A third fundamental concept of systems thinking is that of regulation or control. Other words, really near-synonyms, that you may hear or find used in this connection include *feedback* (particularly negative feedback), *equilibration* (or equilibrium maintenance), *adaptation,* and *self-regulation.* Systems are conceived as having the capacity to maintain key components within an appropriate range of values in the face of external disturbance. According to Tustin (1952), there are at least two kinds of control systems, (1) open-sequence or calibrated and (2) closed or negative-feedback. In open-sequence control, all corrections for environmental variations are built in, and the system cannot compensate for changes not anticipated in the original design. For example, plants that set buds in the spring season of temperate zones in response to lengthening photoperiods may not be able to adjust to a cold snap. In contrast, the closed or negative-feedback control system goes directly to the quantity to be controlled, corrects for all kinds of disturbances to that quantity regardless of their origin, and communicates or "feeds back" information on the degree of departure from the target condition or goal. There are three rules for demonstrating or operationally accomplishing equlibration or feedback control:

> First, the required changes must be controllable by some physical means, a regulating organ. Second, the controlled quantity must be measurable, or at least comparable with some standard; in other words there must be a measuring device. Third, both regulation and measurement must be rapid enough for the job at hand. [Tustin, 1952]

Communication

The fourth major dimension of systems thinking is the notion of communications, which is related to a system's ability to communicate information in order to control what happens within and the forces that come from without. Two in particular are central: feedback, discussed earlier, and *feedforward*. In feedback, a critical aspect of the communications loop is the determination of how quickly information should be transmitted (Churchman, 1979:47). The more rapid the communication, the less extreme the oscillation of vital components. Thus, the desired output of a system may be different from the actual output as compared with some standard or goal. The principal way of reestablishing control of the nature and quality of the output is by looking at or redesigning the feedback controls within the system. This is feedforward communications.

For example, Parnaby (1979) notes that the processes involved in a manufacturing system must be ordered and controlled if they are to meet the system's objectives efficiently. Yet at the same time they must be able to change their structures and characteristics in order for the business to survive. Their ability to adapt comes from the involvement of people in the control procedures and the decision-making strategies to maintain a way to adapt to market and other environmental disturbances. In this instance, feedforward control might involve routinely forecasting such things as sales, market conditions, and input availability and providing that information to managers so that periodic adjustments can be made (e.g., increasing inventory, phasing out a product line, and so on). While feedback is a response to an effected displacement, feedforward is an anticipatory response in relation to a possible future displacement.

Simply stated, feedback pays attention to dealing with things after they have already happened and correcting the situation so that the undesirable kind or quantity of output is improved. In feedforward, attention is directed in advance to predicting and possibly correcting those disruptions of the system that might affect output in the future. Feedback and feedforward are often combined in human activity systems (Carter et al., 1984).

Hierarchy

The fifth major concept is the idea of hierarchy. One way of reducing the complexity of the real world is to construct nested or hierarchical systems models in which smaller units or subsystems are "nested" in larger systems. An example that should be familar to many readers is the biological hierarchy in which the biosphere consists of ecosystems, ecosystems consist of communities, communities consist of local species populations, populations consist of individual organisms, and so on. The notion of a hierarchy of systems (or subsystems) is the systems version of reductionism. It says that *some* properties of systems are reducible to properties of components. We can often effectively use similar logic to impose some order on business firms and their operating units or on political entities such as states and their subdivisions (e.g, municipalities).

Accordingly, control is of interest, not only within the system, but also as it manifests itself among and between the hierarchies of systems present in a particular situation. Basic to the notion of control is recognition that systems exist within environments that influence and affect the system and its ability to control and adapt its activities. Higher-level systems may provide inputs to lower-level systems because they form their environment. A system may influence the environment by its activities, but it cannot totally control the environment. Hence, uncontrolled external forces can influence the performance and outputs of the system.

Emergent Properties

The sixth and last major premise to be introduced emerges from the principle of hierarchy. It is also the expression of the earlier statement that in systems the whole is different from the sum of its parts. That *difference* is the emergent property and, in any given hierarchy, emergent properties uniquely pertain to particular hierarchical levels. If you move from one level to another, new properties emerge while those pertaining to the former level either are absent or are radically changed.

Take the example of the development of new technology and the issue of its appropriateness. A technology that is eventually to be used by a particular group of clients in a specific setting can be thought of as a subsystem within a larger system. If a technician develops a technology (e.g., a new soil management practice or a new post-harvest pest-control process) that does not account for the characteristics and functions that are present within the already established wider system, there is great risk that the new technology will be ignored or rejected. Paying attention to the appropriate hierarchical level and relevant systems is therefore important to systems practitioners. "There is a fundamental limitation on any modeling of a system, that *the system is always embedded in a larger system*" (Churchman, 1979:76). Accordingly, to err in the direction of holism rather than reductionism is the best course.

To summarize, the core of systems thinking is learning to think holistically. Basic to holism is to think in terms of people and other organisms of varying complexity that can survive in a changing environment, showing emergent properties, hierarchical structure, processes of communication, and means of measurement and control (Checkland, 1985). These are a few key premises that you will use if you are trying to think systemically. There are also related premises regarding the approach to inquiry in systems analysis.

Inquiry into Systems

The basic premise of holism comes to the forefront once again in the inquiry approach to systems. In a holistic inquiry, the greatest attention and activity focus on describing the complexity of a situation, with the investigator using techniques to deal with the interactions of things rather than on the things themselves. The objective is to understand (1) interactions among parts and within hierarchies, (2) their emergent properties, (3) the transformations that occur among components of a situation, (4) control processes, (5) communications linking the parts of the system, (6) the objectives and performance measures of the system, (7) the environment, including constraints, (8) resources and inputs, and (9) details of management, ownership, or dominance.

This chapter discusses two current approaches to systems-based inquiry, hard and soft systems. The words *hard* and *soft* may bear unfortunate connotations, since most people give these words meanings that the originators did not intend, such as "difficult" and "easy." Naughton (1984:7) provides a useful summary of the intended meaning of *hard* and *soft*:

> The adjectives "hard" and "soft" do not mean difficult and easy respectively, but are ways of describing two approaches to systems analysis. There are some significant differences between these approaches on such matters as their underlying assumptions about the nature of social reality and of organizations. They also differ somewhat in the intellectual tools which they employ: the "hard" approach, for example, makes extensive use of quantitative models and simulations. To get started, I might say — as a first approximation — that the hard systems approach is one which evolved in the context of machine-based or hardware-dominated systems and is thus concerned with settings in which *goals can be set, performance maintained* and *implementation achieved*. In contrast, the *soft systems approach* evolved to deal with problem-situations in which *human perceptions, behavior or actions seemed to be the dominating factors* and where *goals, objectives* and *even the interpretation of events are all problematic*. [Italics added.]

Hard Systems Inquiry

Figure 3.1 is an outline of a general inquiry procedure associated with the hard systems approach (Jenkins, 1969). This inquiry procedure is found, with minor variations, in most systems engineering, systems analysis, and operations research texts. The process begins with the recognition and quantitative definition of "the problem." The analyst goes on to organize the project, defining its purpose. He or she then designs a system relevant to the purpose and problem. A systems model is then formulated and used to assess the relative efficiency of alternative technologies, policies, or strategies. An alternative solution is selected and validated in relation to the original definition of the problem. This constitutes the solution to that problem.

This procedure is best suited for tackling problems involving those biological, physical, mechanical, and human activity systems *for which it is realistic and appropriate to define clear-cut quantitative goals:* for example, the energetic efficiency of a process in the context of an energy crisis, with the understanding that the analysis may be put aside once the crisis is over.

How models are used after they are built gives a clue to the objectives of those using this type of inquiry process. As discussed subsequently in detail, models may be used to communicate complex ideas, as a tool for discovering new things about a complex situation, and as a means to test alternative approaches to dealing with a problem. Some general considerations about the kinds and uses of hard systems models are presented in the following sections.

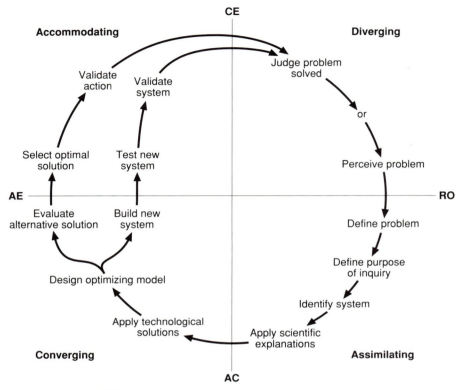

FIGURE 3.1   The process of hard systems inquiry.

On Modeling

Models are created through a process called *formulation,* which is based on assumptions and data. While a model is by definition always a simplification of the real world, the accuracy of the model is tested by reexamining the original assumptions and data. Just the attempt to formulate a model forces one to make assumptions and collect certain kinds of data, which are subsequently reviewed and tested (or debated). To apply the results of the manipulation of the model, one should *not* go directly from the model to the real world, but should go back through the initial assumptions and data to make sure they are not distorted. The results should only be applied when the researcher is satisfied that the formulation is acceptable. This is not always easy to do when one is trying to use models formulated by other analysts. In the case of certain forms of symbolic modeling, such as mathematical models, the assumptions used to create equations are often not articulated for all to examine. Such assumptions must be made explicit. When the researchers are satisfied through debate that the formulation is acceptable, then it should be applied or implemented in the real world. More will be said about this point as the types of models that might be developed are described.

The Purposes of Modeling

Models are used to amplify the human process of learning and, in the context of systems thinking and practice, have four particular uses:

1. To communicate complex interrelationships.
2. To communicate concepts about the meaning of something.
3. As a novel construct for the search for new insights about how a system might be, might work, or might behave.
4. As a test bed or simulation for the evaluation of alternative strategies of change.

Systems modeling is a way of amplifying the human thought process where complexity is involved. The human mind readily recognizes patterns and generates concepts but often stumbles when dealing with complexity. The mind is always receiving other, often diverting, input, resulting in an inability to hold factors constant in time. The continuity of activity in a complex and dynamic situation often interferes with the ability to study and understand it. Modeling offers the analyst the opportunity to interrupt the phenomena that the model represents and reflect on its behavior.

Models may be used as an experimental device for designing new systems. A prototype or several prototypes may be created to be examined and evaluated as a basis for the development of entirely new approaches to doing things. Using such simulations allows the contemplated system to be manipulated and managed beyond the expectations of the final system. For example, a simulated field plot of land may be subjected to excesses of fertilization, irrigation, soil working, and so on in order to assess the nature and severity of the impact of these factors. Soils vary widely in their ability to accept water, nutrients, and machine traffic. What is acceptable for one soil may be unacceptable for another. Simulations offer an opportunity to test alternatives in ways that would be too damaging if conducted in the real world.

The effective use of time may be a reason to use a model to represent a particular phenomenon. The time required for a cycle of particular biological phenomena may be too long or too short to permit useful observations. A plant growth study may be effectively speeded up for observation by using time-lapse photography. The longer the time interval between frames of a film, the shorter the viewing time when projected at normal speed.

It is often easier, less expensive, and more informative to experiment on a simulation model rather than on a real farm or field plot. It may also be ethically necessary to use an animal model to test a new medical technology. Models and simulations may be the only way to test ideas about the possible consequences of proposed social innovations or to train people to work with them.

The consequences of introducing a new material, procedure, or component into a system may be unpredictable. A model allows the consequences to be explored without affecting the actual situation. Introducing a new crop or testing a new fertilizer, such as a nutrient-reinforced sewage sludge compost or a new herbicide, in a farming system can have profound impacts. Using a representative field trial on an experimental farm enables the input to be studied before the decision is made to use it on an operating farm. Insecticides and herbicides are two farm inputs that are rigorously tested and government-approved before they are offered to farmers.

The risk associated with a new or modified system can be evaluated by using a representative model. In the Chatham River case (located at the back of this book), for example, the consequences of reallocating water can be examined by using a simulation model that shows how income, or plant yield, or the abundance of aquatic life are affected as various amounts of water are reallocated. It enables people to assess the proposed changes before they occur.

Another reason for modeling is to convey the mathematical relationship between components. A graph is a symbolic model that displays a functional relationship between two variables (Figure 3.2).

A typical production function in agricultural economics, represented in Figure 3.3, is a graph used to help us understand the relationship between some input and an appropriate output. It is particularly useful for indicating when the application of some input has reached the point of diminishing returns, that is, that amount of a resource that actually begins to reduce productivity.

Several production functions can be combined in a model to measure cost effectiveness, providing a possible mechanism for choosing between alternative strategies, as in Figure 3.4. Note, however, that this is not the sole way to assess options in natural resources management or agriculture.

Nevertheless, a systems model is useful for predicting outcomes of alternative strategies. Thus, one can optimize various parts of a system, such as physical, biological, economic, or logistical factors, to understand better what will happen to the rest of the system. Note that in connection with systems modeling, to optimize means to adopt assumptions about the most *desirable* relationship between factors in a system, typically an input and an output.

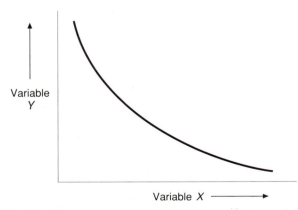

FIGURE 3.2 A graph expressing the functional relationship of X and Y.

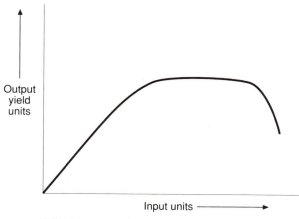

FIGURE 3.3 A production function model.

Some authorities feel that biological, physical, and mechanical systems are easier to model than human activity or socioeconomic systems because their parts are assumed to have less variability. Two counterpoints are worth noting, however. There is a strong reaction among ecologists against hard systems modeling (e.g., Colinveaux, 1973:273ff). The core of this ecological criticism is that assumptions regarding relevant inputs and outputs, such as limiting and tolerance factors, energy budgets, and the like, prove to be unjustified in practice because they reduce complexity in ways that are not analytically useful. Hence, according to this argument, biological phenomena are just as variable and complex as human activity–related phenomena. The other counterpoint is that much original work in hard systems modeling, including some adopted by biologists, was car-

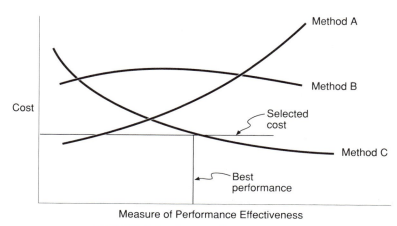

FIGURE 3.4 A cost-effectiveness model.

ried out by economists (e.g., Leontief, 1966) precisely to deal with national economies. This version focuses on the inputs and outputs characterizing transactions that occur between various sectors of an economy. Accordingly, the problem is to optimize all these relationships. This approach continues to be used widely in economics. Thus, there are grounds for questioning the belief that biological and physical systems are easier to model than socioeconomic systems.

Another way in which models can be used is to highlight key environmental influences on a system and to show how the behavior of one part of the system transacts with other parts. Since the model relies on existing data bases, deficiencies in current scientific knowledge may also become better understood. Hence, those who use models for this purpose are seeking new knowledge—basic science inquiry—rather than trying to improve a current real-world situation.

It is common for hard systems studies in economics, biology, and physical sciences to start with a model of a system, rather than with the investigation of a problematic situation. This occurs when modeling is used as a tool or technique in a stage of either technology development or basic science inquiry, rather than in a hard systems inquiry as such.

Some examples of hard systems analysis teaming up with technology development involve a new generation of "smart farm machines" currently under development. Eventually, chugging down rows of crops, a farm manager's new assistant—a computerized, radar-controlled vehicular robot fitted with a video image processor, a pump, and a few gallons of herbicide—may pause at a clump of vegetation to analyze its shape and complexity. If the image differs substantially from that of crop plants stored within the computer's memory, a signal will go to the pump to deliver a burst of liquid to the doomed plant. Irrigation and fertilization systems, automated animal-control devices, and food handling and processing equipment are but a few of the applications now under development. The challenge will be for the designers to consider enough variables, including people factors, ahead of time so that what is in the computer is appropriate to the actual environment and situation in which it will be used. It will also be important that the lower-level employees responsible for the actual care and use of these computers and computer-controlled devices know as much about them as the managers who decided to buy them (Churchman, 1979:75f).

Types of Models

Many different classifications of models are found in the modeling literature. Each classification helps us understand the special nature of and differences among the various kinds of models that one might develop. Some classifications stress the distinction between whether the model is designed to describe something or to prescribe a change. Others help us conceptualize events as if there were causes leading to certain effects, while yet other kinds demonstrate how effects can change causes too! Some of the key distinctions among various kinds of models are described in the following sections.

Physical Models

At their most basic, models are abstractions that represent simple or complex reality. The representation may be what we assume is an appropriate representation of an imperfectly understood reality. Or the abstraction may be a real duplication. Examples are a photograph of a farm tractor, a sketch of a tree, a painting of a landscape, or a sculptured head. They are all perceptions of something real displayed in a manner that endeavors to mimic reality—they try to be structural lookalikes. Thus, Mihram (1972) described modeling of this kind as the "art of mimicry."

Such models are often classified as *iconic* models or physical models (Forrester, 1961). It is important to note that, even though they are often intended to duplicate reality, they all bear the idiosyncratic stamp of the modeler. This issue assumes much greater significance in the next class of models, *symbolic* models.

The primary purposes of developing iconic models are to change the scale of something, to catch basic characteristics of something at a point in time, or to test the characteristics of something under varying conditions. Thus, panoramic views can be reduced down to something much more manageable, while photo micrographs do the opposite. Or in the case of the Chatham River situation (found at the end of this book), a photograph of the river is useful to show the physical extent of the river. Lines can be drawn on the photograph to show the extended flow during flood stage and the minimal flow during drought. For some people, when informed that during the flood stage a river carries 100 cubic feet of water per second and during minimal flow only 10 cubic feet of water per second, there is little recognition of how a river acts. The photograph and sketch provide documentation during the investigative and learning phases of a methodology. A variety of ways of providing documentation in addition to photographs and sketches exists, such as movies and videotapes. The latter can in some cases contribute substantially to the understanding and persuasiveness of the models finally developed.

Scale is an important dimension of a physical model. By manipulation of the scale, the model can be made much smaller or much larger than the actual entity. When studying a farm, the scale can be made very small by using various techniques, such as a drawing or an aerial photograph on an 8-inch-by-10-inch sheet of paper, to portray the geographical features of a multiacre farm. Scale can also be utilized agronomically, using a small-plot trial to predict the performance of a plant species or variety on the whole farm. When studying an individual plant, the scale can be increased by using magnification of several forms to provide more detail and thus information about the plant.

A physical model can be created using analogs. In the modeling sense, an analog is a proportional substitute that behaves or mimics a behavior in a way that is similar to what one thinks will characterize an entity in real life. Analogs can be constructed of components not present in the actual entity studied. They are similar substitutes. Reasons for using analogs are that they may be more manipulable; have a faster response time; be more visual; and, as a result, more informative. A prime example is the analog computer, which uses electrical phenomena to represent physical phenomena.

There are many limitations to iconic models, including the fact that normally they do not illustrate how their components work, only what they look like or how they respond under a range of conditions. Such iconic models can be designed to feature a static condition; others, a dynamic condition. Static physical models describe a relationship that does not vary with time. Dynamic models deal with time-varying interactions (Forrester, 1961).

In recent years, a whole new class of analog models has been developed and used, the game simulation or role play. A game allows participants to play out roles within designed limits or rules, which are based on a model of a situation (Greenblatt and Duke, 1979). Although they possess some of the limitations of other analog models, they have nevertheless proven to be powerful tools for communicating proposals for change and allowing people to learn their way through the alternatives and potential impacts of change. And, like other simulations, they can be run for experimental purposes and the results recorded by observers. More will be said about the uses of game simulations in Chapter 7.

Symbolic Models

Through the use of symbolic models, it is possible to model things that are hard to see or can't be seen by observation, such as our ideas of relationships and properties. This is a crucial issue when it comes to systems thinking, for the characteristics of systems that have been highlighted only exist in the mind of the beholder. Thus, as Peter Checkland has it, "Systems are ideas in the heads of observers of the world which they may find useful in trying to understand it" (1981). They are not "entities out there." Systems aren't real. They are merely figments of the imagination. True, some of the properties that the designer places in a given systems model may be tangible and observable, but the arrangement as visualized and described in the model is best thought of as a conceptualization of what it is we observe or think exists or could exist currently or at a future point in time. For some faculty across the U.S., this notion goes contrary to everything they have been taught. Their mental model contains a world-view statement that "agricultural systems are real things out there."

Symbolic (or conceptual) models are much more idiosyncratic, for they reflect a highly individualistic or specific group's conceptual view of things. Symbolic models are actually quite common, but they are less often recognized for what they are. Symbolic models attempt to portray our notions of how things might function, rather than how they look. We are now concerned with interpretations of our concepts of the way things are or could be. For example, an agribusiness manager talks about the business as a corporation. The use of the term *corporation* implies that the manager expects people to do certain things and relate to one another in certain ways. Such a manager deals continuously with these mental and verbal models of the agribusiness, even though he or she may not recognize it.

As with physical models, many kinds of symbols can be used to model something. Typically, special languages, schematic symbols, and mathematical symbols are used.

While mathematical models are in common use, Forrester (1961) reminds us that they are less easy to comprehend than physical models and are less frequently used in everyday life, including agricultural and natural resource enterprises, than narrative models. Mathematical models rely on a very precise language. Mathematical symbols are less ambiguous than the English language and therefore do have greater clarity than most verbal models, although, if designers of verbal models are rigorous, the same degree of precision can be achieved. Even when the ultimate goal is to design a mathematical model, the designers must begin by designing a narrative model, which is then translated into a schematic and then further translated into another more limited language, mathematical symbols.

Mathematical models are valuable because they can be more easily manipulated than verbal or physical models. Their logical structure is more explicit and precise. Where a word has more than one meaning, a mathematical symbol has only one meaning. But because the language of mathematics is limited in expression, care must be used not to fool oneself into thinking that mathematical models are better or more accurate conceptualizations of a particular situation under study.

The following section reviews the kinds of models and approaches to modeling that various authorities have developed and shows how they are related to some of the key concepts in systems literature. Each kind of model attempts to demonstrate the characteristics of an entity that emerge when one is thinking about it systemically. When one is thinking of a particular set of components as if they were a system, key qualities that make various entities different from one another can be modeled. Some of the key differences among systems are whether they are conceived as stable or unstable, dynamic or static, steady state or transient. Only three features will be reviewed here in order to give the reader a beginning idea of the different purposes for which models are created and what they may attempt to demonstrate. Symbolic models and iconic models are usually further classified based on these features, as well as others.

Modeling Static and Dynamic Features

Symbolic and iconic models can be static or dynamic. Therefore, time is another major feature that is used to classify types of models. Will the model deal with the features and results of time-varying interactions, or is it meant to describe a relationship that does not vary with time? This classification feature of a model is an important distinction, since we sometimes have a tendency to describe things statically when we know by our narrative models that the properties of and interactions between states being modeled change through time. Hence, spending time to develop a mathematical model that holds constant some properties that are known to change out there in the real world should be seriously questioned.

Modeling Linear and Nonlinear Relationships

Another major way models are classified is by whether they are linear or non-linear. While linear models are adequate in much of the work of the physical sciences, they fail to represent the essential features of many biological as well as business and social processes (human activities). As Forrester (1961) points out, "In linear systems the response to every disturbance runs its course independently of preceding or succeeding inputs to the system; the total result is no more nor less than the sum of the separate components of system response. Response is independent of when the input occurs in the case of a linear system with constant coefficients." (It is possible to have time-varying coefficients.) When expressed mathematically, linear models are much easier to express than nonlinear models. "With a few exceptions, mathematical analysis is unable to deal with the general solutions to nonlinear systems. As a result, linear models have often been used to approximate phenomena that are admittedly nonlinear." And the nonlinear characteristics have been lost. One must learn when it becomes foolish to proceed with a particular type of model design or to believe in model results that have departed from reality!

For the past thirty years, modelers have been experimenting with better ways to mathematically design models so that the assumptions would be truer to the way things seemed to be in real life. Boughey (1976) provides a very good and understandable review of the kinds of modeling efforts that occurred and had major impacts on decison making and policy making regarding social, agricultural, and natural resource development and management during the 1960s and early 1970s. Berlinski (1976) focuses on the limitations of these efforts.

Mathematical models usually attempt to convey the relationships between structures rather than the structures themselves, although, as always, there are plenty of exceptions around. The limitation, however, is that demonstrating relationship is not demonstrating cause and effect. Most models only show that as one variable changes, another variable changes in some way. Why there are the kind and rate of change is not determined by the mathematical language, nor is one variable said to cause the change in the other variable. The modeler or researcher must make an interpretation based on the model and his or her **W**.

Sometimes mathematical models are classified as deterministic or stochastic. Deterministic models are characterized by assumed certainty and predictability. Chance does not enter into the formulation of the mathematical formula and the solution that it reveals. Stochastic models are characterized by uncertainty and probability. Chance has an important impact on the solution and is somehow built into the mathematical equations used in the model. Some examples of deterministic models are linear programming, as mentioned previously; network analysis; dynamic programming; and many engineering equations that model natural phenomena. Some examples of stochastic models are queuing, inventory, simulation, and gaming.

Farming is fraught with chance, and the farmer is often identified as a gambler because he or she must constantly evaluate the odds regarding the occurrence of an event. Many models used to represent farming are stochastic. Consider a farmer who is producing alfalfa hay. Three days of continuous drying time (sunlight) may be required to dry a cutting of hay in the field. On the first day of sunlight, what are the odds that two subsequent days will be sunny? They may be one in three or four, depending on the geographical location of the farm and the season of the year. To reduce these odds, he will consult the weather service, farmer's almanac, or other weather information source and integrate it with his own weather experience. All are associated with chance, and at the end of a growing season's production, after multiple haymaking efforts, the farmer will have experienced some hay being destroyed, some being reduced in quality, and some reaching the desired quality. The proportions are based on the chance associated with the decision to begin making a cutting of hay.

Modeling Stable and Unstable Features

Another major feature that has been used to classify the kinds of models developed is whether or not the model portrays stable or unstable features of a particular situation. In those models where the designers are trying to deal with time-varying interactions (dynamic models), the key properties of the model may be conceived as being stable or unstable. In a stable system, the properties are designed to show how a particular set of features changes when a disturbance is introduced and then how it returns to a stable state — in other words, how change is handled, what it does to the properties of the components under study, and how these components react or handle change.

Unstable systems models demonstrate that the properties of the system do not handle a particular disturbance very well or at all, and, as a result of sustained, intermittent, or one-time disturbances, the features of the properties deteriorate over time. Moreover, given the present way things are being handled or function, the model demonstrates that the system will decline or die out. For example, the well-known Club of Rome report shocked the world's leaders by projecting what would happen to the world's food supply, natural resource base, and the like if the world's population continued to grow unmanaged (Mesarovic and Pestel, 1974). While the report and modeling work were widely challenged, it did fulfill the purposes of modeling: It communicated complex interrelationships; it served as a novel construct for the search for new insights about how a system might behave; and it was a test bed, or simulation, for the evaluation of alternative strategies of change.

Modeling Steady-State and Transient Features

Related to the notion of unstable and stable systems is the notion of the steady-state and transient qualities of systems (Forrester, 1961). A system can be designed to show that the behavior of the properties under study maintains the same fundamental qualities over time and, in some cases, even when disturbances occur. The system's character is assumed to remain untouched in light of disturbances. Transient systems models, on the other hand, try to demonstrate that as disturbances, even of a one-time nature, affect the situation under study, the character of the key features in a situation changes. While some physical and biological systems are often modeled as relatively stable systems, social systems are typically modeled as transient systems. And, as discussed previously, since the early 1970s biologists have been challenging the notion that biological systems are stable, as had been assumed in the 1960s.

Many of the models found in the economics and management literature of the recent past are steady-state, stable, and linear models of agricultural and natural resource systems. The practical utility of such models has been questioned greatly in the literature (e.g., Berlinski, 1976). Since the early 1970s a search has been on to find ways to model situations that have some of the following characteristics: (1) features of the system can evolve in response to internal or external changes; (2) effects can react on causes; and (3) the system can be managed through qualitatively changed but acceptable states.

To summarize, *iconic* (physical) models attempt to duplicate the structural realities of things. *Symbolic* (abstract) models, on the other hand, are abstractions of real things, representing someone's thinking about that reality — in the present or in the future. These abstractions can take the form of physical analogs (any substitute that is said to replicate something in a similar way), narrative languages (spelling, syntax, etc.), schematic symbols (representing such features to be discussed here as states, sinks, and rates), or mathematical equations and symbols.

The Languages of Modeling

There are many languages used to model. There have been many attempts to standardize conventions for modeling to reduce the rampant confusion and ambiguity that currently exist. Most of these attempts at standardization have failed to make an impact. Two standardization attempts commonly encountered in systems literature are the dynamic language of Forrester (1961) and the symbolic code of ecology. Simulation flow charts frequently use hybrids of these two. Forrester's industrial dynamics language (1961) is found in Figure 3.5. Odum's symbolic energy language (1971; Odum and Odum, 1976), which is used to model ecosystems, is found in Figure 3.7.

Jay Forrester's symbolic language will be taken to illustrate how one of these languages is used by people in the agricultural and natural resource sciences. A good explanation of how such modeling is done was published by Forrester in *Industrial Dynamics* (1961). His particular kind of modeling involves the development of a narrative model, which is then developed into a schematic using the symbols found in Figure 3.5. Next, each concept represented by a symbol is quantified (translated into the language of mathematics) and rate equations are established. Models using this procedure can be illustrated, computed, and run.

The schematic for one such model, originally developed by Smith and Langlands (1973) for an Australian case study, is shown in Figure 3.6. (See page 86). It illustrates some of the key relationships between the components of a grazing

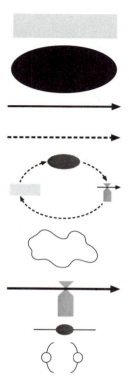

State variable or **system level** (in some applications) An element of a system that changes only as a result of a change in a flow rate.

Auxiliary variable Part of the flow-rate description but separated from that symbol because it is most clearly described independently as a mathematical function of two inputs.

Material flow The path followed by a resource (material or energy) through the components of a system from source to sink.

Information or **effect flow** Information line designates parts of feedback loops controlling flows.

Feedback loop Any closed path through the diagram; must follow the direction of arrows in information lines but need not for material flows controlled by system rates. There are two types of loops, positive, which generate growth, and negative, which tend to seek equilibrium.

Source or **sink** Respectively, the origin of an important resource that flows through the system, the state of any resource whose level is variable within the system, and the fate of that resource in which it is irretrievably degraded; both external to the system.

Rate or **flow valve** Symbol for regulated flows of resources; flow rates depend on system levels/state variables through an information network, as shown by dashed lines and circles.

Initial condition Starting rates and ratios affecting flows.

Driving variables Variable rates and ratios affecting flows; data supplied external to model.

Note: System structure consists only of variables (levels) and rates (flows). Alternative names for symbols reflect variables adopted by a range of users.

FIGURE 3.5 Forrester's industrial dynamics language.

Mathematical relationships have been established for these variables, and the rates of change between them are such that the schematic symbolic model shown can be converted into a mathematical version, entered into a computer, and run as a simulation of what might occur in the real world with the performance of, in this case, grazing animals under a range of conditions.

The modeling language developed by the ecologist Howard T. Odum is somewhat more specialized than Forrester's, but it is useful for looking at a wide range of combined ecological, agricultural, and socioeconomic situations. And because it is based on energy and energy laws, it is dynamic. It assumes that energy is basic to all processes and that change occurs through transformations of energy from one form to another. Odum's approach is mostly concerned with modeling those transformations, how they occur, the pathways through which energy flows, and the mechanisms for controlling that flow. Thus, Odum's modeling language is built around 14 graphical symbols, each representing something that happens to energy. Eleven of the basic ones are presented in Figure 3.7 (see page 87), and discussed here.

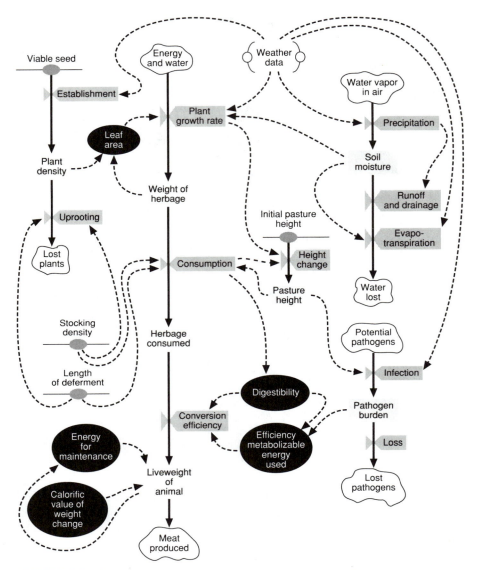

FIGURE 3.6 A dynamic flow model of a grazing system in Forrester's language. (Redrawn from "Model Development for a Deferred Grazing System," © 1973 R. C. G. Smith and J. Langlands, *Journal of Range Management* 6:455, by permission of the authors.)

An application of Odum's modeling language is presented in Figure 3.8, a replica of the subsistence system of certain East African peoples, including the Dodos, Turkana, and Karimojong, who mix livestock operations and plant agriculture (Little and Morren, 1976:58–61). Note that (by convention) energy originating from the sun flows through the system from left to right, while con-

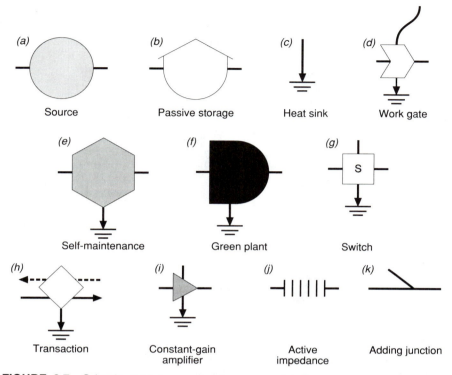

FIGURE 3.7   Odum's modular energy language. *(a) Source* — An environmental source or constraint involving energy or some other factor that is external to the system. *(b) Passive storage* — A variety of forms of stored potential energy. *(c) Heat sink* — Energy loss necessary to all processes that involve transformations. *(d) Work gate* — One form of energy (or factor) applied to the valvelike gate allows the flow of another form of energy or a factor through the gate in proportion to the quantity applied. *(e) Self-maintenance* — A self-maintaining, work-generating component typically referring to life forms, organisms, populations, human communities, and the like. *(f) Receptor* — A type of component capable of receiving radiant energy and converting it to another form, typically a green plant component. *(g) Switch* — A form of work gate (d) with only binary states ("off" and "on"). *(h) Transaction* — Economic exchanges in which symbolic value flows in one direction, facilitating the flow of energy or materials in the other direction. *(i) Constant gain amplifier* — A positive-feedback work gate in which the force applied at the top increases the output by a constant factor. *(j) Active impedance* — A backforce that resists energy flows proportional to their magnitude. *(k) Adding junction* — The addition of flows of similar forms of energy or currencies.

trolling factors, including people's labor, move from right to left in the diagram. This approach to modeling has been widely used by anthropologists, geographers, and ecologists to describe the structure and environmental context of long-established agricultural and herding systems around the world and to study changes in them (e.g., Little and Morren, 1976; Moran, 1982; Morren, 1977).

These and related types of modeling are sometimes carried out in stage 4 of

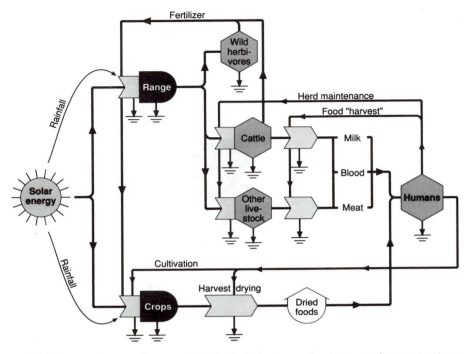

FIGURE 3.8 Energy flow model of East African mixed pastoral–crop system. (After Little and Morren, 1976:58, by permission.)

the soft systems approach, along with a specialized form of modeling (see subsequent text and Chapter 5) to assist people to gain a big picture of a situation and make decisions regarding resource allocations, potential impacts, and other alternatives.

For example, in connection with a river-basin management initiative in Canada, which embraced hundreds of communities and interest groups, a computer model of the drainage was made available to the people involved (Marks, 1984). The immediate purpose was to provide people with a means of *seeing* the entire drainage as an entity and to help them to *understand* how their personal or local interests affected those of others in different localities and situations. The larger objective was to prepare people to participate in the planning process.

Models such as these are not quickly done if the people who need them (see Chapter 5) must start from scratch. Therefore, depending on the time available, it may be expedient to recommend this type of model to appropriate people for future development so that it will be available to make additional improvements in a situation. Some computer-based models similar to the animal-grazing example mentioned previously are available for agricultural groups through the Cooperative Extension Service. Unfortunately, designers don't always provide the narrative and schematic models on which the mathematical model is based.

This makes it difficult for the user to determine what assumptions were made.

Computer resource allocation or decision-making models for the major commodities exist and can sometimes be modified to match the needs of particular clientele, thus saving development time. Models have also been developed for biological systems, such as forests and wildlife populations, and for physical systems, such as watersheds. In addition to models with resource management and allocation applications, many other kinds of decision support aids are available, and these too can be modified to meet particular needs. Such models can assist in forecasting sales, designing procurement systems, monitoring inventories, and performing other functions found in a wide spectrum of enterprises. Several examples are represented in the case materials in the back of this book.

Ask your nearest state university-based Cooperative Extension Service what kinds of software are available. Or check one of the many computer software catalogs available in book and computer stores (e.g., North Central Computer Institute, 667 Ward Building, 610 Walnut Street, Madison, WI 53705; Blackie and Dent, 1978; Bonozek, Holsapple, and Whinston, 1981). Some of these software packages are available free, and others are modestly priced. A new array of user-friendly software packages is available that assist with impact assessment, strategic planning, and resource allocation. EZ-Impact (Biosocial Decision Systems, 1987) is an example. For those interested in developing models of their own, a new software package called STELLA (High Performance Systems, 1987) has recently been released for the Macintosh. STELLA allows you to draw an explicit schematic model onscreen, generates many of the equations automatically, runs your model either as a simulation or in output mode, and produces graphics to use in communicating your model to others.

It should be emphasized that the models presented in this chapter merely represent how modelers arranged things in their own minds. The models are not sacred, carved-in-stone monuments to be accepted without question. Their authors may have made flawed assumptions, put in inappropriate data to represent an effect, left out critical variables or relationships and effects, and so on. So the value of a given model cannot be known until you have carefully examined it, including trying out the information it yields and its results. The question is, does it make sense? This comment is not intended to devalue the benefits of such models, but to make us mindful once again that a model is an abstract conceptualization of reality and not reality itself!

Another language used to model is the schematic. Figure 3.6, which used Forrester's industrial dynamics language, is an example of a schematic. More familiar schematics also are used, however. Charts and graphs are examples. They are constructed of blocks and line connections that identify components, relationships, or activity. Decision trees, activity networks, organizational charts, diagrams of intended and actual communication channels, and maps of formal and informal political power networks are but a few examples of schematic charts introduced in many agribusiness and natural resource management texts. The typical graphs that show the relationship among inputs used and yields produced are a prime example of schematic graphs.

Useful Models of Agricultural and Natural Resource Management

One of the values of models is that they provide a powerful way to communicate to others our conception of what things are like or how things could be associated and could behave. All textbooks provide such models. We learn the content of a discipline through them because they are powerful in helping us gain an understanding of a field. They provide a conceptual base to build on, modify, or change radically our own thoughts. Using conceptually powerful models is one of humanity's major ways to stimulate radical change in thought and also conformity of thought.

While this section is not meant to be an exhaustive review of key conceptual models in agricultural and natural resource management (there are hundreds!), some are provided for those who are beginning to acquire useful frameworks for thinking about the world of agriculture and natural resources. The conceptual models discussed are of different scales and levels of abstraction.

A Farming Systems Research/Extension (FSR/E) Model

The first example is a particular kind of model that is frequently referred to in the farming systems research and extension (FSR/E) literature (Shanner, Philipp, and Schmehl, 1982). As discussed elsewhere in this chapter, hard systems analysts use modeling to describe the present situation. They then move up or down the inquiry spiral to work on basic science, technology development, or resource allocation concerns. This depends on the nature of the hypotheses that are based on their modeling efforts and that they generate during the beginning stages of their inquiry. The FSR/E inquiry process combines some phases of applied science inquiry with hard systems analysis. It is used to address technology development concerns in nations around the world.

> Following the initial data gathering, the team is to analyze the data it collects, develop conceptual models of the system, and hypothesize about ways to improve the system.
>
> Conceptual models in their simpler form identify the major components of the system and the links both among the components within the system and between the system and its environment. Later additional data can be gathered to quantify these relationships and eventually to search for improvements in the system's functioning. [Shanner, Philipp, and Schmehl 1982]

While several different kinds of modeling have been employed in FSR/E, the work of McDowell and Hildebrand (1980) is widely cited and emulated. It appears prominently in the modeling handbooks developed by the Farming Systems Support Project, which was funded by the U.S. Agency for International Development specifically to improve methodologies for economic and technology development work overseas. McDowell and Hildebrand's procedures for model development closely resemble Forrester's, described previously — from narrative

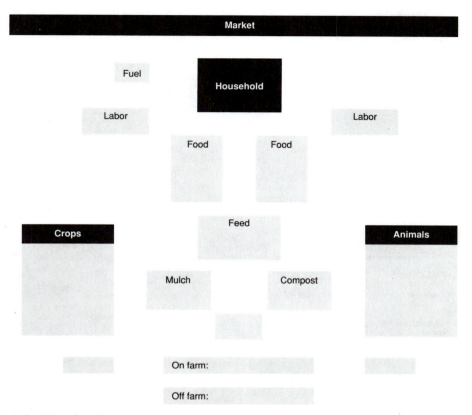

FIGURE 3.9 Outline conceptual map of the components of a typical farm in a developing country. (From "Systems Approach to Livestock Production." © R. E. McDowell, Cornell International Animal Science Mimeo No. 7, Figure 1, p. 3. Reprinted by permission of author.)

model to schematic model to mathematical model. This work is introduced here because these authorities provide a conceptual model of the key components of a typical farm in developing countries. Their conceptual model is a good beginning for readers who lack a conceptual foundation for thinking about the components of a rural farming enterprise in systems terms. Figure 3.9 is the outline conceptual map that they use to determine the "levels of integration of crops and animals and portrayal of the infrastructural dependence of the crop and animal subsystems."

A specific application of this approach is represented in schematic form in Figure 3.10. This diagram is the result of an inquiry into a rural farm situation in Kenya, East Africa. The authors outline the suggested process for completing the conceptual map as follows:

> The box identified as "Market" represents all off-farm activities and resources
> (except land); hence it includes products sold or labor going off the farm as well as

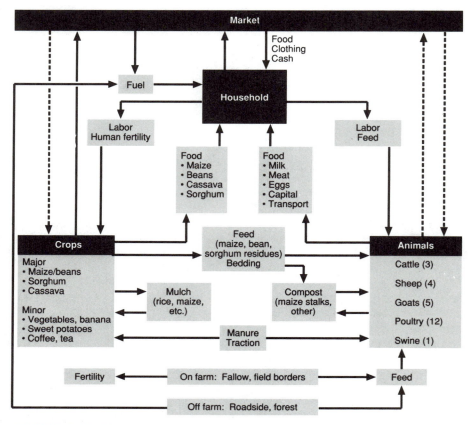

FIGURE 3.10 Model of a low-resource farm in the western Kenya highlands. (From "Systems Approach to Livestock Production." © R. E. McDowell, Cornell International Animal Science Mimeo No. 7, Figure 2, p. 4. Reprinted by permission of author.)

purchased inputs and household items. The "Household" is the core of the farm unit.

Using the model, survey data can be used to characterize a typical low resource farm in the highlands of Western Kenya. In preparing the models of the typical system, labor use, sources of human food, household income, animal feed, and the roles of animals were the main focus. The solid arrows (⟵⟶) portray strong flows or linkages (e.g. more than 20 percent of total income arises from the sale of crops, animals, or household-processed products). Broken arrows (⟵---⟶) are used when sales of crops or animals contributed less than 20 percent of household income, the interchange among functions was intermittent, or there was as yet no routine pattern identifiable; e.g. purchase of seed, fertilizer or pesticides, to support cropping or a member of the household's visits to the market occasionally with no predictable pattern. Family labor applied to crops or animals was identified but off-farm employment or the amount of hired labor was not quantified except generally and is indicated by broken or solid arrows.

For most farm output there is a direct relation to the market, absent where there is little sold or when the household changes the characteristics of the product before sale (e.g. wool to yarn, milk to cheese, or manure to dung cakes). Household modification is shown by solid arrows from crop or animal products to household to market." [McDowell, 1984]

McDowell and Hildebrand intend that the schematic diagram developed through a survey of the present situation be translated into mathematical relationships and then used to "identify features at the subsistence level which [analysts] wish to examine in greater detail."

As models are expressions of "ideas in the heads of observers," there will be an infinite variety of possible models of farming, food, agricultural, or natural resource systems. A number of other examples will suffice to underline this *fundamental* idea.

Models of Agriculture as Managed Ecology

The concept of agriculture as managed ecology and the contingent view of farms as agro-ecosystems have inspired many modeling attempts. Such a conceptual model has been created by the Hawkesbury group in Australia (Bawden et al., 1984) and has influenced many faculty members and other professionals in institutions in the U.S., Asia, and the Pacific. The model provides a basis for thinking through what might comprise an effective education for agriculturalists and natural resource managers as well as for thinking about the nature of agriculture.

In the Hawkesbury group's view, farming is a human activity in which farm family members attempt to manipulate the physical and biological elements of their environments for their own defined (although frequently unclear) purposes. These purposes are markedly influenced by the constraints imposed by the nature of the physical environment and by interactions with a wide range of other human beings. There are also profound interactions among family members and, of course, changes within each individual. All in all, then, farming is an extremely complex activity conducted in ever-changing environments.

An important dimension of the Hawkesbury approach is the particular concern for and focus on farmers in relation to their farm as a human activity system. In attempting to apply Checkland's soft systems approach (which will be introduced later in this chapter), the Hawkesbury group believes that the role of professional agriculturalists is to help farm families to learn to manage change more effectively so that their now-threatened life-style will be sustainable in the future, irrespective of what happens to the environments in which they operate. Sustainability in this context, therefore, refers to the interactive complex between the family and its physical, biological, socioeconomic, and cultural environments. And at base, it is a function of the effectiveness of learning by family members involved in the management of those interactions.

At the center of their model, they put the human "management subsystem," comprised of the three major human activities sometimes attributed to the management function. They chose this particular view because they wished to express the notion that each of the three functions involved a different set of inquiry methodologies. Thus, drawing heavily on ideas of Chaudhri (1969) and of Kast and Rosenzweig (1981), they presented management as a hierarchy of inquiry routines (or sub-subsystems).

> At the STRATEGIC level, we saw people acting to relate their farming activities to the characteristics of the environments with which they needed to interact. This was seen as strategic "purpose setting," deciding on the overall thrust of what it was that the family was attempting to achieve and providing constant adjustments to those plans as conditions changed. The sources of these changes are mandated from within the person, between persons and from "outside" the farming system. Objectives (at this level of human activity) tend to be "satisfycing" and problem solving has a judgmental focus. The strategic level is relatively open to the system's environment and innovations are responses to its dynamic and variable nature. At the other end of the scale we perceived the operations level where the primary task is to carry out those activities necessary to move towards the achievement of the set purposes. Operations are relatively insulated from changes in the environment and hence comparatively stable and fixed. At the allocative level, the primary concern is to integrate the internal operations through resource allocations in ways consistent with the strategies and innovations. The PURPOSES derived from strategic activities therefore set the whole system in motion (as if it were an entity!). [Bawden et al., 1984]

Figure 3.11 presents the management subsystem at the lowest level of abstraction as a component of a stereotyped human activity system.

Having established the "people perspective," they turn to what it is the people are actually trying to manage. Here they slide into an ecological mode. Farming can be easily embraced by the concept of a managed ecosystem. Further, the simplest way to indicate the complexity of natural ecosystems is merely to illustrate trophic relationships between those organisms that transform solar energy to chemical energy (autotrophs, essentially green plants) and the animals that feed upon the former (heterotrophs). The third trophic function to be included is decomposition, whereby nutrient wastes from the other two subsystems can be recycled. This is represented in Figure 3.12.

To put these two submodels together, they add inputs and outputs. All farming systems can be examined in terms of the interrelationships between inputs and outputs: Productivity is the relationship between the amount of output that

FIGURE 3.11 Hawkesbury management subsystem.

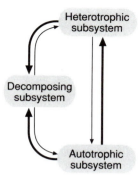

FIGURE 3.12 Trophic relations within a managed ecosystem.

can be transformed from given inputs and is thus a measure of the efficiency of the transformation process. This is represented in Figure 3.13.

Elements of this model concerned with environmental effects caused more debate than all others combined. Hawkesbury group members argued about modeling conventions, terminology, and underlying concepts. They sought to illuminate the forces in the natural and social environments that farmers confronted. The view that finally prevailed recognizes that members of the farming family must account for the potential impacts of their operations on the environment in the face of certain forces and changes.

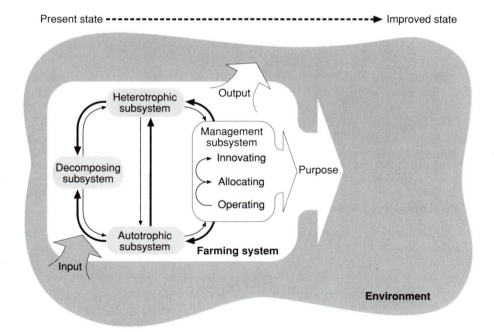

FIGURE 3.13 Inputs and outputs added to subsystems.

Present state --> Improved state

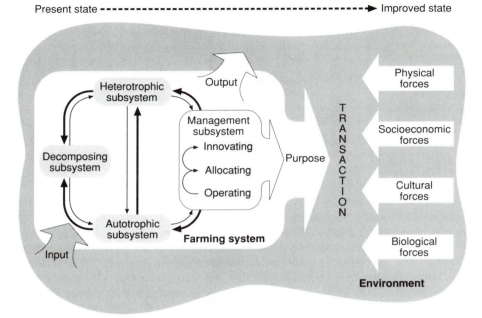

FIGURE 3.14   Hawkesbury model of farming as a human activity system.

The Hawkesbury group believes that farmers must be aware of the impact of these forces on them and, in turn, of the impact of their practices on the environment. There is ample evidence to suggest that in Australia, the principal focus of their concern, these dimensions are ignored to a significant degree and at great personal, social, economic, and environmental cost.

Debates notwithstanding, the completed model has been disseminated in the form presented in Figure 3.14. Many have found it a useful way to begin to think about farming as if it were a human activity system.

The Hawkesbury model builds on the learning cycle introduced in Chapter 2 because it focuses on people in a problematic situation as prospective learners and makes explicit use of the soft systems approach, which is also concerned with learning.

The Spedding Model

Another modeling approach encountered in the food and agricultural systems literature was developed by Colin Spedding (1979, 1980, 1984) of the University of Reading in England. He describes agricultural systems "in terms of their main components and interactions, all related to the central output or purpose of the system." On this basis, the intention is to impose a logic on the modeling process that (1) relates the relative importance of components to the main purpose or output and (2) allows the identification of subsystems. Spedding has developed a standard for representing agricultural systems involving circular diagrams to provide a logical framework for determining what are the components of a particular system. Spedding's procedure begins by putting in the center of the circular diagram the output of the system that is of primary concern. In the case considered in Figure 3.15, the output of primary concern is *profit*. Then, grouped in a ring around this center, are major factors that influence this output. The items on this ring may, in turn, be influenced by yet other factors which are grouped around a third ring. Factors on the same ring that influence each other can also be connected with arrows, and various graphical conventions can be used to indicate influences, relationships, boundaries, and so on.

Modeling of an actual case begins with gaining answers to two questions:

1. What is the system to be improved?
2. What constitutes an improvement?

The answers to these questions, which derive from a dialog with the farmer-client, determine how modeling proceeds — specifically, what it is that is to be modeled.

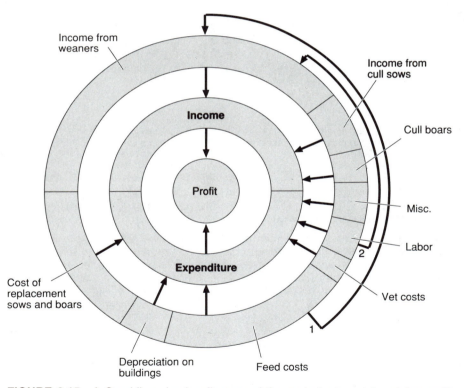

FIGURE 3.15 A Spedding circular diagram of the main factors determining profit in pig weaner production. (Circular arrows: 1, Cost of feed may affect performance and profit; 2, labor costs may have similar effects.) (Reprinted from *An Introduction to Agricultural Systems,* C. R. W. Spedding, © 1979 Elsevier Applied Science Publishers, Ltd., Barking, Essex, England. Figure 2.9, p. 26. Reprinted by permission.)

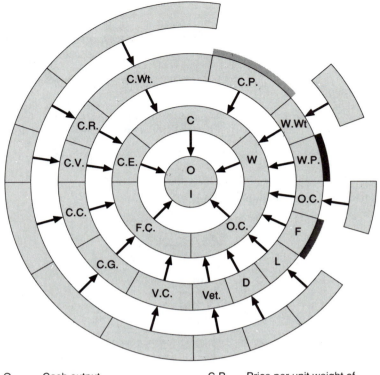

O	= Cash output	C.P. = Price per unit weight of
I	= Cash input	lamb carcasses
C.E.	= Value of cull ewes	W.Wt = Weight of wool
C	= Value of lamb carcasses	W.P. = Wool price
W	= Value of wool output	C.R. = Culling rate
F.C.	= Feed costs	C.V. = Value of a cull ewe
O.C.	= Other costs	Vet. = Veterinary costs
C.C.	= Cost of conserved forage	D = Flock depreciation
C.G.	= Cost of grazed forage	L = Labor costs
V.C.	= Value of concentrates fed	F = Fixed costs
C.Wt.	= Weight of lamb carcasses	

FIGURE 3.16 A Spedding diagram of sheep production showing how firm (dark shading) and tentative (light shading) boundaries may be defined in terms of the independence of the factor (for example, W. P.) in relation to the effect on any part of the production system (for example, C. V. is affected by C. R., and C. P. may be influenced by c.wt.) (Reprinted from *An Introduction to Agricultural Systems,* C. R. W. Spedding, © 1979 Elsevier Applied Science Publishers, Ltd., Barking, Essex, England. Figure 2.11, p. 28. Reprinted by permission.)

This is illustrated in Figure 3.16, which presents a model of a system concerned with the profitability of sheep production. The circular diagram shows "how firm (dark shading) or tentative (light shading) boundaries may be defined, in terms of the independence of a factor; e.g., wool production (W.P.) in relation to the effect of any part of the production system. For example, the value of cull ewes (C.V.) is affected by culling rate (C.R.) and the price per unit weight of the lamb carcasses (C.P.) may be influenced by the weight of the lamb carcasses (C.wt.)." (Spedding, 1979)

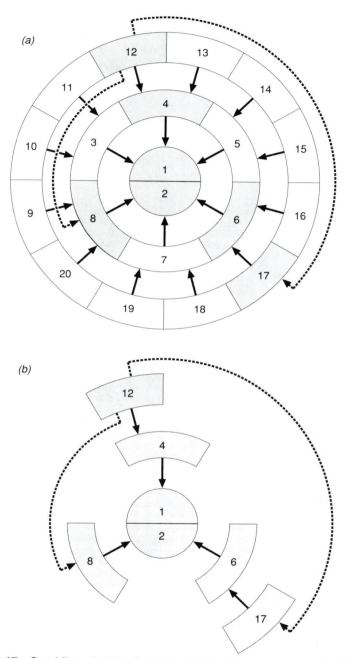

FIGURE 3.17 Spedding circular diagrams identifying a subsystem. (*a*) The subsystem identified by shading. (*b*) The subsystem extracted. (Reprinted from *An Introduction to Agricultural Systems*, C. R. W. Spedding, © 1979 Elsevier Applied Science Publishers, Ltd., Barking, Essex, England. Figure 2.12. p. 29. Reprinted by permission.)

Figures 3.17a and b demonstrate how the diagram is used to "extract" appropriate subsystems. The identification of a subsystem concerned with the effect of component 12 is illustrated. In Figure 3.17a, the subsystem is identified by shading. In Figure 3.17b, the subsystem identified is extracted and visualized separately so that the interactions are more distinct.

The Nieswand–Singley Model

George Nieswand and Mark Singley of Rutgers University have created a conceptual model that is intended to encourage students to reflect on agriculture

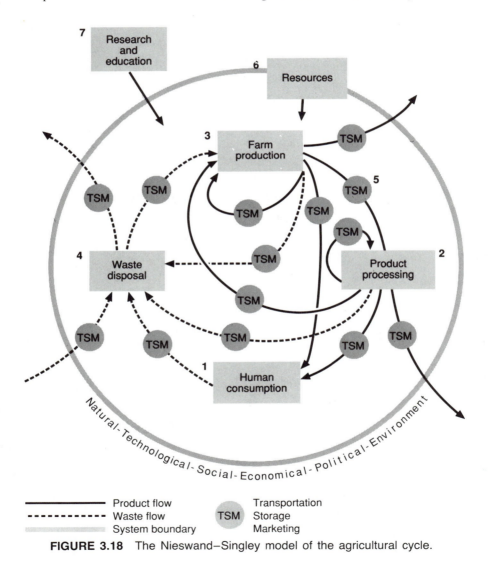

FIGURE 3.18 The Nieswand–Singley model of the agricultural cycle.

as an interrelated complex of activities and flows. It embraces traditional concerns with farm production and also reaches beyond those concerns to include issues such as environmental impacts. And like Spedding, as noted previously, they have adopted a circular convention. The general model, which is presented in schematic form in Figure 3.18, contains seven subsystems or "general activity blocks" whose interrelationships define what the authors see as the essential nature of agriculture. The seven blocks are (1) human consumption, (2) product processing, (3) farm production, (4) waste disposal, (5) transportation, storage, and marketing, (6) resources, and (7) research and education. The order in which the blocks are numbered coincides with a process of discovery appropriate to students with nonfarm backgrounds.

The various activity blocks are connected by streams of product and waste flows. The authors point out that it would be equally valid and interesting to describe interconnections in terms of energy and information flows. The cycle is delimited by a conceptual boundary that provides a basis for considering the complex environment within which it functions; as was true of the Hawkesbury model, here the environment consists of natural, external technological, social, economic, and political factors that constrain or move the system.

In Figure 3.19, the human consumption block is explored from the perspective of a consumable food product. Other possibilities not represented here include (a) the product stream derived from a parent product (e.g., corn → corn on the cob for direct human consumption, corn flour to be used for "indirect consumption," and dye stock used for nonconsumption industrial purposes), and

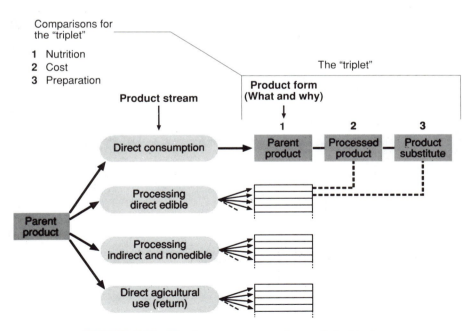

FIGURE 3.19 The human consumption activity block.

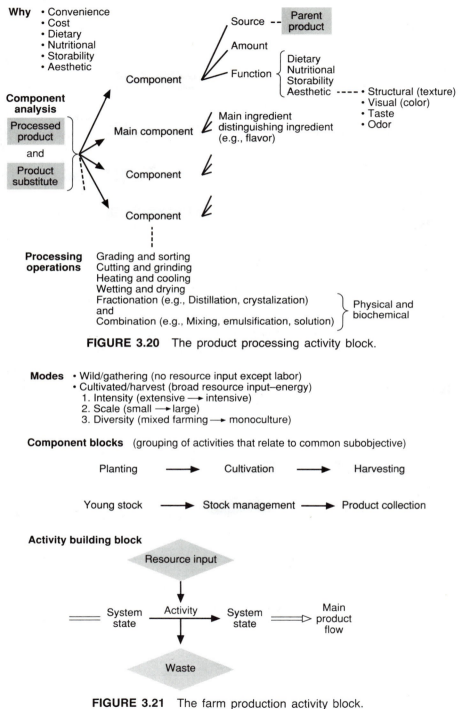

Why
- Convenience
- Cost
- Dietary
- Nutritional
- Storability
- Aesthetic

Source -- Parent product

Amount

Component — Function
- Dietary
- Nutritional
- Storability
- Aesthetic ---- • Structural (texture)
  • Visual (color)
  • Taste
  • Odor

Component analysis

Processed product

and

Product substitute

Main component — Main ingredient distinguishing ingredient (e.g., flavor)

Component

Component

Processing operations
Grading and sorting
Cutting and grinding
Heating and cooling
Wetting and drying
Fractionation (e.g., Distillation, crystalization)
and
Combination (e.g., Mixing, emulsification, solution)

} Physical and biochemical

FIGURE 3.20   The product processing activity block.

Modes
- Wild/gathering (no resource input except labor)
- Cultivated/harvest (broad resource input–energy)
  1. Intensity (extensive → intensive)
  2. Scale (small → large)
  3. Diversity (mixed farming → monoculture)

Component blocks   (grouping of activities that relate to common subobjective)

Planting ——→ Cultivation ——→ Harvesting

Young stock ——→ Stock management ——→ Product collection

Activity building block

Resource input

System state === Activity → System state ⇒ Main product flow

Waste

FIGURE 3.21   The farm production activity block.

(b) the "triplet" of parent product, processed product, and product substitute (e.g., oranges, frozen orange juice concentrate, and powdered instant orange drink).

Figure 3.20 is an outline of the product-processing block and places particular emphasis on the analysis of components, that is, the reasons and the required inputs to manufacture a particular product or product group. The variety of common processing operations is also represented.

Production is the focus of Figure 3.21. Here, the authors' choice of entry point is the spectrum of production modes from simple hunting and gathering of wild food and fiber sources, with no controlled resource input required other than labor, to complex modern intensive agriculture, characterized by a broad spectrum of high technology (e.g., chemicals, machinery) and basic resource (e.g., land, capital) inputs. At a lower level of specificity, the submodel also focuses on basic required human activities such as planting, and at an even lower level, it examines inputs to and outputs of specific activities, the "activity building blocks" in the schematic subsystem model.

The fate of waste products from all the activities of our food, agricultural, or natural resource management system is an issue that is often overlooked or ignored. Waste disposal is taken up by the model in Figure 3.22 from the source/sink standpoint. The main concerns expressed in the schematic are with internal and external *sources* and alternative *sinks*, or modes of possible utilization or ultimate disposal.

The transportation, storage, and marketing (TSM) activity block involves operations that determine the spatial and temporal distribution of inputs and outputs of the other blocks or subsystems. These movements are represented in the form of a matrix in Figure 3.23 and are visualized as a mechanism to match the actu-

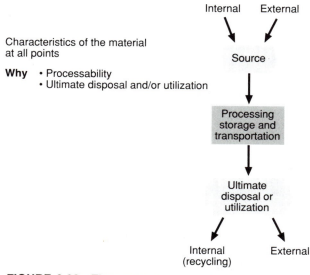

FIGURE 3.22   The waste disposal activity block.

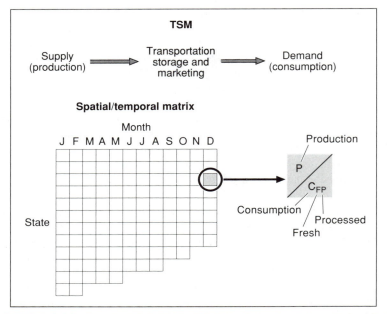

FIGURE 3.23   The transportation, storage, and marketing activity block.

alities of the product input stream (supply) with the spatial and temporal dimensions of needs and wants of the end user (demand).

Strictly speaking, in the Nieswand–Singley model, the resources block is not an activity block in the sense that other blocks have been so labeled. Instead, this is the source of inputs from outside the cycle which, in some way, fuels the activities represented by the other blocks in the system. The example presented in Figure 3.24 offers various categories of natural, artificial, and human resources for review.

The final block in this model of the agricultural cycle is research and education, which are important to the system for their informational, technological, and human resources impacts. The U.S. version, embodied in the land grant sys-

Natural	Human	Created	
Soil Water	Management Labor	Physical and biochemical: Buildings Machinery Electricity and fuel Chemicals Seed and plant stocks Animal stocks	Nonphysical: Information Social Economic ⎱ Programs Political ⎰

FIGURE 3.24   The resources block.

The Land Grant Concept

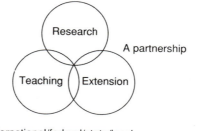

• International/federal/state/local
  institutionalization, implementation, and interaction

• Past/present/future?

• Public and private (industrial and farming) sector activities

FIGURE 3.25   The research and education activity block.

tem of institutions combining teaching, research, and extension, is represented in Figure 3.25.

The main blocks of the overall system, which was presented in Figure 3.18, are tied together by various flows and product streams involving such things as mass, energy, information, time, and, ultimately, measures of performance, which might include assessments of efficiency, productivity, nutritional value, profitability, and sustainability, depending on how the user chooses to employ the model.

Preceding sections of this chapter have provided a wide-ranging overview of the hard systems inquiry process, particularly emphasizing the quite varied approaches to modeling that have been employed in agriculture and resource management. The chapter has also reviewed some more specific models that are intended both to provide analytic frameworks and to tell you something about agriculture and the environment. Some of these approaches are also notable for introducing novel elements into the systems field. For example, the Hawkesbury group and Nieswand and Singley invite attention to the idea that *human activity* is the bottom-line issue in any model of agriculture. Spedding emphasizes that the real purpose is not modeling for its own sake, but *improvement*. The Hawkesbury group and the Farming Systems Research approach emphasize the integration of a variety of basic science, technological, and socioeconomic perspectives. These innovations provide a springboard for introducing the soft systems approach in the next section.

Soft Systems Inquiry

Soft systems analysis is a type of inquiry and a problem-solving approach, not an object of study or improvement. As was previously suggested in both the Hawkesbury and the Nieswand–Singley models, the object of soft systems analysis is the *human activity system* (HAS) (Checkland, 1981). The human activity system differs fundamentally from natural and designed systems. The latter have a degree of objective existence and identity independent of observers. While a human activity system may be defined in seemingly objective terms as *sets of purposeful human activities,* it can be manifested *only* as perceptions by people, along with the meanings they are free to attribute to those perceptions. In other words, while the basic components of human activity systems are drawn from a universe of real, observable *activities,* it is up to an observer, in consultation with the people involved in a situation, to select the activities assigned to a given human activity system. In the case of this book, the focus is on human activity systems in which the activities have something to do with food, agriculture, and natural resources. And many of the specific activities will be familiar to laboratory and field scientists because they are the kinds of transactions between people and the physical world that they investigate experimentally.

In practical terms, if someone were to ask, "What human activity system are you studying?" you might answer, "It depends on whom I last spoke with!" This is in accord with the way Spedding initiates inquiry, as was discussed in a previous section. According to Checkland, "There will . . . never be a single (testable) account of a human activity system, only a set of possible accounts all valid according to particular *Weltanschauungen*" (1981:14). Human activity systems and their problems are messy and ill-defined. This rules out use of a hard systems approach to inquiry, and an alternate inquiry approach is mandated. This is where the soft systems approach comes in.

Chapters 4 through 9 of this text present fully the premises and procedures of soft systems inquiry. Therefore, what follows is an overview of the approach. Following Checkland (1981), the seven major stages of the soft systems approach to inquiry are outlined in Figure 3.26.

Stages 1 and 2 are aimed at making sense out of a situation. Stage 1 involves the initiation of an open-minded inquiry, gathering basic facts from written sources and the people involved to see what makes the situation problematic. This also is when you begin to find out what **W**s are relevant to the situation even as you attempt to establish yourself as a neutral analyst/facilitator in relation to the people who are experiencing difficulties. Being a facilitator involves founding a *mutual learning* relationship with the people such that they are fully involved in the process of fact-finding, planning, testing, and implementation from beginning to end. Chapter 4 presents basic techniques for building rapport with people, as well as fact-gathering and also outlines approaches to describing a situation for stage 2.

Stage 3 begins the application of *systems thinking* to gain insight into what improved systems might be. This initiates a process of design in which the beneficiaries, decision makers, and other actors are identified, focal transformations (what improved systems do) are designed, and features of the environment in which the systems may operate in the future are specified. A key technique is to identify and describe the relevant **W**s of the people involved because these largely determine the relevance of the improved systems tentatively identified. Stage 3 techniques are presented in Chapter 5.

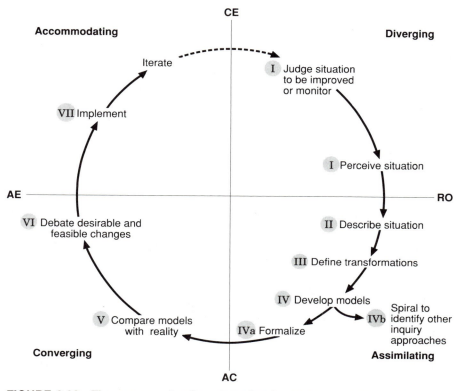

FIGURE 3.26 The process of soft systems inquiry as an application of the learning cycle. (After Checkland, 1981, modified.)

Stage 4 continues stage 3's application of formal systems concepts. The key technique is developing human activity system (HAS) models, and this is also the time for "spiraling" in inquiry to tap useful knowledge, using hard systems, technology development, and basic science approaches. HAS modeling and associated techniques are described in Chapter 5.

Stage 5 requires that we take our new conceptualizations back to reality and compare them with the original situation, particularly with our original description of it that was developed in stage 2. Chapter 6 discusses the procedures appropriate to this vital step.

In stage 6 the appropriate people in the situation are assisted to consider the desirability and feasibility of the proposals for change that emerged from the earlier stages of inquiry. Do the proposals actually address the issues of real concern? And will they work, given the established structure of the situation? Are the proposals desirable? Can people agree on a sense of future improvement, what they want the future to be like? Techniques for facilitating debate in various settings are presented in Chapter 6.

Stage 7 of the approach is the actual implementation of the developed and agreed-upon changes that are aimed at solving problems or improving a situa-

tion. This is not described as the final stage of the approach because it is important that there be some monitoring to validate that the changes have actually
been put in place and that they effectively improve the situation. Indeed, it is
realistic to be ready to go through all or some stages of the soft systems
approach again when a need for modification or revision becomes known.
Chapter 7 discusses approaches and techniques associated with implementation.

Because managers handle complex situations involving groups of people in
which each person has some say in and control of events, they need a new systems-based approach. Even the purposes of an inquiry into a current situation
may vary widely and entail different end results because what constitutes improvement depends on whom you talk to. Thus, the soft systems methodology
is particularly useful when value differences exist and the goals of an improved
state are hard or impossible to agree upon.

Some situations just don't lend themselves to developing clearly defined statements of what an improved state might look like. People's views of an end state
may differ. The food irradiation concern presented in Chapter 2 is an excellent
example. What state department of energy officials, economic planners, university
food researchers and technologists, business venturers, and community members
might believe to be the satisfactory end state may be very different. Indeed, it is
unrealistic to expect clearly defined objectives to emerge from a multiparty situation, at least not comparable to what sole actors in a situation are willing to accept.

The soft systems approach to inquiry has great utility with the innovating and
strategic planning functions of management. It is also well suited to those situations in which multiple organizations are involved, such as farm cooperatives,
shipping companies, government regulatory agencies, and food processors and
distributors, that interact with and depend on each other.

Some observers feel that more and more of the problems we face in food,
agriculture, and natural resources are of this sort — value-laden, full of complexity in terms of the amount and kind of information that must be dealt with, and
where few parties to the situation would be interested in anything more than
finding satisfactory improvements to deal with their own basic concerns. Some
would say that many agriculturalists have approached problems at too low a level
of complexity and that they must begin to account better for the human activity
systems in which their technological concerns rest.

In the past, the tendency has been for these types of problem situations to be
put in the "too-hard" basket or reduced to a particular specialty component (agro-
nomic issues, soil issues, animal husbandry issues, economic issues, or whatever). In doing so, a technology development or applied-science approach to
inquiry is adopted. Many agricultural or natural resource professionals then see
the situation through the reduced perspective of a "technological fix." Furthermore, even the utility of what the technical advisor could offer is often lost because
in the process of inquiry and the application of the principle of reduction, there
is only one particular technical solution that is worked on. These "solutions"
have a track record of being rejected by the community or the agricultural or
natural resources clientele groups for whom they were intended to be useful.

Career Opportunities Involving Soft Systems

What kinds of career areas can and should use a soft systems approach to inquiry? According to the U.S. Department of Agriculture, there are many openings for people who will be able to deal with complex, value-laden situations on a routine basis (Coulter et al., 1986). The job categories in which there will be more openings than trained graduates in the next decade are applied scientists, engineers, and related specialists who will work in the public and private sectors (e.g., landscape architects, animal scientists, agricultural engineers, plant scientists, safety engineers, nutritionists, soil scientists, veterinarians, rangeland scientists, water engineers, weed scientists); management and financial specialists; and technical service representatives within the marketing, merchandising, and sales representative category. Within the social services professions, there will also be more jobs than trained graduates. Good opportunities will exist for personnel and labor relations specialists, recreation workers, youth workers, naturalists, regional planners, and community development specialists with a strong natural resources or agricultural background. In addition, and within this same category, the highest number of possibilities exists for dietitians and nutrition counselors. While location will somewhat determine whether there are jobs or not, those contemplating careers in the agricultural or natural resource production areas will also gain much from learning soft systems approaches to inquiry and situation improvement (e.g., farmers, aquaculturists, forest managers, game ranchers, fruit and vegetable growers, food processing managers, nursery product growers, ranchers, tree farmers, turf producers).

Systems Hard and Soft: A Comparison

Comparisons of these two basic approaches will improve our understanding of when one or the other will be useful to our problem-solving activities. First, the approaches differ regarding the point in the inquiry process at which they employ systems thinking. The hard systems approach uses systems concepts and thinking during the problem identification stage, when the problem is defined and a model of the relevant system is described. The problem is not reduced in the manner of the scientific method. Rather, the problem is sculpted to fit the requirements of the optimization assumption (see subsequent text). In the soft systems approach, the focus is on problematic situations, and modeling involves improved human activity systems. An improved system is conceived only after careful descriptions of the current situation are developed. In summary, hard systems start with a systems model, whereas soft systems defer modeling to a much later stage.

The two approaches also differ on whether or not systems actually exist. The typical hard systems view is that agricultural and natural resource systems exist in the real world. The soft systems view emphasizes the heuristic value of looking at proposals for improvement as if they were a system of parts interacting with each other and describes them as such. The hard systems thinker tends to say, "Out there in the real world are agricultural systems that can and should be described." The soft systems analyst says, "It is better to recognize that systems are actually a function of our ability to think abstractly. Hence, we will look at what is happening and describe, analyze, and synthesize what we see. We then will conceptualize improved situations as if they were systems, but we will do this only after using other techniques that help us understand the situation."

Modeling is the sine qua non of hard systems inquiry. In fact, if one were to ask many systems agriculturalists to describe their preferred inquiry technique, the majority would say that they were interested in developing models. Soft systems inquiry emphasizes encouraging those involved in a situation to propose improvements. Models serve the secondary role of stimulating debate, or else they are used to examine the consequences of particular proposals for improvement.

In both the hard and soft approaches, systems models will include a recognizable boundary, inputs, outputs, essential transformations, and some measures of performance. Subsystems and their interactions will be identified. In hard systems inquiry, there is a definite preference for examining quantitative features of situations, and mathematical relationships are sought wherever possible. The mathematical model is often built by relying on data collected using the scientific method of inquiry (or reductionist approaches) and economic and demographic data available from such sources as state and federal Departments of Agriculture, the U.S. Census Bureau (a division of the federal Department of Commerce), the U.S. Department of Labor, and other governmental bodies involved in regulating environmental and economic affairs. Once the mathematical model has been constructed, it is used to simulate some aspect of the real world. Some proponents of the approach come close to forgetting that the model is just an approximation whose value rests on its realism and the amount, kind, and quality of data fed into it.

While the hard systems approach utilizes mathematical modeling almost exclusively, the soft systems approach may use mathematical modeling when appropriate, but only after first developing one or more human activity system models. The premise here is that all food, agricultural, or natural resource enterprises can be thought of as human activity systems. That is, they might be conceptualized along the lines of the Nieswand–Singley model, as for example, people-managed systems comprised of major activities needed to produce, process, distribute, market, or consume products of varying kinds. The basic level of hierarchical complexity handled is people and their interactions with other life forms and physical factors present in the environment. Hence the focus of soft systems inquiry is on human management of agricultural, and natural resources and the resultant effects and interactions.

Yet another major difference in methodology between hard and soft systems inquiry is whether or not improved end states or goals can be defined clearly and unambiguously at whatever organizational level is being investigated (e.g., at the farm-family level or above, such as the agribusiness level, cooperative level, state level, national level, etc.). Hard systems analysts believe that one need not expend too much effort in understanding the whole of a situation. Rather, they focus on defining the desired goals. To a large degree, the definition of goals is biased by the single most important premise of the hard systems approach. Known as the optimization assumption, it equates effective system performance, and therefore the character of desired goals, with maximum efficiency or productivity. It thereby also restricts the scope of inquiry to system properties that are quantifiable. The approach also ignores qualitative properties such as resilience, flexibility, and sustainability.

In contrast, the soft systems approach recognizes that goals or desired end states are often ambiguous, conflicting, and constantly shifting. Moreover, they will be seen differently by the key individuals and groups in a situation, so that a varying sense of improvement is also part of the problem to be described. Because of these philosophical differences between respective proponents, the corresponding inquiry processes also differ.

You may gather from this discussion that you must choose between hard systems analysis or soft systems analysis. In fact, this has been the prevailing tendency, although systems scholars and practitioners are increasingly recognizing that it need not be an either/or choice. Nevertheless, the hard systems approach has been the dominant paradigm in economic analyses of enterprises, whole industries, and sectors, in risk assessment, natural resource management, and wildlife management, to name a few important applied fields. According to critics, proponents of the hard systems approach have expressed a metaphysical adherence to the optimization assumption, and users of simulations have been attracted because of the prospect of "automatic decision making," untrammeled by concerns for fickle human nature. At a minimum, one must learn to match the kind of approach and inquiry process to the nature of the problem one is asked to handle.

Conclusions

There is an unquestionable need to generate fundamental principles to explain the natural phenomena underlying agriculture. Those phenomena, however, must be put into ever wider contexts and with a particular emphasis on relationships to human activity. Furthermore, many of the phenomena cannot be explained in such fundamental terms. If they were, then agriculture and natural resource management would present no challenge to its practitioners. Every little aspect would have a predictable outcome, and that clearly is not so!

As Chapter 1 tried to demonstrate, America's agricultural and natural resource enterprises are having difficulty coping with natural, as well as social, political, and economic forces and their growing interdependence and competition for use of limited land and water resources with other sectors of society. The very integrity of our food, agricultural, and natural resources sectors is viewed by many to be at risk. Our educational institutions that prepare professionals for the future have tended to plod on oblivious and undeterred, ignoring crucial interactions and dynamic forces affecting them, not explicitly providing the inquiry processes needed to cope with the range of complexity that a person will have to deal with on the job.

All four approaches to inquiry — the scientific method, applied science/technology development, hard systems, and soft systems — have their place in tackling problematic situations in food, agriculture, and natural resources. More emphasis must be placed on systems thinking in these areas; in learning about them, we should start by viewing food, agricultural, and natural resources situations holistically and then we should think about them as if they were systems, and finally reduce them, *not* the other way around. In other words, an urgent re-look at the fields of food, agriculture, and natural resources from a systems perspective is as essential as are new ways of learning about food, agriculture, and natural resources in terms of those systems.

Spiraling Through Several Methodologies

Chapter 2 presented a model of learning that is relevant to our responsibility to inquire into the problems and situations we will encounter in our careers. It was stated that learning is a process of transforming experience into meaning. Furthermore, it was indicated that not only do we need to find out about situations, but we will be expected to translate meaning into action. Inquiry or problem-solving processes help us to take action by transforming experiences into knowledge. The finding out → taking action process was said to be cyclical and iterative, driven dialectically from one pole to the other on four critical dimensions: apprehending experience to comprehending experience (prehending); reflectively observing to actively experimenting (transforming); reducing or sustaining a holistic view of a situation (methodology); and trusting intuition or relying on rational, systematic, and systemic thought (insight).

This model of learning also recognized four interrelated learning activities: diverging, converging, assimilating, and accommodating. All methodologies of inquiry utilize the competencies involved in these activities. In order to complete all the phases of any known methodology, one must diverge, assimilate, converge, and accommodate. While the four methodologies introduced differ in the range of complexity they attempt to handle during the inquiry process, all four rely on all of the investigator's competencies.

Members of the Hawkesbury Agricultural College faculty in Australia have visualized the connection between learning styles, the type of methodology used, and the dimension of reductionism and holism as a spiral. Accordingly, we need to learn to spiral up and down, depending on the kind of problem or situation we face. That is, we need to learn to know when to reduce and when to view things in their entirety. We need to choose the methodology that is appropriate to the question we are asking (or that others are asking) and to the extent that the situation is definable.

Figure 3.27 is a spiral that shows the major phases of each of the inquiry processes described in Chapters 2 and 3 and the kinds of learning-style competencies that are needed in each phase. It also gives you a picture of how the

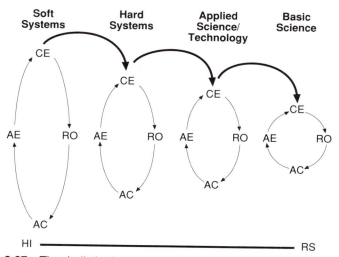

FIGURE 3.27 The holistic integration–reductionist separation axis: a spiral of interconnected learning cycles.

reductionism–holism issue is currently seen, what methodologies help us reduce problems, and which are more useful in helping us retain a picture of the wider context within which themes of concern are expressed. It is particularly relevant to stage 4 of the soft systems approach.

The next chapter begins a detailed explanation of what is involved in using the soft systems inquiry process. The learning competencies required at each phase of this process will be described. In addition, an overview of the premises and key inquiry techniques to use in each phase will be provided.

In many of the succeeding chapters, you will find practice exercises at the end, which you should do either individually or in small groups. Inquiry procedures cannot be learned very effectively by just reading and talking about them. One becomes competent only through practice. Therefore, reading this text will give you conceptual frameworks and suggested practices, but it does not insure that you will be at all proficient once you have read it. That will only come with practice, either under supervision or with other individual learners who learn new things by finding out and taking action!

REFERENCES

Bateson, G. *Mind and Nature: A Necessary Unity*. New York: Dutton, 1980.

Bawden, R. J., R. D. Macadam, R. G. Packham, and I. Valentine. "Systems Thinking and Practice in the Education of Agriculturalists." *Agricultural Systems* 13: 205–225, 1984.

Berlinski, David. *On Systems Analysis: An Essay Concerning the Limitations of Some Mathematical Models in the Social, Political and Biological Sciences*. Cambridge, MA: MIT Press, 1976.

Biosocial Decision Systems. *EZ-IMPACT: The Judgment Based Systems Modeling and Decision Analysis Program.* College Station, TX, 1987.

Blackie, M. J., and J. B. Dent. *Information Systems for Agriculture.* London: Applied Science Publishers, 1978.

Bonozek, R. H., C. W. Holsapple, and A. B. Whinston, *Foundations of Decision Support Systems.* New York: Academic Press, 1981.

Boughey, Arthur. *Strategy for Survival: An Exploration of the Limits to Further Population and Industrial Growth.* Menlo Park, CA: W.A. Benjamin, 1976.

Carter, R., J. Martin, B. Mayblin, and M. Munday. *Systems, Management and Change.* London: Harper & Row, 1984.

Chaudhri, D. P. *Education, Innovation and Agricultural Development.* London: Croom Helm, 1969.

Checkland, Peter. *Systems Thinking, Systems Practice.* New York: John Wiley & Sons, 1981.

Checkland, Peter. "Achieving Desirable and Feasible Change: An Application of Soft Systems Methodology." *Journal of the Operations Research Society* 36(9):821–831, 1985.

Churchman, C. West. *The Systems Approach.* New York: Dell, 1979.

Colinveaux, P. *Introduction to Ecology.* New York: John Wiley & Sons, 1973.

Coulter, K. J., M. Stanton, and A. Goecker. *Employment Opportunities for College Graduates in the Food and Agricultural Sciences: Agriculture, Natural Resources and Veterinary Medicine.* Washington, DC: USDA, Higher Education Programs, 1986.

Darwin, C. *The Origin of Species.* New York: Mentor, 1958; original 1859.

Forrester, J. W. *Industrial Dynamics.* Cambridge: M.I.T. Press, 1961.

Greenblatt, Catherine, and R. D. Duke. *Games-Generating Games: A Trilogy of Issue-Oriented Games for Community and Classroom.* Beverly Hills, CA: Sage, 1979.

High Performance Systems. *STELLA.* Lyme, NH, 1987.

Jacobs, Jane. *The Economy of Cities.* New York: Random House, 1969.

Jenkins, G. M. "The Systems Approach." *Journal of Systems Engineering* l(l), 1969.

Kast, F. E., and J. E. Rosenzweig. *Organization and Management: A Systems and Contingency Approach.* New York: McGraw-Hill, 1981.

Leontief, W. *Input–Output Economics.* New York: Oxford University Press, 1966.

Little, Michael A., and George E. B. Morren, Jr. *Ecology, Energetics, and Human Variability.* Dubuque: Wm. C. Brown, 1976.

Marks, James V. "Vetting Long-Range Planning Options with the Public: The Case of the South Saskatchewan River Basin Planning Program." *Social Impact Assessment* 90/92:8–15, 1984.

McDowell, R. E. "Systems Approaches to Livestock Production." Cornell International Animal Science Mimeo. Ithaca, NY: Cornell University, Department of Animal Science, No. 7, November 1984.

McDowell, R. E., and P. E. Hildebrand. *Integrated Crop and Animal Produc-*

tion: *Making the Most Available to Small Farms in Developing Countries.* New York: Rockefeller Foundation, 1980.

Mesarovic, Mihajlo, and Eduard Pestel. *Mankind at the Turning Point.* New York: E.P. Dutton, 1974.

Mihram, Arthur G. *Simulation: Statistical Foundations and Methodology.* New York: Academic, 1972.

Moran, Emilio. *Human Adaptability: An Introduction to Ecological Anthropology.* 2nd ed. Boulder: Westview, 1982.

Morren, George E. B., Jr. "From Hunting to Herding: Pigs and the Control of Energy in Montane New Guinea." In *Subsistence and Survival: Rural Ecology in the Pacific.* T. Bayliss-Smith and R. Feachem, eds. London: Academic, 1977.

Naughton, John. *Soft Systems Analysis: An Introductory Guide.* Milton Keynes, England: Open University Press, 1984.

Odum, Howard T. *Environment, Power, and Society.* New York: John Wiley & Sons, 1971.

Odum, Howard T., and Elizabeth Odum. *The Energy Basis for Man and Nature* New York: McGraw-Hill, 1976.

Parnaby, J. "Concept of a Manufacturing System." *International Journal of Production Research* 17(2), 1979.

Shanner, W. W., P. F. Phillipp, and W. R. Schmehl. *Farming Systems Research and Development: Guidelines for Developing Countries.* Boulder: Westview Press, 1982.

Smith, R. C. G., and J. Langlands, "Model Development for a Deferred Grazing System." *Journal of Range Management* 6, 1975.

Spedding, C. R. W. *An Introduction to Agricultural Systems.* London: Academic Press, 1979.

Spedding, C. R. W. "Prospects and Limitations of Operations Research in Agriculture — Agrobiological Systems." In *Operations Research in Agriculture and Water Resources.* D. Yaron and C. S. Tapiero, eds. Amsterdam: North Holland, 1980.

Spedding, C. R. W. "Agricultural Systems and the Role of Modeling." In *Agricultural Systems: Unifying Concepts.* R. Lowrance, B. R. Stinner, and C. J. House, eds. New York: John Wiley & Sons, 1984.

Tustin, A. "Feedback." *Scientific American.* Scientific American Offprint No. 327. Salt Lake City: W.H. Freeman, 1952.

CHAPTER 4

Making Sense out of Situations: Stages 1 and 2 of the Soft Systems Approach

This chapter is about how professionals in food, agriculture, and natural resources can learn to see problematic situations in new ways and from many perspectives. The perspectives of the principal people in a situation are particularly important. The relevance of scientific and technological perspectives remains an open question. To accomplish this objective, you must not only understand and assimilate the approach and the required ways of thinking, but also the action competencies that, depending on your preferred learning style, may go against the grain. At this stage of the inquiry process, the strengths associated with the divergent learning style come to the fore.

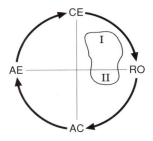

To recapitulate the argument in Chapter 2, the competencies needed to explore a situation initially are associated with divergent learning. We reach for meaning by moving into the situation to talk with people who have differing perspectives and by suspending our own biases about the situation. We concentrate on hearing people and summarizing accurately what they say: their values, interpretations of cause and effect, senses of meaning, beliefs about who is doing what to whom, views of what needs to be done, and expectations about what will happen if the situation goes uncorrected. In short, we learn by grasping meaning through direct experience with people in a situation.

In addition, we reflectively observe a situation by seeing people interact with each other and the environment and watching for patterns to emerge. We test our tentative interpretations by bringing a sense of meaning to that situation. Does our explanation fit? Does it account for what is happening? Does it seem to provide a guide for future action to improve the situation? In short, we build an adequate map of a situation by reflectively observing whether the meaning we tentatively impose on people's transactions with the environment and with each other adequately explains a situation and provides a comprehensive base for determining courses of improvement.

> To *reflectively observe* means to watch something and think about it without drawing conclusions. It is the learning activity that accompanies the question "What does this mean?" Reflective observation is a precursor of assimilation when we impose some kind of conceptualization. Thus, reflective observation leads to a preliminary ordering of elements, a tentative identification of objects, a proposed definition of relationships, and so on.

Many readers will say, "What's new in that? I do it all the time." To the contrary, most people don't use these competencies very well, but many people can learn to do so. Because this book is aimed at future professionals who are still in training, it is important that you start your careers by learning how to enter into situations fully, openly, and without bias. Many people in the applied sciences run into trouble because they jump to conclusions before they have reflected upon other possibilities. Too often, reflection is seen as a waste of time, whereas a little time spent here may avoid future problems.

The soft systems approach commences when we engage in what Checkland calls real-world activities (1981:163), that is, encounters with the concrete experience of people in their contexts. Activities of the real world are distinguished from systems-thinking activities, which pertain to the abstract conceptualization side of our learning cycle. Here, the first two stages of the approach require that we assemble the richest possible picture of a situation in which people express uneasiness about a problem or an opportunity. Our immediate task is to display the situation in all its detail and complexity so as to reveal a range of possible and relevant choices regarding the identification of themes of concern and changes desired in the structures and processes of a current situation. Here, too, the stance of the analyst is neutrality regarding the direction of change and the merits or priority of one theme of concern relative to another. Accordingly, you will draw on your abilities to learn, first divergently, and then assimilatively.

This chapter introduces key concepts related to the tasks of Stages 1 and 2 of the soft systems approach. It also presents several inquiry techniques for rapidly identifying and summarizing information you discover about a situation. During stages 1 and 2, your principal task is to collect and synthesize people's accounts of current events and past history and their opinions regarding people, tasks, processes, and things as well as documentary information such as organizational charts, statistics, financial data, government reports, rosters of individuals or groups, and specifications of products or services involved.

No matter how experienced (or inexperienced) we are, all of us take a stab at giving meaning to new situations. Inevitably, we use the results of past learning about similar situations as a guide to interpret new situations. Yet, a key point of this chapter is the need for professionals to see situations afresh and without preconceptions. How can we avoid seizing an interpretation too quickly? How might we put aside advice giving or prescriptive approaches to inquiry? How can we resist proposing solutions before we have adequately understood the problems and issues involved?

Research indicates that we tend to talk to people who are similar to us in values, attitudes, beliefs, and action styles (Rogers, 1981). Yet we know that situations are viewed differently by different people. It's easy never to talk to all people involved. While it is not common practice, it is common sense to try to talk to everyone in a situation. Most situations we face will not have a sense, value, or meaning common to all parties. These disparities are part of the situation you are supposed to uncover.

Your Role as Facilitator in the Inquiry Process

The soft systems methodology discussed in this book originally developed in the course of hundreds of management consultancies. Within the past decade it has been applied to the agriculture and natural resources sector. In the course of that development, two central questions regarding the behavior of consultants and analysts had to be dealt with. First, its creators adopted the role of facilitator rather than that of expert. It is suggested that the facilitator role is particularly appropriate to the demands of searching for improvements to complex situations. A facilitator's competence rests on his or her ability to guide people through a process of inquiry. A facilitator may also have expertise in one or more aspects of the situation, but the importance of the role is not attached to how much knowledge he or she has about some problem's technical aspects. Rather, it is the ability to assist people to think rigorously about their difficulties and to see their situations in new ways so that beneficial change is possible. The alternative role, which was rejected by the soft system approach's developers, is the common one of the "expert" who prescribes remedies and simply tells people what to do. In short, your role is that of a *change agent* or catalyst. You're there to help others to bring out and build upon their own perceptions of the dilemmas they face in given contexts (Naughton, 1983:19). In an attempt to reflect the special nature of the complex role you are being called upon to assume, the book usually uses the titles *analyst, facilitator,* or the two together, as in *analyst/ facilitator.*

Second, when you accept the responsibility to become involved in the improvement of any situation, you are no longer detached. You become part of it the minute an inquiry begins. In other words, *the intervention process affects the dynamics of the situation.* As this chapter discusses, there are ethical as well as procedural reasons for being open about your own values, goals, and objectives in the inquiry process; how you plan to get involved; and what you intend to do with the information you obtain. If you were hired by a group, then some of these questions are partially answered, but if you are a service representative, extension agent, or public official, it might not be so clear. These issues will be returned to later in this chapter and also in Chapter 9.

The role of client is at least as important as the role of facilitator. Throughout the book, the term *client* (or *client group*) is used to refer to the person or people who commissioned the inquiry process (Checkland, 1981:294) and the *people who are involved* for those (including clients) who want to know about or do something related to a food, agricultural, or natural resource situation. Clients and others involved have *feelings of unease* about a situation that range from profound fear to a sense of mismatch between what is and what might be to a vague feeling that things could be better if something were done about it. In the terminology used in this book, the *situation owner* is the person or persons among the clients who is the current or possible future decision maker in a situation. These people have various kinds of power that may be very influential in either maintaining things the way they are or contributing to any changes that may occur (for better or worse). As some people said in the 1960s, "They are either part of the problem or part of the solution!"

Key Inquiry Activities

The techniques you select to carry out the first two stages of inquiry should clearly demonstrate the values of participation, mutual learning, and empowerment of the people involved; of wanting to describe (rather than reduce) complexity; and of initially suspending the tendency to view the situation from only one perspective. There is also the premise that, no matter how technology-oriented a situation appears, it is better to start the inquiry process by fully understanding the human context in which, say, a particular food, crop, soil, animal, or plant concern resides.

Contrasting with the practice of hard systems analysis, the facilitator should avoid using systems terms during stages 1 and 2. Since the objective of these two stages is to describe human activity, its context, and the divergent interpretations people give it, using systems terms would be seriously misleading. For example, drawing systems boundaries and objectives (in the strict sense of these concepts) is impossible because rarely will all relevant actors agree on a set of concerns and objectives of improvement or believe that they can control the same subsystems. Systems terms are best left to later stages of the inquiry process when they can be applied with technical rigor *on a firm basis of experience.* Prematurely applying systems concepts will lead to false and inadequate understandings. Therefore, *in the initial stages, build the richest possible picture of the situation.*

Stage 1: Inquire into the Situation

The task objectives of the inquiry process during stage 1 are as follows:

1. Capture the perspectives of people in a particular context. Any technique used to do this should help people describe a complex situation rather than isolating "the" problem or prioritizing problems.

2. Capture the way that people associate their activities with time, place, impact, and outcome.
3. Grasp an understanding of a situation by engaging people and events.
4. Capture people's expressions of concern, opportunity, and hope and relate them to the structures and processes of the ongoing situation.
5. Assist analysts (yourself and colleagues) and the people involved in the situation to avoid the development of a restricted viewpoint from which to look at the situation (for example, we are agriculturists, so we will look at and hear only agricultural problems — or animal-related problems, or resource-use problems, or production issues, or whatever specialization is involved).

The conventional approach to inquiry into problematic situations is to start by identifying *the* problem. In the soft systems approach, that is exactly what you do *not* do. Why is this? First, the overwhelming majority of the situations you will face do not have one problem, but rather are characterized by a complex mess of problems that interrelate with one another. Moreover, these problems are not easily understood at the beginning of inquiry. If one's orientation is to look for *a* problem at the outset, the process of reduction is usually too fast and lacks rigor. Second, as was pointed out before, various people involved in a situation will view it differently. These differences of perception may be a determining feature of a situation. To look for *the* problem implies that all parties see the situation in the same way, which most often will *not* be the case.

Therefore, the inquiry process starts by looking *at* problematic situations rather than *for* a problem. In a problematic situation, people believe that if things were done differently, the situation would improve. A situation is comprised of people as individuals and in groups, themes of concern, a historical context that bears on the present, key human activities, decision-making structures, physical and biological environmental factors, the political–economic and social context, and relational climates.

In order to begin to understand a situation, you must enter it in a formal way. This means to take care that everyone is informed of your presence and the purpose of your involvement with them, including who you are working for (who is paying for the activity). Everyone is familiar with the adage "He who pays the piper calls the tune." In addition, you must let them know up front that the purpose of your involvement is not to "solve" their problems, but rather to facilitate their own development of alternative courses of action and their understanding of the consequences of taking those actions.

The procedures of stage 1 assume that at the beginning of inquiry, the situation is unstructured in your mind and in the *organization* embracing the people involved in the situation (the latter means slight group consensus but lots of individual ideas). While each person will have opinions about what a situation is like and what his or her sense of unease is, the situation will be unclear, complicated, and ambiguous to you. Avoid the temptation to prematurely label *the* problems, even though people in a situation will look to you as the expert who is there to identify and solve their problems — and will even pressure you to do so.

You advance in stage 1 by accumulating information. The tangible results may consist of news clippings, interview tapes and transcripts, personal notes and journals, organizational records and documents, and so on; all of it will be unclassified and disorganized (except possibly for retrieval purposes). In practical terms, you cannot avoid thinking about these materials as you gather them, and you will almost imperceptibly slide into stage 2. What you should be thinking about, however, is not closure on a problem or some other neat and final characterization, but rather the richness and variety of alternative ways of looking at things.

Stage Two: Describe the Situation

The following are the objectives of your work in stage 2:

1. Help participants display the situation so as to reveal a range of possible and relevant choices for improvement.
2. Fully describe the present and necessary structures and processes of a situation and the climate resulting from their interactions.
3. Fully describe the principal themes of concern or issues and the primary tasks associated with the current situation.
4. Prepare a synthesis report documenting the foregoing in written and graphical form.

The information gained from all sources is synthesized in a comprehensive oral or written report on the situation. Such a synthesis will cover a range of features of the situation, including a description of the individuals and groups involved, their themes of concern, and their corresponding **W**s; the historical context that bears on an adequate understanding of the present; key human activities involved in and related to the group's themes of concern; decision-making structures and processes; environmental factors related to the present situation, including the constraints; climates related to structure and process; and other qualitative and quantitative data on the physical, biological, economic, and demographic features of the situation that bear on various groups' themes of concern.

These reports are often written and shared with key people. For example, an executive summary—that is, a concise, digested version of the total report—can be prepared for circulation at a cooperative meeting, or to share with your employer, or as a "backgrounder" for a staff assembly, association meeting, or legislative hearing. These are mentioned as representative of the kinds of contexts in which you might be working. While the people involved should be fully engaged in gathering and synthesizing information on their situation, a designated person must take the lead to compile and present the actual synthesis report. Often the analyst/facilitator drafts the report working along with a small task group of representative people. (An example of the first draft of such a report will be found in the appendix at the end of this book—the Mucho Sacata Ranch case. The summary in the Chatham River case is somewhat more developed.) The presentation of a report to the people involved is critical to the process because only they are in a position to identify misleading, inaccurate, or missing information. It is particularly important that the holders of a particular **W** be the ones to comment on the adequacy of the representation of that **W**.

In summary, the synthesis report highlights at least five features of the situation:

1. The quantitative and qualitative aspects of structure, process, and climate.
2. The acknowledged mission or principal function of the various groups or organizations in relation to each other and to current themes of concern; this is called primary task analysis.
3. Themes of concern other than those associated with primary tasks, e.g., those relating to structure, process, or climate.
4. The significant features of the **W**s of the individuals and groups involved with the foregoing themes of concern.
5. Various proposals for improvement that the people involved present.

The importance of describing these essential features is underlined because they relate directly to succeeding stages of the inquiry process. This synthesis will be used in stages 3 and 4 to identify relevant systems and to design models of improvement to structure and process. Get it all down now, for you will need it later.

The following sections discuss in depth the concepts of structure, process, climate, primary task, and primary issue. Note that the meaning of **W**, *Weltanschauung*, has already been extensively discussed in Chapter 2.

Structure, Process, and Climate

The *structure* of a situation consists of the relatively durable physical, biological, and social patterns and organizations associated with a situation in a particular place and time. This can be thought of as the slow-to-change context within which people and other dynamic elements act. Structure may include aspects of the natural or modified environment; the "built" environment of buildings, roads, and other infrastructures; and social patterns of political, economic, and social institutions, corporate and community organizations, and associated beliefs and sentiments. *Process* is how things are actually done and by whom within the constraints of structure. Features of process are more changeable. *Climate* is the quality of the relationship between structure and process, how well things work together, *and* the resulting emotional response.

In order to identify relevant structural elements, we must develop an understanding of the setting and the relatively fixed background relationships, including aspects of history before a situation developed. If you think of a stage-play, you may see more clearly what is implied. In a play there is a setting (e.g., auditorium, stage, scenery, props, lights, curtain) and a set of relatively fixed background relationships (e.g., script, cast of characters) that reflect relevant history prior to the current staging of the play. All of that is structure. So structures are aspects of situations that support and contain its more dynamic aspects, including its constraints and limits. Against this background, the current action takes place. In a sense, a story will unfold within a historical context. In the stage-play metaphor, the director and individual actors have considerable latitude regarding how they actually put on the play, and they may even modify it from performance to performance. That is process. When the audience applauds, remains silent, or boos in response to their view of the interaction of structure and process, that is climate.

Structures in agricultural or natural resource situations include physical and biotic properties and the organizations and institutions that have been a part of a local community for a long time, laws and political institutions, established reporting relationships, traditional formal leadership patterns, and past alliances or alienations that may have arisen due to past events and that influence dialogue between groups in the current situation. The identification of crucial roles in a situation is central to understanding a situation's structure. The norms of behavior expected of people in those roles should become clearer, and the values people use to judge performance will also emerge. How power is acquired, exercised, retained, and passed on should surface also (Checkland, 1985:824). The structural features of a situation are elements that the people involved believe are slow or hard to change. In the Mucho Sacata Ranch case, one important structural feature — but not the only one — might be that "Old man Bell owns the whole lashup (but the banks own him)!" An environmental feature of that same case might be the rainfall regime.

Carter and associates (1984) identify types of social structural relationships: (1) causal or logical relationships, (2) personal or contractual relationships, and (3) inclusion or membership relationships. Examples of causal or logical relationships are A is necessary for B, or A has influence over B, or A is connected with B, or A is the controller of B. Examples of personal and contractual relationships are A dominates B, A sets the price that B must abide by, A is the only shipper for B products, and A owns the water that B needs. Examples of inclusion and membership relationships are A is part of B, or A is of a higher order than B, or A is a member of B, or A is *not* a member of B.

Within the structural frame is the action currently going on in a situation. Processes are the activities that occur within, through, or in spite of the structure. Analyzing information you have collected in order to understand processes involves finding out what kinds of physical interactions and transformations are occurring, how these are controlled (including how decisions are made), and what kinds of interchanges occur between key actors.

The *climate* of a situation is the match (or mismatch) and sense of ease (or unease) arising from interactions between structure and process in a situation. The slow-to-change features and the fast-changing features of a situation often produce tensions. Hence, some authorities refer to the overriding emotional tone of a situation as the climate. Your task here is to understand the emotional charges and motivations, why they exist and what conflicts are occurring over what or whom.

Before you move on to stage 3, note the climate features of the situation and discuss them with key people. Strategies for action, which will be designed later in the inquiry process, have to account for the climate. If people feel helpless or defeated, certain action plans may have to be included (and others precluded). In the fruit-pest control case and the suggested action of using ionizing radiation, originally cited in Chapter 1, the emotional climate for some actors could be characterized as scared and threatened and for others as annoyed because "these people just don't know the technical information or they would accept our way of doing things." Another possible source of distrust involving the mismatch between structure and process might be the failure of the authorities involved to consult all the interested parties before rushing to the planning and even implementation phases.

Often people describe concerns that are, by their nature, proposed solutions developed by *other* people whom they question. For example, food irradiation may be mentioned as a concern. It is useful to help people focus on what their concerns are with a proposed solution or line of action (food irradiation is a line of action intended to solve the problem of food spoilage). In many agricultural situations, there will be themes of concern regarding the management of the natural and social environments. Both will have to be identified. For example, issue statements from a fruit grower's perspective might be.

> We have got to find a way to kill the pests in fruit so that it is acceptable for market. It must be safe to people who process the fruit and eat the fruit and not hurt the environment. Furthermore, it must not require harvesting the fruit at critical moments because members of our families have second jobs and we can't afford to take our children out of school and skip work just because the fruit has to be picked within the next two days.

As field researchers have talked with agribusiness people, government officials, community members, and consumer advocates, certain types of concerns are mentioned again and again. A few observations and insights from this experience is sketched out here.

People may be concerned about a situation that results from a person, corporation, or agency of government doing something to them. Or they may have concerns that are a routine part of living or that occur in the normal course of managing an agricultural or natural resource enterprise. In practice, the two types of situations are often intertwined. When they arise, people tend to express at least one of five different kinds of concerns:

1. *Basic well-being or survival concerns* are perceptions of threats to safety, health, or well-being. A few examples are concerns over the use of too much pesticide on vegetables; the possible effects of spraying a farm or forest area on neighbors' health, foliage, or pets; and pollution of the water supply due to agricultural activities.

2. *Project-related concerns* are expressed by one or more groups when an agricultural or natural resource (including land) development project is planned

for an area without people having been engaged effectively in the planning and development process or when a development is perceived to have negative consequence on people, families, ways of life, or property. For example, a proposed housing development may compete for land presently used by small-scale vegetable growers.

3. *Long-range planning concerns* are expressed when a particular facet of agricultural and natural resource development or management is so problematic and complex that most people involved recognize that it will take a long time to handle and that no immediate solution or even improvement is forseen. Sometimes it is described as an intergenerational equity concern—"What kind of future are we wishing on our children or our children's children?" There is usually some sense of unease as to whether the current structures are able to handle the concern. These concerns usually involve outside groups and may require a combination of state, national, and/or international resources in order to work on improvements. They can become crippling to local people and groups who feel they have no say or control over what is happening. Examples include the implementation and impacts of policies favoring agricultural diversification in a state; commodity price changes; the development of new water sources; stopping soil erosion; import and export policymaking; and changing price support programs and procedures. The siting of hazardous facilities also comes in here.

4. *Due-process concerns* are expressed when people feel left out of or are denied a voice in a policy-making, legislative, or planning processes. Usually they are concerned that adequate public meetings were not held, that governmental hearings violated "sunshine" laws, or that community or industry sentiments about changes in agricultural or natural resource activities were not listened to. For example, the board of water supply decides to change water rates charged to agricultural users without adequate public hearing or disclosure of vital information on the need for an increase.

5. *Generic concerns* are issues that, in some sense, go with the territory; that is, they are generally associated with a class of groups, organizations, and enterprises. Processors worry about materials, cash flow, markets, and so on. Municipal governments express concerns about their tax base (technically called *ratables*), spiraling costs, services, and so on. For families, concern for the well-being and future of children is universal. In addition, certain concerns are generic to managing agricultural or natural resource enterprises. These include the onset of new diseases, pests or parasites attacking crops, departures from normal weather patterns such as drought or high rainfall and floods, new government policies affecting the viability of established production methods, and changes in capital markets. In short, any change outside of the manager's control causes new concerns. These external forces may be biological, physical, economic, political, social, or cultural. Still others are due to changes internal to the enterprise and may call for improvements in management practices, training, cultural methods, labor, or more extensive changes. Due to the dynamic and interdependent nature of agricultural and natural resource systems, new themes of concern regularly surface and have a mixed character.

The Life Cycle of an Issue

Themes of concern also tend to have life cycles related to their magnitude, duration, and underlying causes. The cycle described here deals with the way concerns are often discussed by the people involved in a problematic situation. Chapter 7 describes similar issue cycles from the perspective of different kinds of situation owners and other concerned parties in the context of debating change.

Emerging concerns usually are not found in newspapers, but rather are heard as one talks with various client groups. People involved will be heard to say "This is terrible!" "Someone (or we) should do something about it!" or a combined "This is terrible! What do you think should be done?" People are not yet demanding action. Rather, when people confront a new issue, they may begin talking with each other, sharing information, finding out who potentially may be able to help, and speculating about the ramifications on their current activities and products. Note, however, that there is an alternative scenario in which some affected people view their situation as isolated, internalize it, do not communicate with others (including you, the analyst–facilitator), feel a burden of personal blame and guilt, and may thus also lack the capacity to take action.

If a situation persists, it enters an existing-concern stage. The tone and substance of people's conversations change, and the press may begin reporting the story. The latter may be important also in breaking down the psychic isolation referred to earlier, fostering the realization that people are in fact *not* alone in a situation. Conversations among affected and interested parties switch from "Something should be done about it" to "You do something about it." Power holders, decision makers, organized client groups, and other interested people begin demanding action to improve their current situation. They now see themselves as part of a group that is expected to take action, but they will also hold others accountable, based on their view of who is responsible for managing various kinds of resources on their behalf. Often these resource managers are public servants of some kind, such as extension agents, university researchers, or government officials from such departments as health, environment, or agriculture. Some situations move toward improvement during this stage, but others move into a conflict phase.

During a conflict phase, the ones who are able or allowed to manage a situation repeatedly change. If those who are held responsible for taking action fail to do so, then often a new group of managers takes over some or all of what the current managers were supposed to handle. "We will do it ourselves!" is often heard at this stage. Also, various parties to the situation have split apart according to polarized value positions and in support of different emergent leaders. They form firm beliefs regarding the "opposition's" beliefs, values, and wants, as well as expressing their own. Polarized situations are often very heated. For example, as one talks to various people, the polar opposite statements "We are for food irradiation" and "We are against food irradiation" are often heard.

Experienced intervenors and mediators advise that the best time to handle problematic situations is during the emerging to existing phases of the issue life cycle. Then more options are open for getting people together who have a stake in the situation but who represent divergent perspectives and interests. In these phases, people are usually able and willing to talk to one another because they are still formulating positions or senses of improvement. The longer the situation goes on, however, the greater will be the polarization and development of rigid, uncompromising beliefs, values, and attitudes.

It is best for facilitators of change in agricultural and natural resource situations (including the employees of public agencies) to handle themes of concern quickly. Procrastination eats up time and fritters away options. You cannot wait for more information or the results of more technical studies before taking action. During the time when the facilitators aren't taking action, those involved in a situation are. If you plan eventually to enter a career path in the public sector, this information is particularly relevant. There are numerous situations in which agricultural and natural resources officials have been declared to be irresponsible, incompetent, or worse by a constituency, and the management of portions of their missions has been given to another agency or official. Alternatively, their very tardy initiatives have been ignored or tied in knots by an aroused public. Those of you who are moving toward careers as technical information service representatives in business and industry should also take heed. If you are seen as unable or unwilling to help out, the viability of your company's products and services can decline overnight.

Experience also shows that the more a situation is seen as life-threatening or seriously endangering to the health or safety of a group, the faster it will heat up. This also occurs when the viability of a product is threatened. A recurrent example that may be painful to some readers is the contamination of the milk supply that has occurred in several states over the past decade. Skills in pulling concerned groups together quickly and openly discussing the situation are essential. Unfortunately, there is a tendency on the part of some professionals to favor particular groups and leave out others or to minimize the magnitude and scope or overall seriousness of a problem, particularly if legal action is in the offing. Delay in taking action is sometimes a reflection of wishful thinking or unrealistic optimism that a situation will clear up by itself. Indeed, sometimes it does!

In the case of the milk contamination situations as they unfolded in several states, there is evidence that favoring the agribusiness party and ignoring or excluding others, such as consumer groups or dairy farmers, led to actions and consequences that might have been avoided by establishing a more open dialogue. No matter how messy the situation, how inadequate the available information and uncertain the results, *it is wiser to get people talking to each other, collectively gathering and sharing information and working through a sense of improvement than to stand back and await developments.*

The number of people, groups, or businesses involved in a particular situation varies. Trying to determine who is, or potentially could be, involved is not as easy as when we are dealing with one clearly defined group. Because agricultural and natural resource industries involve many businesses interacting in order to get a product or service to customers, many people and organizations are also involved. Many situations will engage government regulatory agencies (e.g., agencies and departments responsible for agriculture, environment, or health); producers; processors; distributors; retailers; consumers and the community groups affected by such things as site locations for waste disposal; and production, processing, and distributing activities. And each of the groups involved in a particular situation will initially talk about a situation differently. Those differences are often the source of the issue-based concerns you attempt to describe in stage 2.

Primary Tasks

Another type of concern expressed by people in a situation focuses on the mission, or primary task, of the group, enterprise, or organization in which they participate. For a food, agricultural, or natural resources enterprise, a primary-task statement will summarize *its reason for being,* its basic nature, what it does, and its niche or role. These statements are expressed by people in operational terms and summarize what an enterprise or organization does or should do. In systemic terms, what you are attempting to do is to capture people's sense of the *central transformation processes* that are occurring or must occur if an enterprise or other organization is to be sustained. Resources or other raw inputs are taken in, to be transformed into something else. Your summary focuses on this transformation: hence the label *primary task-based expressions of a situation.*

Sometimes the people involved will represent their primary task as a general concern. In an example of this mode, Bruce, a member of a growers' cooperative, says, "Our co-op is supposed to collect our crops, grade them, and give us the money we rightfully deserve according to the quality of the crop we turn over to them. Instead, everything is being dumped together and Joe, whose produce is inferior, is getting the same price as I am. What good is this outfit, anyway?" Embedded in this statement is the speaker's sense of the proper mission of the cooperative and what it is supposed to do to realize his expectations. Several rounds of discussion with the same person might reveal that, to him, the co-op is a quality-control and monitoring enterprise that rewards growers who have high standards.

Superficially, this looks straightforward, but in practice it can give rise to prolonged discussion: "What is this enterprise all about?" "What functions are central to our mission?" Similar questions will arise in other settings: What are the primary tasks of a co-op? A state department of agriculture? A farm? A slaughterhouse? Of other organizations connected with the identified concerns? What primary tasks are being overlooked, thus leading to the inability of an organization to produce certain desired outputs?

It follows that engaging people in a discussion about the very essence of an enterprise will develop useful observations. Defining the primary tasks of an enterprise helps everyone to understand clearly what is at stake. If there are major differences in **W**s represented, several different definitions of the primary task will have to be created. For example, Bruce (the speaker in the earlier passage) sees the co-op as a quality-control mechanism, while Joe (another co-op member) sees it as a distribution and wholesale marketing enterprise. *Always use verbs to describe primary tasks;* e.g., *controlling* quality of produce put on the market.

A few words are needed regarding the distinction between *process*, which was discussed earlier in this chapter, and *primary task*. Both label critical features of the situation. On the one hand, *primary tasks* are part of the structure, and refer to specific activities related to the mission or reason why the organization or enterprise exists. *Process,* on the other hand, refers to how a wide variety of tasks is actually done or should be done. As you read up on a situation and talk to the people involved, note all the key activities described, whether they are working well or not. As you begin your synthesis, connect these activities with corresponding structural features and ask which ones people link to the central missions of an enterprise or organization. Those are the primary tasks. The ones that are not so linked are key human activities that may also need correction but, as features of process, are of a different kind.

Gathering Information

This section discusses ten key issues regarding the collection of information during stages 1 and 2, and particularly our responsibilities and tasks in talking to people in a situation. First a few useful tips on getting started in the inquiry process are presented. Second, some ethical issues involved in formal and informal interviews and surveys are discussed. Third, a distinction between going into a situation with an open mind and a blank mind is drawn. Fourth, advice on how to establish rapport with people you don't know is offered. Fifth, the types of questions you might ask and what they yield will be reviewed. Sixth, ways to increase your observation and inference-making skills are suggested. Seventh, the kinds of information discrepancies you are apt to find are described. Eighth, some guidelines on notetaking are provided. Ninth, the critical issue of whom to talk to is discussed. This section concludes with suggestions on where to find and how to use written information related to the situation you are approaching.

Getting Started

Your first plunge into a situation is driven by information sources and written documents of various kinds, but especially by the people involved. Written and other hard-copy material includes reports, newspaper articles, scholarly works, records, files, and the like. A variety of tools for retrieving, gathering, and even generating these materials is also available. These range from data base searches to remote sensing. Some of us are strongly tempted to depend on documentary materials alone. The more schooling you have, the more likely you are to orient yourself by means of extensive reading of printed information. While you need to consult written materials, it is suggested that you *start* by talking with key people in a situation.

If you are new to a company, organization, or situation and don't know anyone, naturally you will not immediately hit on key people. The first people you meet, however, can identify some of the key people, allowing you to move on to them. It is particularly important to talk with those who are voicing concerns about something in their food, agricultural, or natural resource management activities. As you talk to people, note who else is mentioned, particularly other key figures. One aspect of the rich picture you are building is to develop a roster of all the people involved and the nature of their involvement. As you read documentary materials, key people also will be named in news articles and other sources. Make a point of seeking them out and discussing the issues with them.

As you talk with people, you will also get a sense of who will be affected if the situation continues or if particular courses of action are pursued. List the names of these people, and talk with them too. Get their perspectives on the story, and record their concerns.

Usually, as part of the story they tell, people reveal the following pieces of information, which can also be supplemented with written material. These six items constitute a minimum data set for getting started, the kinds of things people are likely to talk about that can also be used to form key questions that will help guide your discussions with situation owners and clients:

1. Descriptions of a situation as seen by various people and their accompanying senses of opportunity, hope, unease, or concern. Sample questions: "What opportunities do you see for your enterprise?" "What concerns do you have about your enterprise?"
2. What various people think should be and/or can be done about a situation. Sample question: "What should (or can) be done about these opportunities (concerns)?"
3. The consequences of a situation continuing as it is. Sample question: "What will happen if things continue as they are now?"
4. The key people and groups involved in the situation from the standpoint of (a) causing/contributing to the current state and (b) doing something about it. Sample questions: "Who (or what group or organization) is causing these concerns?" "Who should be doing something about this?"
5. Current approaches to coping with the situation. Sample question: "How are you dealing with these opportunities (or concerns) now?"
6. The associated values, usually expressed as "shoulds" or "oughts," and rationales for why people think a particular line of action is appropriate or inappropriate. Sample questions: "Why are you doing it this way?" "Why do you think that would be a good thing to do?"

Many of us are not used to initiating conversations, particularly if we feel we are not on top of a situation. Yet if you can pose only one question to a subject, you are underway. Start by saying something like "What are your concerns regarding _____ (for example, producing beans/potatoes/corn)?" As a matter of fact, often you won't have to worry about asking that first question because people will freely talk about their concerns. Note that formal structured interviews and survey questionnaires are *not* appropriate at this point because you are only beginning to find out what the questions are.

If a situation has persisted for a while — involving a chronic or endemic problem — or it involves people's sense of well-being or safety (or both), usually the situation will have been reported by the news media. Some big-city and state-wide newspapers have indexes, which can be consulted in the library. The *Readers Guide to Periodical Literature* and on-line computer services in some libraries can direct you to national and regional magazine articles, including those in so-called trade publications serving particular industries. Many university libraries serve as depositories for federal and state government publications and documents — meaning that they get them all! Some librarians keep issue folders of news clippings on a variety of themes. You may want to check there. It never hurts to get a librarian to give you a crash course in library research techniques and information retrieval. Municipal offices and government agencies often keep files of news stories on important issues affecting their areas. They provide good sources for gathering the story quickly. While newspaper accounts are not always totally reliable in terms of getting their facts straight or quoting people's perspectives accurately, they do provide a running chronology of events, and they will give you important insights and people's names to add to your contact list. Editorials, columns, and letters to the editor in newspapers also convey important perspectives in themselves, since their viewpoint is being read by so many.

In summary, the way to take action on a situation is to begin talking with a variety of people and groups connected with the situation. Earlier it was suggested that formal interview and questionnaire techniques are not appropriate to this phase. These techniques are best used when we have a clearer understanding of a situation. Since the quality of the questions in a questionnaire or interview guide depends on how well we at least tentatively understand a situation, these techniques are best left to later iterations of stages 1 and 2. In addition, when done well, questionnaire or interview information often is descriptive, when what is really needed at this early stage is information that will help develop a more appropriate course of action toward improvement. Therefore, a more fluid and informal approach to information gathering is recommended at this stage. The risk in suggesting this approach is that you might infer that rigor is not required, and that certainly is not the intended message.

Ethical Responsibilities

Due to abuses of research subjects and unethical conduct by investigators in the past, the federal government has imposed rules regarding the involvement of people in a wide variety of research projects. Today all institutions receiving federal support must follow guidelines aimed at protecting people who participate as subjects in research from harmful psychological manipulation, physical trauma, violation of confidentiality and privacy, professional misconduct, jeopardy of legal and criminal sanctions, and other risks. Moreover, in response to both the history of abuse and the regulatory initiative, some professional

organizations have developed ethical guidelines shaped by their historical experience and molded around their own standards and practices. Two examples of
guides to professional conduct are reproduced in Tables 4.1 and 4.2.

In accordance with regulations, most universities have institutional review
boards empowered to scrutinize research proposals involving people as subjects.
They can turn down projects or impose changes of protocol in order to insure
that research subjects are protected.

Survey questionnaires and structured interviews are particularly scrutinized
when they involve such topics as sex, crime and other illegal or illicit behavior,
personal finance, and business practices. Typically, the bottom line for institutional review panels is that the data are collected in such a way as to (1) assure
that participants are advised of the purposes of the research, the fate of the data,
any risk they may be exposed to, and their right not to participate (this is known
as informed consent), (2) make it impossible to identify individual respondents
or, (3) if the foregoing would violate a research protocol, as when there is a
need for monitoring or followup, store the data securely.

Not only should you behave ethically because it is morally right, but damaging people by failing to adhere to professional standards may expose you to a
lawsuit and professional disrepute. Here are ten specific rights that we must accord the people with whom we talk in pursuit of our professional duties. They
are the right

1. Not to participate.
2. To quit at any time.
3. To have information they provide held securely and in confidence.
4. To demand anonymity.
5. To have a "fair return" on their contributions.
6. To be free from harm.
7. Not to have data used against them and their interests.
8. Not to be placed in jeopardy of the legal or criminal justice system.
9. Not to have their personhood violated in any way.
10. To be fully advised of any aspects of their participation in a project that
    many involve any of the foregoing.

Just because a project touches one of these rights does not automatically kill
it off. Number 10 is the critical one here; if (potential) subjects fully understand
the issues involved, they have a right to participate and indicate their acceptance
by signing a form you provide.

The protection of human subjects guidelines and statements of professional
standards, such as those reproduced here, were developed to inform basic and
applied research in the social and behavioral sciences, medicine, and other
fields. They are now also applied to agricultural research sponsored by the
U.S.D.A. and other governmental research organizations. So it is legitimate to
ask: What do they have to do with soft systems analysis in agriculture, natural
resources, and related industrial contexts? Are they applicable to a well-trained,
ethical employee of or consultant to, say, a farmer cooperative, regulatory
agency, legislative committee, or agribusiness firm?

TABLE 4.1 Professional and Ethical Responsibilities:
Society for Applied Anthropology

This statement is a guide to professional behavior for fellows and members of the Society for Applied Anthropology. As members or fellows of the society we shall act in ways that are consistent with the responsibilities stated below irrespective of the specific circumstances of our employment.

1. To the people we study we owe disclosure of our research goals, methods and sponsorship. The participation of people in our research activities shall only be on a voluntary and informed basis. We shall provide a means throughout our research activities and in subsequent publications to maintain the confidentiality of those we study. The people we study must be made aware of the likely limits of confidentiality and must not be promised a greater degree of confidentiality than can be realistically expected under current legal circumstances in our respective nations. We shall, within the limits of our knowledge, disclose any significant risk to those we study that may result from our activities.

2. To the communities ultimately affected by our actions we owe respect for their dignity, integrity and worth. We recognize that human survival is contingent upon the continued existence of a diversity of human communities, and guide our professional activities accordingly. We will avoid taking or recommending action on behalf of a sponsor which is harmful to the interests of a community.

3. To our social science colleagues we have the responsibility to not engage in actions that impede their reasonable professional activities. Among other things this means that, while respecting the needs, responsibilities, and legitimate proprietary interests of our sponsors we should not impede the flow of informations about research outcomes and professional practice techniques. We shall accurately report the contributions of colleagues to our work. We shall not condone falsification or distortion by others. We should not prejudice communities or agencies against a colleague for reasons of personal gain.

4. To our students, interns or trainees we owe non-discriminatory access to our training services. We shall provide training which is informed, accurate and relevant to the needs of the larger society. We recognize the need for continuing education so as to maintain our skill and knowledge at a high level. Our training should inform students as to their ethical responsibilities. Student contributions to our professional activities, including both research and publication, should be adequately recognized.

5. To our employers and other sponsors we owe accurate reporting of our qualifications and competent, efficient and timely performance of the work we undertake for them. We shall establish a clear understanding with each employer or other sponsor as to the nature of our professional responsibilities. We shall report our research and other activities accurately. We have the obligation to attempt to prevent distortion or suppression of research results or policy recommendations by concerned agencies.

6. To society as a whole we owe the benefit of our special knowledge and skills in interpreting socio-cultural systems. We should communicate our understanding of human life to the society at large.

This table has been reproduced from *Human Organization* and by permission of the Society for Applied Anthropology.

TABLE 4.2 Code of Ethics for Members of the Society of American Foresters

Canons

1. A member's knowledge and skills will be utilized for the benefit of society. A member will strive for accurate, current, and increasing knowledge of forestry, will communicate such knowledge when not confidential, and will challenge and correct untrue statements about forestry.
2. A member will advertise only in a dignified and truthful manner, stating the services the member is qualified to perform. Such advertisements may include references to fees charged.
3. A member will base public comment on forestry matters on accurate knowledge and will not distort or withhold pertinent information to substantiate a point of view. Prior to making public statements on forestry policies and practices, a member will indicate on whose behalf the statements are made.
4. A member will perform services consistent with the highest standards of quality and with loyalty to the employer.
5. A member will perform only those services for which the member is qualified by education and experience.
6. A member who is asked to participate in forestry operations which deviate from acceptable professional standards must advise the employer in advance of the consequences of such deviation.
7. A member will not voluntarily disclose information concerning the affairs of the member's employer without the employer's express permission.
8. A member must avoid conflicts of interest or even the appearance of such conflicts. If, despite such precaution, a conflict of interest is discovered, it must be promptly and fully disclosed to the member's employer and the member must be prepared to act immediately to resolve the conflict.
9. A member will not accept compensation or expenses from more than one employer for the same service, unless the parties involved are informed and consent.
10. A member will engage, or advise the member's employer to engage, other experts and specialists in forestry or related fields whenever the employer's interest would be best served by such action, and members will work cooperatively with other professionals.
11. A member will not by false statement or dishonest action injure the reputation or professional associations of another member.
12. A member will give credit for the methods, ideas, or the assistance obtained from others.
13. A member in competition for supplying forestry services will encourage the prospective employer to base selection on comparison of qualifications and negotiations of fee or salary.
14. Information submitted by a member about a candidate for a prospective position, award, or elected office will be accurate, factual, and objective.
15. A member having evidence of violation of these canons by another member will present the information and charges to the Council in accordance with the Bylaws.

Adopted by the Society of American Foresters by member referendum, June 23, 1976, replacing the Code adopted November 12, 1948, as amended December 4, 1971, and November 4, 1986. Reprinted by permission of the Society.

Legally and bureaucratically, probably not, *unless you belong to a professional group that has adopted standards of professional practice* such as those reproduced earlier. As those guidelines indicate, from a moral and ethical standpoint, the situation should give you pause. This is because such guidelines

Having a code of professional ethics does not absolve one from the need to act morally and with good common sense. In an open situation, you have the burden of thinking about the issues and setting your own standards. The down side is that if you attempt to apply such standards in certain bureaucratic or business settings, *you* are in jeopardy of being labeled a whistle blower and dealt with accordingly. A whistle blower is an employee of an organization who exposes or acts to prevent wrongdoing within it. Some jurisdictions accord limited protection to some kinds of whistle blowers, but this area of the law is relatively undeveloped. So this vital question, of whether or not you can carry out necessary investigations in accordance with ethical standards, remains open. The only advice that can be offered is (1) be true to yourself, (2) think about the issues, and (3) plan ahead along the lines of "What would I do if ... ?"

As indicated, your first obligation is to protect your clients and others you talk to professionally. How realistic is it to warn you about ethical dilemmas? Is it really likely that you, a professional in food, agriculture, natural resources management, or related industries will run across situations where your investigatory activities will somehow place people at risk? You might pause here, backtrack a few pages, and reread the earlier passages in this section. While you do so, try to *imagine* situations like that. What might farmers, or users of resources, be up to that would be *useful* to know about, but if known, could subject them to legal sanctions or other harm? What kind of information do farmers like to hold close to their chests (but that agricultural economists would love to know about)? Later this chapter will have something to say about establishing rapport with people. Trust (and distrust) is a major issue. What are sensitive topics that you would do well to skirt around *at least* until you have achieved an appropriate level of acceptance? Will you share information gained from one person with another?

Pause and see what you come up with in trying to answer these questions.

Do farmers ever engage in illicit or illegal activities? Both in the U.S. and in other nations, farmers have been known to produce high-value crops such as opium, coca, and marijuana to sell to processors and distributors just like any other commodity. One of the very first internal disturbances facing the new government of the United States in 1794 was the so-called Whiskey Rebellion, involving farmers in northern Appalachia who forcefully objected to the imposition of taxes on their highly convenient and valuable technique of processing some of their field crops into alcoholic beverages. And moonshining is still a rural industry in some areas. Farmers have also been involved in illicit disposal of hazardous wastes on their land. In the resource area, there are numerous kinds of illicit activities, ranging from poaching game to water and timber theft. Whether or not ethical issues arise depends on *your role* and *your objectives*.

It is conceivable that some of you will take jobs in environmentally related law enforcement, as conservation-oriented rangers, or with the pollution squad of a police force (increasingly common). In those cases, you will be governed by an entirely different set of professional, ethical, and legal standards and will be intensively schooled on them to boot. Others might become journalists, a profession increasingly interested in agriculture and the environment, but subject to yet another kind of professional code. Here, in addition to protecting sources of information, a significant concern is avoiding libel. This book's concern remains the behavior of the applied-science professional employed by organizations where it may be feasible (and certainly desirable) to advise superiors of the existence of a professional code of conduct.

Even legitimate exploiters of natural resources hold the equivalent of trade information that they consider to be sensitive, such as their land and mineral leases and related information. One of the authors has encountered a situation in New Jersey where a lot of rented land is cultivated, in which farmers will not discuss the land-rental market. They are locked in competition with neighbors for rental land that is being held by speculators. If you were investigating, for example, soil conservation practices, this kind of information might be vital to test the observation that farmers are more conservation-minded on their own land than on rented land.

Cindy in the Woods: A Scenario

Cindy is a professional employee in the forestry division of a state environmental conservation department. She is attempting to gather information on a situation of concern to her agency, the impact of "unregulated activities" on state forests and other land tracts that the department supervises. She has decided to use state-of-the-art methodology, the patch dynamics approach (Pickett and White, 1985) in forest ecology, combined with the soft systems approach, in order to develop proposals for improvement. Her first task of experiencing the situation requires that she conduct an inventory of forest gaps and other disturbances; measure their size; describe their causes, including their intensity and frequency; identify and talk to the people involved, and assess the impacts of the facts she discovers. Remote sensing, using satellite data and imagery, turns out to be a valuable tool, although it requires "ground truthing" — Cindy must survey selected sites revealed by the satellite data in order to document their characteristics and permit them and others of the same type to be identified in the imagery.

While visiting different areas for this purpose, Cindy also makes it a practice to consult local people. After all, they are good sources of information about the origin of particular forest gaps, and they provide information that she could never gain by other means. For example, one long, narrow break in the forest is identified by an old timer as being the result of a tornado that touched down in 1934. People are intrigued by the computer-generated survey maps and readily share all sorts of information.

Cindy approaches people in a variety of public places, including general stores and restaurants, presenting herself in a nonthreatening way. She tells them that she is an employee of the State Department of Environmental Conservation and explains her interest in forest gaps that might contribute to environmental degradation. She is only rarely confronted with the suspicion that she is involved in enforcing conservation laws.

So far, things are straightforward and uncomplicated. So, as the story continues, complications emerge in the form of two alternative scenarios:

Scenario A. One night in a roadhouse at a small country crossing, Cindy gets in with a younger, laid-back crowd, has a few beers, blows a little weed, and tells people about her forest study. Everyone thinks it's cool and several of her new friends volunteer to show her some sites she has already seen in her satellite pictures. The next morning, Cindy goes with them, only to discover that the gaps are actually small marijuana fields and her guides are the growers.

1. So what's the problem?
2. What did Cindy fail to do that might have helped her deal with her ethical bind?

> Stop and consider possible ways in which she could have avoided the problem.

Scenario B. Cindy follows her usual procedure of walking around a small country town and approaching people who look like they might be helpful. After she has explained her job and objectives and promised that their contributions will be anonymous, some local people volunteer the fact that several of the gaps she is curious about are marijuana fields cultivated by "hippies." They warn her not to try visiting the sites herself, but are very positive about their ability to identify the fields in her satellite images and do so.

1. What is the problem here, if any?
2. From the standpoint of professional ethics, how does Scenario B differ from Scenario A?

> Stop and consider the implications of this information. What should she do now?

In summary, with a little care, it is not difficult to develop a protocol for assuring anonymity and/or confidentiality and gaining people's confidence so that they will cooperate and you can use their input. You are responsible for guaranteeing that this protocol is followed. Little details sometimes matter. For example, do not leave your notes of conversations lying around for others to read, and do not share them unless you have rendered them unidentifiable. You may want to develop a master sheet of those you have talked with and give each name a number. Number your notes so that if they are read by others, participants are not identified. And remember, these ethical responsibilities apply, whether the information is gathered formally or informally.

An Open Mind Versus a Blank Mind

Critics of research techniques often use the phrase *blank mind* or some equivalent in discussions of why informal surveys are not as useful or reliable as more structured interviews or questionnaires. They imply that advocates of the less time-consuming, informal discussion approach actually believe that an interviewer can enter a situation with a blank mind and have no preconceived ideas. They correctly point out that the blank-mind approach is impossible, but they go on to claim that an interviewer will inevitably rely on his or her beliefs and impressions of a situation to steer the line of questions. Allegedly, the interviewer may bait the conversational hook and move the interview in directions that, either consciously or unconsciously, reinforce his or her established views of the situation.

These cautions should be taken seriously. Technically oriented interviewers have been observed who claim they are able to listen effectively to people, but who actually selectively screen important information and move people to discuss areas that they believe are important. Specialists are particularly prone to do this. For example, an extension specialist with a soil science background hears a family farmer's complexly intertwined set of concerns regarding the effects of current fertilizer use as a cry for help in reducing soil degradation.

So if it is not possible to enter a situation with a blank mind, what might we do to protect ourselves from our own biases and prejudgments about a situation? How might we suspend immediate judgment about the meaning of a situation that is like something we have handled in the past? It *is* impossible to approach situations with a blank mind because we immediately conceptualize and try to impose meaning. Indeed, our minds are far from blank! They are full of conceptual frameworks that have been useful in the past in sizing up and taking action on a variety of situations.

In learning to think divergently, we must recognize that we have prior conceptualizations of the world to deal with. "Forewarned is forearmed." Instead, one can develop an *open* mind by adopting ways to handle one's own beliefs, values, attitudes, and opinions so that they don't dominate or direct interviews.

Communications researchers remind us that this competence is not typical or common. Apparently, the ability of Americans to engage in *reflective listening,* to listen to others and to summarize accurately back to them what was said, is fairly low. Have you ever heard a heated conversation in which one party hurled at the other the charge "You're not listening to me!" — and the other shrugged it off? Well, next time take it seriously. More importantly, this competency can be learned! Developing the skill of having an open mind involves learning *how to listen.* Practice by listening to family members or friends. Try your hand at summarizing back to them what was just said. Ask them to be honest in assessing how much you actually got right—from their perspective.

Typically, when confronted with the experience, most people are surprised at how poorly they do. Often this poor performance is due to the personal concern to say what is on one's *own mind* rather than some inherent inability to focus on what others are saying. Most people can improve their reflective listening competence with sufficient practice.

Many believe that it is possible to listen to people talk and treat the information they give objectively. What is implied is that you can take the factual information of a message at face value and ignore the emotional content. Egan (1979, 1982, 1985) and others remind us that each message sent has an accompanying emotional climate. It is a mistake to believe that we can and should ignore the emotions that are conveyed with the verbal content of a message. What is said will also evoke emotions in you. Better that we find ways to accurately record the emotions expressed as well as the content of people's stories about a situation. Just as you practice verifying that you correctly heard the content, so also you should *practice verifying your notice of the accompanying emotions.* As discussed subsequently, it is possible to ask clarifying questions about feelings, as well as about facts. So a simple check-up question such as "Does this whole thing make you angry?" will reveal whether or not you have understood the emotional tone. Be prepared in the beginning to be told that you are off target. Also, if you are one of the situation owners, how a particular situation makes you feel may well be different from how it makes others feel. You may be angry, while someone else is frustrated and another is unconcerned. Some may feel hopeless, others hopeful!

Establishing Rapport

When we start a new job, we often don't know other employees or clients. As we talk to fellow employees, our employers, and clients, themes of concern will emerge. People are constantly adjusting to the ever-changing environment. Finding a path toward adjustment is a routine human activity. There will be times when you will want to step back and take a fresh look at the array of themes of concern or senses of opportunity expressed by your clients and other people in an organization. There will be other times when you are assigned to tackle a particular problem, or when your clients ask you to become involved in a particular concern. How wide you diverge in your survey of themes of concern or opportunity, therefore, will depend on the circumstance and demands of a particular job.

If you are new to a job or place and a stranger to your clients, the following guidelines, which were developed by a social scientist engaged in international agricultural research (Rhoades, 1982), may help you to establish rapport with the people with whom you talk.

Of course, you start by greeting people according to local custom. Be sure to make a phone call in advance if this is the way things typically are done. Use acceptable forms of address. Don't talk to a client from a vehicle. Get out and observe up close what is being talked about. Tell people exactly who you are, what organization you represent, why you wish to talk, and the nature of your work. Avoid rambling or launching off into your own stories. Remember you are there to hear their stories, not to tell your own. Be free with information they request, however. They will want to feel you out, too, and it is a matter of basic reciprocity. Be careful not to promise things that can't be delivered. Talk to more than one person in a household, organization, or group; these others will usually have additional pieces of information that will help you to understand the situation.

Our minds tend to wander as people talk to us. Think of times when you felt that someone was not listening to you. It doesn't encourage a person to continue a conversation. In order to avoid terminations of conversations, practice your *attending* behaviors. *Attending behaviors* signal others that you are paying attention to them and also help you actually to pay attention. Attending behaviors are both verbal and nonverbal. The verbal component of attending behaviors involves an occasional word or "mm-hmm" to indicate to the other person that you are still with him or her. This should come naturally, so watch that you don't overdo it. On the nonverbal side is our ability to make appropriate eye contact, to establish appropriate distance, and to avoid distracting gestures and facial expressions that throw the person talking off track.

Appropriate eye contact means *a culturally appropriate level of directness.* In some parts of the U.S., looking a person in the eye is considered rude or aggressive; in other places, it signals sincerity and directness. Take your cue from how other people look at you and follow what they do.

Similarly, in some places, a three- to six-foot separation between seated or standing speakers marks a personal conversation. Getting too close to a person suggests intimacy, while standing too far away signals aloofness. If you are new to a place and job, watch how close people stand to each other as they carry on a conversation and try to match that distance when your turn comes. Ask someone to check you out on your ability to attend to people as they talk; test both your verbal and nonverbal competencies.

Types of Questions and What They Yield

While you always encourage people to voice concerns in their own terms, there are times when you need to ask *clarification* questions. Learning the art of asking effective questions requires gaining an idea of the kinds of questions you can ask and the kinds of information they yield. Listed here are categories of questions with examples (Norem, 1986):

1. "What" questions are fairly straightforward and yield factual information. They tend to be relatively nonthreatening and are good starting questions. Examples include: "What are your major cash crops?" "What do you use the river water for?" "What types of fertilizers are you using?" "What livestock are you having trouble with?" "What kinds of trees are you cutting down?" "What aspects of food quality concern you the most?"

2. "How" questions are good open-ended questions. They yield factual information and usually are nonthreatening. This type of question usually asks for more information than do the "what" variety. Examples include: "How do you prepare your fields?" "How do you plant this crop?" "How many times do you fertilize?" "How do you weed?" "How many times do you weed?" "How do you harvest your prawns?" "How do you prepare this food?"

3. "When/Where" questions are not particularly open-ended, but yield important factual information. Examples include: "Where do you plant this crop?" "When do you harvest this crop?" "When do the pests tend to hit this crop the worst?" "When are you short of laborers?" "When/Where" type questions do not stimulate conversation, but rather turn the discussion into a question-and-answer session.

4. "Who" questions also evoke short answers, but give useful information on who is involved in a particular situation. Examples include: "Who makes the decisions about what and when to plant?" "Who does the planting?" "Who monitors the animals' health?" "Who currently regulates the water flow?" Answers to this type of question should be checked against what you actually observe in an area. Sometimes who is said to do something and who is observed doing that thing are different! As discussed subsequently, "who" questions of the form "Whom else do you think I should talk to?" are important to getting started and selecting a representative and/or useful group of respondents.

5. "Why" questions help to reveal a person's world view and reasons for what he or she does. This type of question can be threatening if you are seen as a stranger or the object of suspicion. "Why" questions may be interpreted as a challenge to what has been said or done or as a demand for justification of words and actions. You may need to think of ways of asking "why" questions indirectly. Examples include: (Direct style) "Why do you plant this crop in this manner?" "Why do you use this amount of water?" "Why do you plant this kind of tree for your timber needs?" "Why do you not hire field hands?" (Indirect) "Can you explain how you decided to plant beans on this plot?" "Can you explain how you decided how much feed to give to these cows?"

The what, how, when/where, who, and why questions can be stated in either an open-ended or closed manner. Open-ended questions allow people to explore their ideas. Closed questions limit the required response and often need only a "yes," "no," or brief answer. Examples include: (Open) "How do you manage when just your family does the harvesting?" "What kinds of concerns do you have regarding your farming efforts?" "What kinds of opportunities do you see?" (Closed) "How many bags of fertilizer do you use?" "Do you want to add another crop?" "Are you satisfied with your yields?" "Do you have any concerns?"

Open-ended questions can also be asked directly or indirectly. Examples include: (Direct) "How do you handle periods of drought?" "How do you manage when the shipping schedules are not regular?" (Indirect) "It must be difficult when the shippers don't show up on a regular basis. I wonder how your family manages during those times."

The what, how, when/where, who, and why questions can also be followed with *probe* questions. These questions explore in depth the themes you or another person have already exposed. You can probe for more details so that you add to your understanding of information previously given. One can also probe for clarification of a response that was not understood the first time. Sometimes information offered seems to contradict other information, or you simply can't understand what has been said. Clarification probes are useful in capturing a person's perspective. Another kind of probe asks someone to elaborate a point he or she just made. Probe questions are important to gain the depth and detail you need to understand a situation and people's perspectives fully. When more than one questioner is involved, they should practice following each other's line of questioning so that the conversation doesn't degenerate into a battery of unrelated questions that yield scattered, superficial information and so that respondents do not become confused or alienated.

Turning Observations into Inferences

As you circulate and talk with people, you may also have an opportunity to see what they are actually talking about. People have a tendency to observe their environments and their own behavior selectively due to their cultural background or specialized education. If you are an advanced student, your observation skills related to your discipline will be highly developed. This can bias your observations. It is common to miss important clues because they are unrelated to your specific expertise or other ingrained experience. For example, professionals have been observed who completely overlook important crop and animal management practices because they focus on the animal/plant/soil nexus rather than on the *person* dealing with the animal/plant/soil nexus! In summary, the people involved in a situation may overlook important features, and you may too.

Instructors who train people destined for overseas assignments provide some basic tips to sharpening your observation skills (Wilson, 1987). First, try to see as many things in the environment as you can. Second, when you go back out for a second look, concentrate on seeing even more new things, even though you are naturally inclined to verify what you already saw or to prove to your partner that what you saw was correct and what he or she saw wasn't! Your mind will do that automatically, so a better strategy is to look for new things. Third, make a practice of recording what you see. Such records will help later when you discuss a situation with colleagues, clients, and other people.

Sharpen your ability to record observations and inferences accurately. Inference-making is what you do when you are internally making sense out of what you see. We are constantly making inferences. Yet our inferences can be very inaccurate, particularly in situations where many things are new or outside our expertise. We may make logical jumps based on old conceptual frameworks that are not useful to explain the current situation. They become hasty generalizations that cannot be verified based on what was actually seen or heard. It is often better to start out describing a situation in terms of observed behaviors and phenomena. For example, which of these two statements is a better way to state initial observations and inferences? (These are from actual field notes where two people observed the same thing but recorded it very differently.)

I. There are three women and two children in the orchard with backpack sprayers. Insecticide containers are set by the side of the field, tipped over with lids off. The women and children are wearing no protective gear. Do only women and children do the spraying here? What health problems are they having because they are breathing insecticide spray? What are the amounts of spray being used and how often?

II. Women and children in this area do the spraying. There are backpack sprayers available to farmers in this area, and they can afford insecticides.

The first example makes fewer hasty generalizations and records more accurately the details seen in a particular setting. It demonstrates an ability to describe what is seen and begins to make tentative inferences about a situation, stated in the form of questions. Contrastingly, the second example rushes to several conclusions and leaves little room for coming back and seeking more information.

Sharing Experiences and Information

As you talk with people, they may ask you what you know about something, particularly if you are seen as an expert. Briefly sharing experiences and information can help to establish rapport. Avoid getting into an advice-giving mode, however. Once some people start giving advice, it is difficult to turn them off. Practice learning how to switch it off. Often it is difficult because the people you are talking with keep asking for more. You can redirect the conversation by asking open-ended "how" or "why" type questions and centering the conversation back on the subject rather than on yourself.

Another tendency is to want to share experiences you have had that are in some way similar to what you are discussing. Lengthy elaborations of your own experiences sidetrack the conversation from people and their concerns to you and your experiences. Practice with friends and ask how you do on this point.

Focusing

The person who initiates a conversation largely determines when and how much focusing occurs. Whether or not to focus on a few points or to get an overall picture of a situation depends on a number of factors. First, time may be limited, in which case you might opt for depth of understanding of a few key areas. Second, people coming and going or noise may be distracting, in which case it might be better to hold off discussing sensitive issues and go for a wider view of what the situation is about. Third, if you have already talked with the person before and need clarification, elaboration, or additional understanding, then focusing may be in order.

Handling Discrepant Information

Discrepancies in the information you collect from printed sources, through your conversations, and from observations will inevitably surface. Inexperienced analysts often find this confusing and frustrating. Depending on your orientation, you will either conclude that you are poor at talking with people and accurately recording information, reject a portion of the information, go back to verify whether you heard someone correctly, or recheck your written sources. The latter two approaches may be the most fruitful, open-minded learning practices. Discrepancies will always occur and are useful data in and of themselves. They point to areas needing further exploration and clarification. They may also mark differences of world view and experience in the people you have interviewed. Several types of discrepancies are common:

1. Analyst/respondent. You see one thing and are told another.
2. Group 1/group 2. Due to differences in their involvement in the situation, members of two or more groups give different descriptions of who they think is doing what to whom and with what results.
3. Men/women/children. People of different social status may give different stories regarding the same situation, including discrepancies in who is involved, what is happening, how often something happens, who does what, when, and how often.
4. Group/individual. You should be prepared to be told one thing in a group (e.g., a co-op meeting) and another story when you talk to people individually.
5. Discussion round 1/discussion round 2. The same person may tell you one thing the first time you meet and something else the next time. Be prepared to seek clarification. Remember we aren't always consistent, and some people might perceive you as a threat and therefore falsify parts of a story or withhold details due to uncertainty regarding your status, objectives, and the like.
6. Analyst 1/ analyst 2. If you are working with other analysts, be prepared to hear divergent reports of the same person's statements, including differences in what you think that person directly said to you. Remember that at times even trained researchers make inaccurate inferences or are inconsistent with others. If you find discrepancies, check with the source to determine what is the more accurate version.

Taking Notes

Experienced field workers disagree about the usefulness of taking notes while talking with someone. Some say that you shouldn't take notes because it may make the person you are speaking with uneasy or be distracting and thus prematurely terminate the interview. On the one hand, if a situation is sensitive, notetaking may make people suspicious and unwilling to talk with you. On the other hand, there have been occasions when a farmer has said, "Why aren't you taking notes? Isn't what I am saying important?" Others say that because it is very hard to capture a person's perspective, it is imperative to take some kind of notes so that more of the information given can be retained and hasty generalizations or faulty interpretations can be kept under control. Notetaking may also function as attending behavior, and most people quickly become used to it. There is wisdom in all of these suggestions and cautions. Take heed and, after you have gained experience, do what is best for you.

A frequent complaint heard from students who are learning this approach is that they cannot take notes fast enough or keep their mind focused on the person talking while also generating questions and taking notes. We all face similar concerns. In response to those of you who are saying, "I'll just use a tape recorder," be aware that you must get a signed statement in which the person you intend to interview agrees to the procedure. Taping a discussion is also labor-intensive. For each hour of conversation, it usually takes two to three hours to transcribe or go back to the tape and take notes so that the information is useful. Most workers in the field do not go back to their tapes and therefore rely on what they can remember. A combined approach that some have found useful is to tape *and* take notes. The notes can then serve as a guide or index to the tape.

Generally, you need to take some notes anyway because recalling information accurately is difficult without them. Here are a few tips:

1. Don't start the discussion with pen and paper in hand. Ease into notetaking. Ask people if they mind if you take a few notes so that you can be more accurate in remembering what they say to you.

2. Don't try to write down every word a person says. It will prevent you from listening and often becomes distracting to the other person. Write down the nouns, verbs, and short phrases that will serve as a memory aid. More will be said on this point subsequently.

3. If working with a partner, you may want to share tasks, including who records what kinds of information.

4. Organize your notes as you go. Some interviewers divide their paper into two columns. They put what the person interviewed says in the left column and their own interpretations, or keys to things they want to follow up, in the right column. This avoids confusion in your notes later on between whether the respondent said something or a particular note is actually your impression or interpretation.

5. If the person you are interviewing comes out with a statement that you

know you will want to return to for clarification or elaboration, you may want to leave space so that you can add information based on followup probe questions.

6. Good notetaking during a discussion needs to be followed by recording additional information *immediately* after you have concluded the interview. You are strongly advised *not* to go from one person to the next without taking time to record additional thoughts and information. To keep who said what straight is not as easy as it sounds! Several authors suggest post-discussion report-writing techniques that they find useful, including Frankenberger and Lichte (1985), Franzel (1984), and Rhoades (1982).

How Many People to Speak With?

As they begin to use this approach, students with no background in field research frequently ask, "How many people must I talk to in order to gain an accurate representation of each group's point of view? Will one person from each group do it? Two? How many must I speak with in total?" There are no hard and fast rules because the situations you might encounter are too varied. Some guidelines can be offered, however.

In social science research, you strive to obtain data that are accurate and representative of the population you are studying, subject to such constraints as time, money, and the accessibility of research subjects. You want and need a representative sample, that is, a selection of people with whom to talk who are likely to respond the same way the whole group would. More often than not, the statistical tests that accompany presentations of social research findings are intended to assess the sampling procedures followed by investigators. The same general principle of *representativeness* is applied in the soft systems methodology, although it may not be necessary to use statistical tests if the groups you work with are relatively small and you intend to interact intensely with group members rather than administering a survey. If you are trying to extract the dominant **W**s of the groups involved in a situation, then you must contact a number of people for each group. But precisely how many? While there is some debate in the literature, a rule of thumb that is often used by field researchers and program evaluators is that you should talk to *all* members of small groups or 20 to 30 people affiliated with larger groups.

How do you select which 20 to 30 people with whom to talk? Two kinds of techniques are used most frequently. If you have a total population numbering in the hundreds or even thousands, then a random sample or stratified random sample is necessary. The other common approach is called network or snowball sampling. It is based on your initial explorations of a situation, when you routinely ask each person you speak with to give you the names of people *they* think you should contact (and don't forget to ask "why" questions here). Thus, you initially select 20 or 30 people because they are frequently mentioned in the newspapers or by those you have already spoken with as being highly knowledgeable, authoritative, prominent, and/or deeply involved in a given situation. And you continue to ask for the names of further contacts.

Do not lose sight of the fact that your understanding of the situation depends on whom you talk to, which, in turn, at least partly depends on how many you talk to. A fundamental principle in selecting respondents is to talk with *the people you seek to understand*. This sounds embarrassingly obvious, but it is the principle most often violated in all sectors of research. For example, don't expect that agricultural scientists will be able to tell you much about the concerns of farmers. If you want to know about the concerns of farmers, then talk to the farmers. Specialists such as extension agents and researchers never talk about farmers' concerns in the same way. The divergence in their expressions of themes of concern is linked to how often they routinely associate with each other.

Related to the foregoing is the point that people who attend meetings of regional, commodity, and national agricultural organizations are not necessarily representative (in the statistical sense described earlier) of their farmer constituency. They may adequately represent the views of current key decision makers, as distinguished from the rank and file. Therefore, if one of your objectives in using the soft systems approach is to gain a feel for a commodity group as a whole in order to help channel requests to the state legislature, don't rely on people attending association meetings as your sole or even primary sources of information.

You must also be careful in making generalizations from the information you have gathered. An important feature of your synthesis report will be an open and honest description of who and how many people you spoke with and how you selected them. For example, you cannot say that vegetable farmers in a certain area share certain themes of concern if you have only talked with three people! In presenting your findings, err on the side of modesty. Present them for the purpose of debate and comment rather than as final conclusions.

Finally, the primary use of the soft systems approach is to help people improve their current situations. It is action-oriented. Stages 1 and 2 are done in order to move in the direction of developing proposals for improvement, not just to describe what is there. The inquiry process is designed to move rapidly, first getting a preliminary understanding of a situation and, as time permits, building on the established information base. The approach is normally used in a professional context in which decision makers or other people involved have requested help or initiated the inquiry themselves. Once you have been asked to intervene in a situation by a particular group, you need to avoid approaching the situation so narrowly as to rule out some significant groups whose viewpoints also need to be known. You need to be concerned also with groups that are interdependently linked with the folks who initiated the inquiry and called on you in the first place. In other words, just because you have been called in by the asparagus growers does not mean that you speak only to them. One of the objectives of your initial inquiry will be to find out about their links with other groups — commodity groups, marketing, canneries, transportation, and so on — until it appears that you have expanded the context widely enough to describe your clients' situation adequately (Vayda, 1983). Your population and scope of inquiry may be much larger than you originally thought. Holistic thinking requires that you begin searching for clues to linkages to wider and wider contexts.

Finding and Using Written Information

So far, only the information that can be obtained by talking to people has been discussed. As you contextualize a situation, it is often useful also to consult published reports and other works, including books, articles, statistical collections, scientific studies, and government documents. Here are a few pointers to help you identify and locate such materials.

Many regions, and even specific communities, of the United States and foreign countries have been the sites or focuses of physical, social, and historical investigations carried out in the past by scientists, scholars, contract consultants, and journalists. The results are available in published books and monographs, professional journals, research reports, and popular publications such as newspapers and magazines. Historical, ethnographic, and geographic works as well as community studies and in-depth journalistic explorations may provide a rich source of observations and understanding regarding the character of particular places and their experience of problems and opportunities in the past. You may also gain information and insights about resources, transportation, agricultural patterns, population change, conflict, ethnic differences, labor, and many other topics that will provide vital context. A college librarian should be able to get you started. Once you hit a good source, check for other references in the author's bibliography at the end or in the footnotes.

As noted earlier, all major universities and state libraries have a special section with current and past state and federal government reports. These reports tend to organize their information around themes (e.g., commodities, health issues, goods, or services) or places keyed to the missions of the agencies that produced them. The U.S. Census data and reports related to your situation may be of particular importance. Monthly and annual reports produced by such departments as commerce, agriculture, and health are often relevant. State and federal labor statistics will tell you about employment in the area that is of interest. Land-use maps of many types are available, usually at no or low cost. These are helpful in visualizing the locations and spatial relationships of people, things, and resources. The land grant university of your state will have extension publications, which may provide useful technical information.

Start by charting statistical data over a 15- to 20-year period in order to get some indication of what changes are going on relative to a given situation. This will give you a feel for stability or change in the situation, along with the magnitude and direction of any changes. The accuracy and reliability of data should always be scrutinized. It may be important for you to determine how the raw data were originally gathered. Suspect data may be used as an indicator of what *might* be present and characteristic of a situation. Remember that most statistical reports are just organized data. Data must be interpreted in order to become useful information. It is also suggested that you work with people in the situation to interpret the data based on their understandings of the interactions among people, objects, and resources. For example, in the late 1980s many areas of the northeast United States have reported very favorable improvements in job markets and employment. The perspective of local people *might* be that the best new jobs have gone to outsiders.

The kind of information available to and used by people in a situation is also important. You will find that they freely share information among themselves. For example, one of the authors was amazed at the amount of technical literature that was being passed around among people concerned about the food irradiation issue in one county of Hawaii. This literature was different from that produced or circulated by Cooperative Extension, the State Department of Planning and Economic Development, and the U.S. Department of Energy. As you talk to people, *ask what they have read on the issues being discussed.* You will be surprised at what you get! Such information networks also give clues to the divergent "facts" of a situation as perceived by various people and groups.

Other Techniques

This section reviews several specific inquiry techniques appropriate to the synthesis task carried out in stage 2. First are some techniques for developing "rich pictures" and "situation summaries." The techniques selected for use should support the premises and objectives of the first two stages of the approach. These premises and objectives were summarized earlier in this chapter.

Pictorial Techniques

Peter Checkland and his associates at Lancaster University and Open University in England (Checkland, 1981; Naughton, 1983; Carter, Martin, Mayblin, and Munday, 1984) suggest that, as you talk to people and see what the situation is all about, you will be faced with an abundance of information, some clear and some unclear. They use a simple *cartooning* technique to capture this sometimes ambiguous information in a representation of a situation that they refer to as "a rich picture." This simple pictorial technique helps to highlight the major activities and the issues involved, who is involved, and what differences of world view or images of the situation exist. Nothing is fixed about the appearance of this picture other than that it contains the actors and their beliefs and roles relative to the tasks and issues of a particular situation. Many different approaches are possible, depending on your personal preference and what features of the situation you are attempting to capture. As you learn more about a situation, new information is added to the sketch. The picture quickly aids analysts and participants to understand the issues more fully and readily. A given situation in food, agriculture, or natural resource management has many issues or themes of concern as well as tasks — human activities related to the issues — that are also

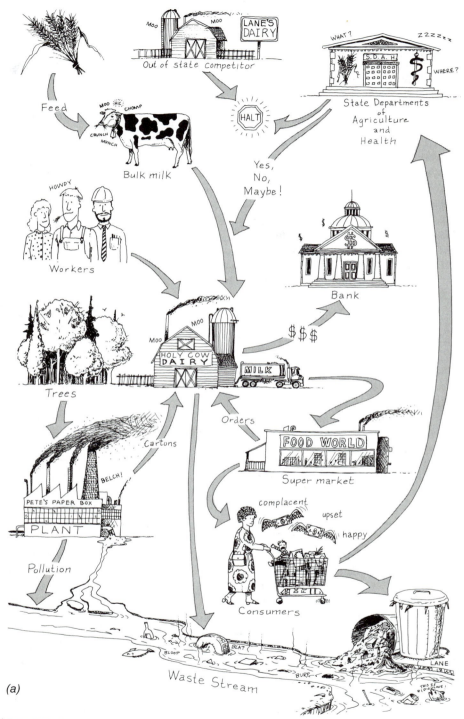

Labels within image: Out of state competitor · LANE'S DAIRY · MOO · MOO · Feed · MOO HIC CHOMP CRUNCH MUNCH · Bulk milk · HALT · State Departments of Agriculture and Health · S.D.A.H · WHAT? · ZZZZzz · WHERE? · HOWDY · Workers · Yes, No, Maybe! · Bank · $ · $$$ · MOO · MOO · MOO · HOLY COW DAIRY · MILK · Trees · Orders · Cartons · FOOD WORLD · Super market · BELCH! · PETE'S PAPER BOX PLANT · Pollution · complacent · upset · happy · Consumers · LANE · Waste Stream · BLOOP · BLAT · BURP · THIS IS IMPOSSIBLE! · (a)

FIGURE 4.1 Alternate pictorial representations of situation and operations of a dairy firm.

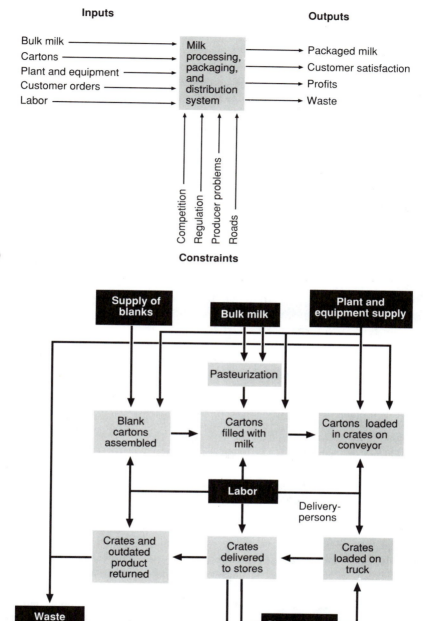

(b)

(c)

in question. Then a listing of key activities and issues is made based on the picture(s) developed.

Figure 4.1a, b and c presents alternate ways of representing the situation and operations of a dairy firm serving consumers in an urban region of the country. The three frames reflect different styles or representations as well as possible successive attempts to get at different elements of structure process and climate. Another example of a Checkland-style pictorial situation summary involving the concerns of important actors associated with the Mucho Sacata Ranch in Texas can be found in Figure A.3, accompanying the case material in the appendix. While Figure A.3 is a "first-generation" pictorial description of key features of the overall situation, Figure A.4 is an initial representation of key structures and processes linking people in the situation.

Mind Mapping

Another useful procedure is a variant of Tony Buzan's (1983) mind-mapping technique. While reviewing brain-function literature for clues about how to teach people to use both sides of the brain, Buzan discovered that most of us tend to develop the left-side functions while neglecting the right side. Right-side functions are related to complexity and divergent learning. Mind mapping is one of the techniques Buzan developed to stimulate right-brain use. Buzan's technique is also a useful notetaking device because it assists us in recalling what people say. In addition to stimulating us to think in images or pictures, and to recall the conversations of various people involved in a situation, it also aids us to organize information in a holistic rather than in linear or sequential ways.

Because this technique is useful and easy to learn by people of all backgrounds, some time will be spent describing it. To reiterate, however, other techniques can also be used as long as they help you to achieve the objectives of the first two stages of the soft systems approach.

Mind maps help build a rich picture of a situation in visual form. Words and ideas of relationships between people, events, and objects are used to develop a visual map. People's sense of impacts and effects are recorded using ordinary speech rather than technical language. The idea is to record accurately the views of the people involved in a situation, not the analyst's views.

As the analyst records conversations, he or she keeps the words used to define a situation free from analyst-imposed terminology, hierarchy, and cause-and-effect relationships. Although this technique is called mind mapping, what is really being recorded are the participants' words, primarily nouns and verbs. Arrows are drawn between nouns to demonstrate cause, effect, or associations between them. The arrows usually are labeled with verbs or short phrases. These arrows or links are generated solely from the participants' discussions of a situation. When finished, the visualizations are maps of various participants' images of a situation, reflecting their experience modified by *Weltanschauungen*. The maps are thus also powerful aids to recall conversations.

Buzan (1983) says that the key words people use in conversations regarding a situation are "multi-ordinate." By this he means that while the dictionary definition of a word is static, how people actually fit words together in speech varies, and this variation determines significant differences of meaning. Mind mapping helps the analyst to capture this by linking all the words used in specific relationships. A key recall word or phrase channels a wide range of important images and, when read at a later time, offers the original meaning back to the analyst and involved people alike. Our minds are triggered to recall lines of conversations and the associations made among ideas, people, events, and objects.

Buzan makes the point that any thought-organization technique should use not only the words, numbers, order, sequence, and lines associated with linear functions of the left side of the brain, but also the visual functions of color, shape, size, pattern, and symbolism unique to the right side of the brain. Consistent with the purpose of stages 1 and 2, mind mapping facilitates the rapid capture, recall, and presentation of information and calls on the right side of the brain to view and reflect on the *whole* of the situation.

During stage 1 of the methodology, a situation is unstructured because our minds have not yet given it meaning, i.e., conceptually structured it. This is particularly so if you are an outsider to the situation. If you are part of the situation, you have a conceptual structure, but it is based on your own **W**. Your task as an outsider is to capture people's current views of a situation. If you are an insider, your task is the same, but with the additional burden not to allow your own **W** to get in the way of looking at others' understanding.

If you are inside a situation, then you should first mind map *your own* understandings of it. Visualize them in as much detail as possible. Then use your self-understanding to suppress the temptation to assess others' perspectives by comparing them with your own. Practice setting your perspective aside and working hard to capture only others' images of the situation.

Mind mapping can be used in two ways, each producing different visual forms. The first is a substitute for traditional notetaking. The second is a tool to form a general picture of the perspectives of many individuals.

With practice, mind mapping a conversation is easier than taking ordinary notes. When someone launches into a story, he or she usually talks very fast, and the logic — the associations being made — isn't always obvious. You *concentrate on capturing the nouns, then draw arrows between them, labeling the arrows with the verbs or verb phrases* used. Rather than starting your notes at the top left corner of a page, *start in the middle!* This will allow you to skip around to new ideas and themes that are introduced as the person tells his or her story. Put the first noun or central idea that occurs in the middle of the page. Simply follow your subject's discourse. Figures 4.2 and 4.3 are examples of mind maps that were drawn during actual conversations. (They have been cleaned up for publication. In real life, these drawings are not neat because they are done so quickly.)

Students, members of community groups, extension agents, and farmers who have been taught mind mapping tend to have several patterned reactions to this technique as they start using it. At first, some are so linear that they simply capture a few words and record them down the page, one after another in a list. Others produce a snakelike diagram going from left to right, then winding back

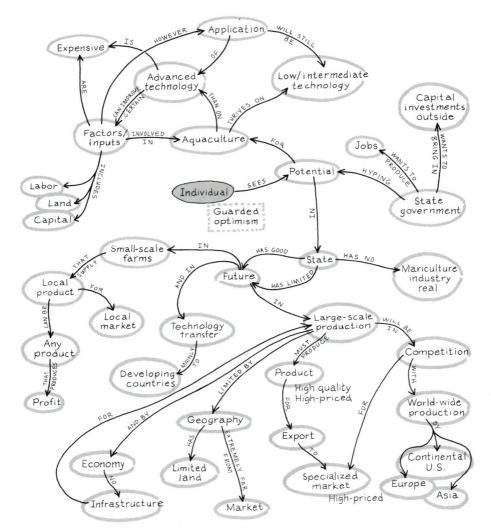

FIGURE 4.2   Mind map: conversation with a U.S. federal government researcher.

right to left. Yet others immediately launch into the "spray" effect that is more characteristic of people who use the technique frequently.

   Any technique used to capture a conversation will reduce your ability to hear what is said, particularly when first learning it. Improvement comes with practice. Just using it a few times and then judging it is not a proper test. It is just like learning to ride a bike. The first thirty tries can be a real downer! If you pledge to use the technique ten times before giving up on it, it is assured that you will begin seeing its power and utility.

   Two conversations with farmers are reproduced at the end of this chapter.

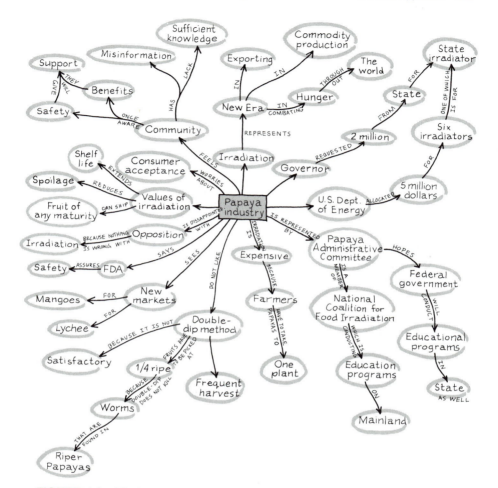

FIGURE 4.3   Mind map: conversation with a papaya industry representative.

You should try mind mapping by having someone read the conversations to you as if he or she were the original speaker. You can mind map written material, but it involves different abilities than those needed to listen to someone and accurately record what you hear.

Mind maps of individual conversations can be posted on a wall and, looking across mind maps, patterns of association can be found. In this way you can discover, in the key concerns of various actors, differing views of who is doing what to whom or what, how, and why. You can also spot contradictory information and determine the emotional tone or climate of a situation. In short, a series of mind maps provides wonderful visual records of conversations that are useful to the inquiry process in ways that linear notes are not.

If the discussion follows a question-and-answer format, then using the mind-mapping technique for recording information is inappropriate. This is because the *questions* follow the analyst's perspective, however formative or exploratory. Other forms of notetaking described elsewhere in this chapter are better.

A second application of the mind-mapping technique is particularly useful during stage 2 of the methodology. Then the emphasis of the inquiry shifts from thinking divergently about a situation to reflecting on alternate meanings of our information in order to express a rich picture of a situation. Drawing a *composite mind map* advances this assimilative process. It is referred to as a composite because the visualization is assembled from several people's understandings of a situation. This is possible if they reflect related concerns and activities. Sometimes two or more composites will have to be drawn because the **W**s are so different. This is particularly the case if a situation has become polarized and conflict-ridden or if you have interviewed members of groups occupying different structural positions in a situation.

To make a composite mind map, start in the middle of a page by circling a brief description of the people or groups whose perspective you have selected as the basis for your composite summary. Also place in the circle a word or two describing the situation's climate. From this center begin to draw lines outward, each representing a different theme of concern repeatedly described to you. Initially it looks like a spider's body and limbs. (Hence, it is sometimes called a spider diagram.) More lines resembling fingers are drawn out from each limb. On the "fingers" write a few more key concepts to help people to understand related details of a particular concern. A few examples taken from various sources are included in Figures 4.4, 4.5, and 4.6 so that you can see what these composites look like. Another example, Figure A.2, accompanies the Chatham River case in the appendix.

Summary

Chapter 4 has discussed stages 1 and 2 of the soft systems inquiry process. Stage 1 emphasizes gathering information by talking to people and collecting related documentary materials. The overriding goal is to think divergently about a situation and place it progressively in wider and wider contexts. The central task is to identify and describe the situation in all its richness and complexity rather than defining *the* problem. During stage 2, the emphasis shifts to beginning to assimilate what has been found into useful alternative conceptualizations of the meaning of the situation. Chapter 4 presented a series of premises on which these two phases of the methodology rest to guide the selection and use of appropriate inquiry techniques. Guidelines regarding the role and ethical responsibilities of the analyst in the process of finding information were also presented. The information collected is synthesized; structures, processes, and climates are analyzed; the issues and primary tasks of a situation are identified; and the preliminary description of the situation is summarized in writing and/or orally and pictorially. Several graphical techniques, including cartooning and mind mapping,

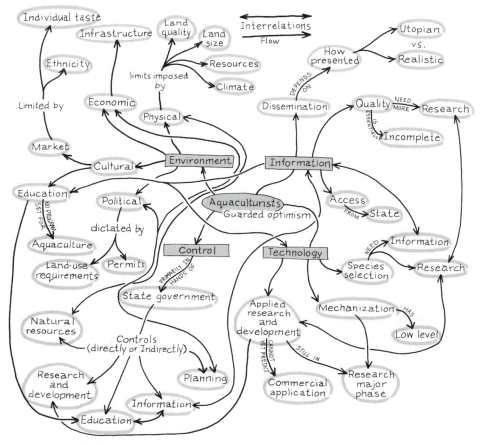

FIGURE 4.4 Composite mind map of optimistic aquaculturists.

were described in detail. Note that stages 1 and 2 may be repeated at any time in the inquiry process if new information becomes available or if circumstances change.

Practice Mind Mapping Conversations

CASE 1: RANCHER DENNIS

SCENARIO

In April 1986, the city of Thornton, Colorado, a growing Denver suburb, announced a $142 million plan to acquire control of the Water Supply and Storage Company of Fort Collins. That firm currently supplies irrigation water to hundreds of farms along the Front Range in the north of the state.

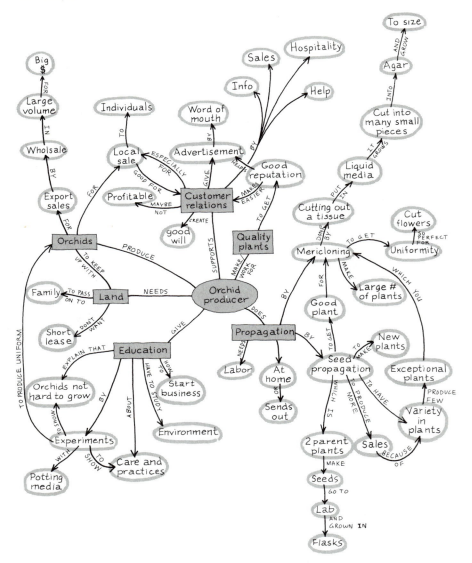

FIGURE 4.5   Composite mind map of concerned orchid growers.

Thornton intends to divert to municipal uses approximately half of the total 70,000 acre-feet the water company controls. To this end, the city has been quietly purchasing options on thousands of acres of farmland to secure the water shares it needs. While Thornton has gotten commitments from many farmers, it faces opposition from others, who have been joined by environmentalists, local northern Colorado officials, the present management of the water company, and other parties. Rancher Dennis is one of hundreds of

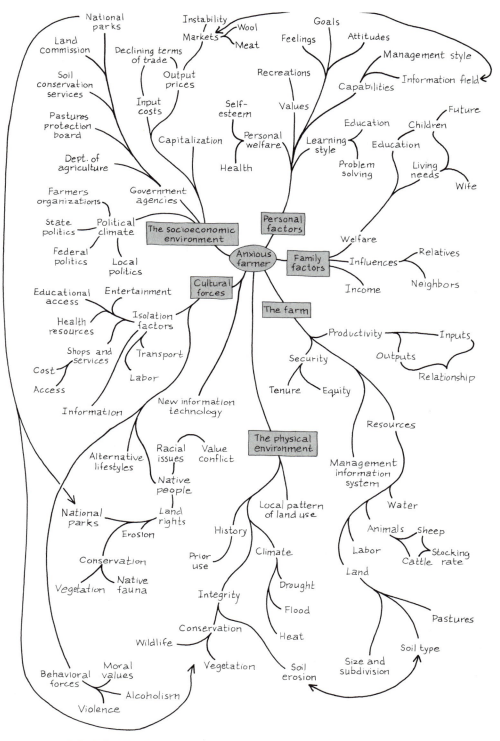

FIGURE 4.6 Composite mind map of anxious Australian farmers.

farmers in the area who may be affected by Thornton's plan and has agreed to talk to you about his concerns.

Bought the ranch in 1977: it's been a long dream of mine. Ranching is a life-style that suits me perfectly. I have two sons. I hope to be able to leave them something so they can be ranchers too. My hired man lives on this spread because it's where I got the feed lot. Altogether I farm 27,000 acres — lease 11,000 and have a deed for 16,000 — not a lot but enough to keep me going. Me and my family live on the other farm and I'm leasing a third, but I'll lose it. The man I've been leasing from has sold out to them Thornton city managers for the water rights.

My main business is cow-calf operations. I'm not one of those efficiency experts so I may be against your grain. . . . You are from the university? I just work through the extension man. He weighs the yearlings and the calves as close to birthing as he can get out here. I guess you university types use the data for something. I use the information for sales records and to keep a history. I'm a market man, keep an eye on beef prices and run my program and timing on that. Also, I am self-sufficient in feeding my 515 head. I raise pinto beans, corn, alfalfa, and stock beets. That's what I use when I bring my herd back down in October from the open range in Wyoming.

My operation is the only feedlot or farm you folks saw when you drove down County Line Road that hasn't sold water rights to them Thornton people. These urban problems just wear you out. A farm sold out to a developer who built a lot of suburban houses next to Monfort's feedlot. Now Monfort has gotta go to court over air-quality problems. We farmers got a historical right to the water. The ditch company built all these ditches with horses at the turn of the century! I got some right to the ditch because I got a share. But now Thornton can change all that. I got faith in our lawyer for the company, but at the last meeting someone said that Thornton has bought enough shares to control the flow. Thornton says that after they pump the water south to their city, they'll return it to the Platt or maybe pump it back up. I just don't know; you hear so much and most of it just don't make no sense. This sector by the feedlot grows corn and I don't even have a ditch share. Between my two pumps from my own wells and the seepage out of the ditch, I got enough. The wells, you see, are partially recharged through the ditch. But if the water gets diverted, it'll screw up the farming. Thornton says they won't exercise their water rights options, but they don't farm and they want the water so I think it's just words. If they do draw the water south, I don't know how I'll recharge the wells on this sector. The other farm is on a ditch share, but that's not enough land to leave to my boys.

Ranching is what I do best. I'd sell out to Thornton too, but only if they'd trade me land in Wyoming. I'd like to get out of living so close to Fort Collins and all the new tax plans they keep talking about. This is the richest cattle county in the country, but we're in for a change and things seem pretty uncertain about how it's all going to fall out.

CASE 2: ZEKE THE GRASS SEED FARMER

SCENARIO

Zeke is a farmer in the Pacific Northwest, trying to make a go of it on 500 acres of land in a fertile valley. To date, he has been a grass seed farmer. This has been his sole crop and has supported his family farm for 40 years. The grass seed industry is both extensive and traditional in the valley. Zeke has been beset by problems, however, and has agreed to talk with you.

Well, as you can see, I am a grass seed farmer and this, for better or worse, is my farm. The equipment shed ain't pretty but does the job. I built all the outbuildings myself, in the winter. That white tractor over there was bought two years ago and has done a good job with the plowing. No, I don't own a harvester. They're too costly and the elevator rents them out with the grain trucks. I just can't afford a good International Harvester on only 500 acres so I rent them and the drivers. When the grass is ready, we got to cut it quick, and after it gets dry, we got to pick it up quick. I usually have three harvesters and four trucks going at the same time. My two boys and I do all the other tractor work ourselves, though. Mostly that is getting the field ready in the early spring, pulling the fertilizer spreader, and plowing the boundaries for the burning. Yeah, I rent the spreaders too, but they come with the lime and fertilizer and don't cost too much. The co-op can load them up at the plant and that saves me a lot of trouble.

The land ain't bad, pretty good for grass seed as long as it doesn't get too wet in the winter. That spot over by the road can look like a lake — did last winter and didn't produce too well this year. Yeah, I fertilize a lot but so does everybody. This here valley has been good to us seed farmers, mostly dry summers after a wet spring. Last year was tough, though, with a wet harvest time. You see, if it rains after we cut the grass, we have to rake it over to get it to dry out. If you rake too much, the heads and seed fall off and you lose out. The elevator people demand that the seed has less than 10 percent water in it. And of course, if it does rain, like last year, then the renting schedule gets all fouled up and you have to wait even longer.

But you see, in a good year, the seed price is good. Mostly it goes to new housing developments and golf courses around the country. Price should go up when the blasted housing industry gets back in high gear. Straw goes to feed and fodder, mostly over on the coast to all those dairy farms. If it stays free of weeds, then eastern Washington and Oregon will take it. I don't have that problem here in the grass fields like the cattle ranches do who want to sell a little of their straw. Everything gets sold through the elevator over in Monmouth. In a good year, I get enough to cover costs, but last year was a killer. The rain cost me plenty. I ain't big enough to get the harvesters over here to pick up the seed when I want, so it laid for two weeks and got wet again. So I lost a lot by raking it twice. I ended up paying full for the trucks and harvesters, but didn't get a full crop in. I wish seed prices went up as fast as the cost of that elevator's crew and equipment.

The buyer over in Monmouth says that as soon as the country gets lower interest rates, then the construction companies will start demanding more seed and the price will go up. I just hope I can hold on. I'm going to need a couple of perfect years just to get back where I was six years ago. But with the cost of harvesting going up each year, I just don't know. Everything is going against the small farmer these days. I'm stuck renting harvesting equipment and crews from the elevator and with all the blasted politics involved with scheduling their equipment, I'm left low on the list.

And if I don't have troubles enough, there has been noise from those environmentalists saying that burning is bad. So they got the State Department of Environmental Quality setting up yet another schedule for us farmers. That's right, now they post "burn" and "no-burn" days. So now we can burn only after it has been dry *and* when some upper air thermals are just right. More delays, more costs. And lately there has been talk of banning open-field burning and forcing us to use a burning shed. I don't see what the problem is. For years we've been burning; it's fast, cheap, and the smoke just goes straight up when it's done right. Nothing clears the field better, and I'm convinced it kills all the pests. Shed burning means lots of propane and tractor time. Locals around here don't mind. They know how important it is to us farmers. It must be those people in the big cities. Yeah, I complained to the state; me and all the other farmers sent our association representative over to the capital to have it out with them. So far, we're holding the line, but some folks seem bent on driving the farmer out of business.

I don't mean to complain, but I'm caught between a rock and a hard place. If prices don't rise soon, at least enough to cover all my costs, I'm going to have to kick it in. Seed growing is all I know and am set up for, but I'd grow anything profitable to keep this farm.

REFERENCES

Buzan, T. *Use Both Sides of Your Brain*. New York: E.P. Dutton, 1983.

Carter, R., J. Martin, B. Mayblin, and M. Munday. *Systems, Management and Change*. London: Harper & Row, 1984.

Checkland, P. *Systems Thinking, Systems Practice*. New York: Wiley, 1981.

Checkland, P. "Achieving 'Desirable and Feasible' Change: An Application of Soft Systems Methodology." *Journal of Operations Research* 36(9):821–831, 1985.

Egan, G. *People in Systems: A Model for Development in the Human Service Professions and Education*. Monterey, CA: Brooks/Cole, 1979.

Egan, G. *The Skilled Helper: Models, Skills, and Methods for Effective Helping*. Monterey, CA: Brooks/Cole, 1982.

Egan, G. *Change Agent Skills in Helping and Human Service Settings*. Monterey, CA: Brooks/Cole, 1985.

Frankenberger, T. R., and J. L. Lichte. "A Methodology for Conducting Reconnaisance Surveys in Africa." Farming Systems Support Project, Gainesville, FL, Networking Paper No. 10, 1985.

Franzel, S. "Comparing the Results of an Informal Survey with Those of a Formal Survey: A Case Study of Farming Systems Research/Extension (FSR/E) in Middle Kirinyaga, Kenya." Paper presented to the fourth annual Farming Systems Research/Extension Symposium, Manhattan, KA, Kansas State University, 1984.

Naughton, J. *Soft Systems Analysis: An Introductory Guide*. Milton Keynes, England: Open University Press, 1983.

Norem, R. H. "Basic Interviewing and Note-Taking Skills for the Informal Survey in Farming Systems Research and Extension." *Proceedings of the 6th Annual Farming Systems Symposium*. Manhattan, Kansas: Kansas State University, 1986.

Pickett, S. T. A., and P. S. White, eds. *The Ecology of Natural Disturbances and Patch Dynamics*. Orlando: Academic Press, 1985.

Rhoades, R. E. "The Art of the Informal Survey." Lima, Peru: International Potato Center, 1982.

Rogers, E. *The Communication of Innovation*. New York: Free Press, 1981.

Vayda, Andrew P. "Progressive Contextualization: Methods for Research in Human Ecology." *Human Ecology* 11(3):265–281, 1983.

Wilson, Kathleen. "A Walk Through Sitiung: A Simulation Game to Increase Observation and Inference-Making Competencies." Washington, DC: U.S. Peace Corps, 1987.

CHAPTER 5

Five Developing Models of Human Activity Systems: Stages 3 and 4 of the Soft Systems Approach

In stages 3 and 4, your objectives shift from the open description, analysis, and synthesis of the present state of affairs to using systems thinking to design and describe proposed future improvements. There are four key inquiry activities in stages 3 and 4. First, *transformation statements* are created, which designate basic features of an improved situation related to the themes of concern and primary tasks identified in stages 1 and 2. Second, each transformation statement is expanded, using the CATWOE outline, to become a *system definition*, a blueprint for a *conceptual model* of the critical human activities envisioned to be in operation in an improved *future* state. Third, conceptual models of *human activity systems* are formulated. Fourth, based on stages 1, 2, and 3, initiatives for *basic science* research, *technology development*, and *hard systems* analysis are taken by appropriate parties. This chapter discusses these activities in detail and presents several illustrations.

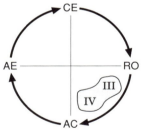

A major shift in learning activities also occurs in stages 3 and 4. The competencies associated with the assimilative learning style is emphasized here. These include the ability to form propositions (deductions) based on your survey of the situation's complexity. In addition, you must be able to use and communicate appropriate technical knowledge and specific techniques.

A Focus on the Future

Typically, once a situation is understood, people are impelled to begin implementing changes without taking the time to decide what the improved way of doing things might look like. The majority of those who first try to carry out stage 3's activities fall into a trap. This is the tendency to say, "Here is what we are doing now; here is how we are going to change," rather than saying "Here is what we are doing now; here is *what we want to be doing in the future*." The distinction between these two approaches is clear-cut. You and the people involved must enter stage 3 with *a vision of what you want a future state to look like*. Think future: To carry out stage 3's activities, you must think in terms of improved conditions and future states. This key perspective is illustrated in Figure 5.1.

You will discuss how to implement your sense of an improved future, but only after you have developed several transformation statements into models. Note also that a full discussion of implementation approaches appropriate to the ends defined in stages 3 and 4 does not occur until stages 6 and 7. Only in stage 6 will you help people engage in a wide-ranging debate about the desirability and feasibility of specific changes. And a detailed implementation plan does not emerge until stage 7 (Chapter 8). To reiterate, people are more inclined to generate ideas about *means* than about *ends*, that is, about specific changes rather than where those changes will take them. Several examples reflecting this distinction may be needed in order to help you and the people involved to assimilate it.

Key Inquiry Activities

Stage 3

The major activities of stage 3 involve the definition of relevant systems using Smythe and Checkland's (1976) CATWOE (see subsequent discussion), a mnemonic that lists the six items to be included in it.

1. Develop a transformation statement for each primary issue; primary task; and the structural, process-related, and climatic concerns identified in stage 2.
2. Further develop each transformation statement so that the following minimum features of an improved state are described: What will be the central transformation process that is in operation? Who will be managing and responsible for the improved operation? Who will benefit from this altered way of doing things? Who might be negatively affected if such alter-

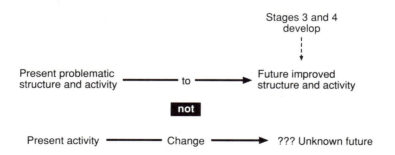

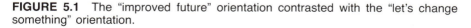

FIGURE 5.1 The "improved future" orientation contrasted with the "let's change something" orientation.

ations were made? What constraints will managers have to work around in order to implement the transformation? Who will have the power to alter or stop the transformation process and, hence, the group's sense of improvement? What are the **W**s, the values, and the assumptions on which this statement of improvement rests? The answers to these questions define the nature of each model.

3. Conduct discussions with actors, owners, and clients involved in the current situation to determine which relevant system definitions will be further developed into models.

4. Formulate recommendations regarding the kinds of basic science, technology development, and hard systems research needed to work on parts of the improved state of affairs. Communicate these recommendations to the appropriate parties for action.

Stage Four

1. Develop at least one model for each *system definition*, using systems concepts.

2. Develop or obtain hard systems models and other aids if they are needed to help you think through or project further potential changes in or among components of a situation; e.g., if we changed this component, how would that component change?

Stage 3: Defining Human Activity Systems

This section discusses the key systems concepts used to develop ideas about improvements. It builds on what was introduced and discussed in Chapter 3. The focus is on how you go about applying systems thinking to the task of defining *what needs to be changed* if the *future state* is to be an improvement over the present state.

Every definition of a human activity system (HAS) model has six essential elements. Checkland's research group from the department of systems at Lancaster University in England has developed a useful aid for remembering these elements. It is the mnemonic CATWOE, referred to earlier, in which each letter stands for one of the system features that needs to be taken into account (Smythe and Checkland, 1976). In the case of HAS models, the CATWOE forces one to articulate assumptions, particularly the world view of all the people involved, including the analyst. For the time being, the meaning of each letter is only summarized briefly so that you will have them "on-line." The following sections will expand on these meanings and, in particular, show how to use the CATWOE.

Developing Transformation Statements

The *T* of the CATWOE is taken up first because the *transformation* describes the central activities that you and the people involved believe will represent an improved future. After you have the transformations defined, you proceed to build the rest of the CATWOE.

Inputs ➡ Transformation ➡ Outputs

FIGURE 5.2   The transformation process visualized.

Developing a sense of improvement involves identifying what transformations will be operating in a desired future state. Because your typical situation is complex, it will be necessary to draft several transformation statements. When you have summarized a central human activity that will have to occur in an improved state, you have developed a transformation statement. A conceptual map of this concept is found in Figure 5.2. Inputs come in, something happens to change them (transformation), and (if the transformation process is working correctly) there is an output or result.

A corresponding transformation statement can be created for each primary issue and primary task identified in stage 2. Central to every transformation statement is a *verb* describing the operation of coordinated activity in an improved state. The statement is action-oriented and expressed with the verbs that best summarize the entire operation. At this point, a sample case is introduced that will be used to help you work through the techniques involved in stage 3 and assimilate the idea of a transformation statement.

Composite mind maps developed in stage 2 are applied directly in stage 3 when the key themes of concern presented in them are converted into corresponding transformation statements. In the same way, a cartoon developed in stage 2 can be used to create concise transformation statements related to structural, processual, or climatic features of the situation being studied. Finally, the primary tasks identified in stage 2 are restated, and corresponding transformation statements are created for them.

From the standpoint of working with the people involved in a situation, there are at least four ways to develop transformation statements:

1. Ideally, transformation statements should be developed collaboratively with the people involved in the situation. People engaged in thinking through improvements will be better motivated, empowered, and committed to undertaking the hard work necessary to make change. Also, if people work through the thinking involved in making changes in a situation, the overall results will be superior to the results of their merely accepting an expert's recommendations. There are two reasons for this. First, people in the situation will help make sure that key factors are not overlooked, and the models of improvement will be better designed. Second, people will know why and how certain proposals came into being. Thus, debates tend to be more productive; and ownership, responsibility, commitment, sense of control over the process and the results, and satisfaction with the proposals for change all tend to be higher. Research on small-group dynamics supports these premises (e.g., Johnson and Johnson, 1987). You can use a technique with the people involved such as brainstorming or the nominal group technique (both discussed in an addendum to this chapter) to quickly develop a list of possible transformations.

2. Alternatively, transformation statements can be generated by the analyst and then discussed with all the people involved or a representative group. This is not the best approach, but it is an option that depends on the circumstances, such as the number of people involved, travel distance, and the time *they* are able to devote to the project.
3. Transformation statements may also be developed during a working meeting with a representative sample of the people.
4. Particular transformation statements may be developed by working with the specific people who originally presented the corresponding theme of concern during stage 2's inquiry.

All of these techniques have been used successfully by analysts who use the methodology in their work. The main point is to engage the people who are in the situation meaningfully to the maximum extent possible in the task of developing transformation statements.

Often issue and task statements can be collapsed or amalgamated to create one transformation statement that addresses several concerns. So, when possible, you should attempt to cluster concerns, senses of unease, and related issues, or else you should closely examine transformation statements and merge those that seem to address the same or similar concerns. Typical food, agricultural, or natural resource situations tend to have several concerns, so clustering whenever possible will make the rest of stage 3's tasks, as well as later ones, more manageable.

Once the transformation statements are created, then the remaining key features of each human activity system are identified according to the CATWOE outline. (Looking ahead, keep in mind that when all features are identified and all the human activity systems are defined, then the modeling process of stage 4 begins.)

In applying the CATWOE, develop each transformation statement in turn. You will recall that these statements identify key activities people believe are necessary to transform their current situation. Each of these statements can potentially become a separate HAS model in stage 4 of the approach. The word *potentially* is used because some of the transformation statements might eventually be seen as subactivities (hence potential subsystems in a HAS model) rather than as a central activity characterizing the whole of an improved system. For now, treat each transformation statement as if it were eventually going to be developed into a HAS model and apply the CATWOE outline to each.

The CATWOE emphasizes that each transformation needs people to carry it out (actors), has impacts on people (customers), will be influenced by powerful interests and decision makers (owners), will operate with various resources and constraints (environment), and will be subject to the owners' and other actors' views of the world (*Weltanschauungen*), which is implied in the group's sense of the transformation. Each part of the CATWOE can be discussed by posing a question to the people involved in thinking about improvements. Conduct these discussions in a creative manner without overemphasizing the development of any given feature.

A situation involving a specific group of floral growers will be used to take you through the facilitative techniques required for developing transformation statements and other elements needed for defining human activity systems.

The Floral Growers' Associations Case

A year ago, representatives of a particular floral industry in the state of Bliss (the name used to ensure confidentiality) requested assistance from their local college of agriculture. In order to pursue stages 1 and 2, an analyst/facilitator spoke with large- and small-scale floral growers, officials of the Bliss State Department of Agriculture, the county land-use commission, the city council, and related floral commodity associations. Based on these discussions, a synthesis of the current situations was created. The floral growers of this state expressed six major concerns: (1) lack of land to expand or to establish new farms; (2) insufficient political clout to influence government policy affecting the industry and agriculture in general; (3) competition between the major floral associations, making it difficult to set prices for their particular commodity; (4) competition between large- and small-scale growers, resulting in seasonal gluts and low flower prices; (5) lack of relevant research by the university (as perceived by the growers) aimed at providing the floral industry with a wider selection of varieties that flower year around and at increasing the variety and availability of adequate cultivars; and (6) inadequate information from the university and associations.

The analyst/facilitator and representatives of the floral growers initially created five transformation statements based on these six themes of concern. (To emphasize the *future action* orientation of these statements, the *verbs* have been italicized.)

1. A system *to share* information so that communications are facilitated among growers regarding production and marketing developments.
2. A system *to expand* production to allow for a greater volume, making a packing house economically feasible.
3. A system *to allocate* land that will devote increased acreage to the crop.
4. A system *to coordinate* government and private-sector efforts to plan and promote development of the industry.
5. A system *to develop* techniques for year-round production and supply of floral and nursery products.

What meaning can these statements have to the growers? To answer this question, the significance of the first transformation statement on the list is discussed. This transformation statement envisions a future in which the growers freely and readily exchange information on production and marketing. If the transformation is adopted, they will be doing several things in the future that they are not doing now, and they probably will also be organized differently. Currently they do not have an organized way to gain access to or to communicate vital information. Five years from now, they will be sharing certain kinds of information in order to stay abreast of technological developments; respond to and predict supply, demand, production, and marketing developments; and be informed of regulatory changes. By explicitly stating their vision of improvement in this transformation statement, the aim becomes clearer to all of them. And so it is with the rest of the transformation statements that define the essence of improved human activity. (Note that the specific "hows" of achieving specific aims are not at issue until later stages of the approach. To make this discussion more real to readers, however, some hints of concrete proposals that could develop later are provided. For example, the floral growers might merge competing organizations and publish a newsletter and technical bulletins.)

The other elements of the CATWOE add flesh to the bare bones of the transformation statements. A few preliminary tips are called for. First, remember that, as with the transformation statement, the focus of the specification of the elements of the CATWOE is on *the future*, not the past or present. "Future" means *that period in time when the improved human activities that are now only being defined will actually be functioning*. Second, the order in which you discuss the CATWOE characteristics determines how smoothly the discussion and thinking proceed. *The suggested order is TACOEW*: that is, transformations, actors, customers, owners, environment, and the **W**s. Third, the modeling procedure will be enhanced and expedited if you refer to the knowledge base developed in stages 1 and 2, consisting of your preliminary documentation of things people said they wanted to improve or change and other observations. Often, when you go back to this valuable material, you will find that some transformations have already been defined or at least indicated plainly.

Actors (A). Here your task is to determine who *in the future* will be responsible for managing and carrying out the key human activities described in your transformation statement. So, during stage 3, develop a preliminary list of actors. Later, as you develop a particular model, additional actors may be identified because critical functions become clearer to the designers. It is also possible that actors initially identified in stage 3 will be deleted in stage 4. For A the question posed to the people involved is: "Who do you think could be (or should be) involved in operating the proposed improved system summarized by the transformation statement?" Remember that *who the actors are* is directly linked to *what activities they will carry out*.

Facilitators must be sensitive to the divergence of thought accompanying this phase and carefully lead discussion in accordance with the needs of the group. As indicated earlier, brainstorming is an effective technique for helping a group to identify key actors. Writing ideas on a chalkboard or flip chart can help to visualize links between the actors identified and the functions for which they might be responsible.

Customers (C). Customers are people who potentially will either benefit or be adversely affected by a change defined in your transformation statement (Checkland, 1981; Naughton, 1983; Wilson, 1984). The question you pose to the people with whom you are collaborating on the design work is: "Who do you think might benefit from the proposed system and who do you think could potentially be adversely affected by the proposed change?" Identifying potential bene-ficiaries is easy. People are much less comfortable identifying possible victims. Since there is no such thing as a free lunch, most improvements carry with them some negative impacts. Therefore, the people involved in stages 3 and 4 are ethically bound to think through how the change envisioned might affect various individuals, groups, businesses, communities, and so on. How can you adjust your thinking to reduce potential negative impacts? What will you do if the im-pacts are potentially very serious to people or the environment? Sometimes a sober look at negative effects results in a change in the transformation statement. Other times, the development of a model continues, but potential victims are identified again in stage 5 or 6, and concerns are dealt with in detail then.

Owners (O). The next step is to identify the prospective decision makers in an improved state. Depending on the range of **W**s represented among the people involved, this discussion is tricky. People often have different views on ques-tions of power and legitimacy. What power sources are considered to be givens and what ones are possible to change? The challenge is to think through who should be the powerholders in an improved state of affairs. These powerholders are called owners (O) by the Checkland group. Who currently isn't but should be? Who currently is and shouldn't be, in an improved state? And who currently is and should continue to be in the future? *Owner* means those people or official positions that the designers of the system want to have enough influence, deci-sion-making authority, and responsibility to be able to cause the proposed system to be altered; maintained under adverse, as well as normal, conditions; or cease to exist in an improved state of affairs. Once again, note that any specification of owners carried out in stage 3 need not be final. During all subsequent stages, there will be ample opportunity to discuss and clarify the distribution of power, decision-making flows, responsibilities, and limits.

Environment (E). Next, identify important features of the environment, both resources and constraints, affecting the activity defined in a given transfor-mation statement. These are all things that the designers of the improved system take as givens rather than subject to direct control. As such, their impacts must be managed and accommodated by the system that you and the people who are involved design. Discussion of this design feature is challenging. Often, for example, farmers take certain things as given, even in an improved state, that other people involved in the design process do not think should be taken for granted. The central question here is: What can be changed or controlled and what can't be? And what activities do we want *not* to control or be responsible for, preferring to see them carried out by some outside group or agency? Some will see existing laws as environmental constraints, while others will see that changing them must be a part of the system's activity or a part of the overall plan of activity necessary to model and implement the proposed system.

A technique called *force-field analysis* (Ford, 1975), described later in this chapter, can be used by facilitators to help farmers, other enterprise managers, and their organizations to examine and think through questions about environmental factors and the limits they are placing on themselves. Do they really have to be givens? Can we envision ways to reduce some constraints or their impacts? Should we think of removing some current constraints? Rather than accepting the constraints and designing them into our view of an improved condition, could they not be thought of as a subsystem that will have to be worked on? As with all the rest of the CATWOE characteristics, environmental constraints need to be articulated precisely and clearly. It is easy to become conceptually sloppy here.

Weltanschauungen (W). The final step in identifying features of a HAS is to clarify the values and assumptions held by those developing the transformation statements and the rest of the CATWOE. What is your **W** and that (or those) of the people involved in this work that made a particular transformation meaningful and "good"? Just as you identified the **W**s of all the different groups involved in the situation during Stages One and Two, now you must reveal the **W**s behind proposals for change.

To aid in remembering the discussion and development of ideas here, Table 5.1 summarizes key questions related to the development of the CATWOE. Once these characteristics have been defined, then a narrative is created which puts them together in complete sentences. When this has been done, you have succeeded in creating a system definition for *one* HAS model.

TABLE 5.1 Summary of CATWOE Questions: Six Items Covered in a Well-Formulated Systems Definition

Customers *C*	=	Who could benefit from this altered way of doing things (as summarized in our transformation statement) — "beneficiaries"? Who could be adversely affected — "victims"?
Actors *A*	=	Who would manage and be responsible for the improved operation as summarized in the transformation statement?
Transformations *T*	=	What could be a central transformation process that characterizes an improved operation? How might we best summarize the operation of coordinated activity represented in an improved state? (Formulate a separate system definition for each transformation.)
Weltanschauung *W*	=	What is the outlook, mental framework, or image that makes this particular transformation meaningful? What values and assumptions are explicit or implicit in our view of improvement as described in our transformation statement? (If there is more than one associated with a given transformation, formulate a separate system definition.)
Owner *O*	=	Who has or could be granted the power to alter or stop the proposed transformation process, and hence the group's sense of improvement, even in the future?
Environment *E*	=	What environmental factors might constrain and assist our improved operation in the future? And what are the *internal or organizational* constraints?

A HAS Definition for the Floral Growers

The case of the floral growers can now be taken up again to show you the kind of output from such a CATWOE-driven discussion. The example illustrates how one group developed a system definition of a HAS model.

Based on the themes of concern described earlier in the situation summary, the growers tentatively created five transformation statements. These were listed earlier on page 169 and you should go back and review them.

After much discussion, the growers felt that the transformation statements could be merged; numbers 2, 5 and 3, 4, while number 1 could be slightly revised, thus providing a blueprint for three human activity systems. As discussed earlier, the technique of combining transformation statements into more comprehensive ones whenever possible streamlines all subsequent tasks, including those of stage 4. Here are the revised transformation statements:

1. A system *to share* information and *to facilitate* communications among growers regarding technical, production, and marketing developments.
2. A system *to expand* production in order to allow for a greater production volume, to make a packing house economically feasible, and to provide for year-round production and supply of floral and nursery products.
3. A system *to coordinate* government and private-sector efforts in order to plan, promote, and expand the industry, including increasing the acreage available to growers.

To further illustrate the process of creating a human activity system model, the expanded production transformation statement (number 2) will be taken and carried through the definition process. (In actuality, all three were developed more or less at the same time so that the people involved could examine alternative future states.) The definition of the model that was developed using the CATWOE aid is as follows:

Transformation. A system *to expand* production in order to allow for a greater volume, to make a packing house economically feasible, and to provide for year-round production and supply of floral and nursery products.

Actors. The production expansion system is operated and managed by the following actors and groups:

1. The three major growers' groups, namely:
   a. the Bliss Floral Growers Association (BFGA).
   b. the Bliss Floral Promotions Council (BFPC).
   c. the Floral Growers Association of Bliss (FGAB).
2. The Bliss State University, College of Agriculture, and the Cooperative Extension Service, namely:
   a. the researchers and extension agents of the department of horticulture.
   b. the directors of research and extension for the college.
   c. the college dean and department chair.

3. The State Department of Agriculture, namely:
   a. the floral and marketing researchers and regulators.
   b. The secretary of the department.
4. The State Department of Land and Natural Resources, specifically staff assigned to agricultural land-use and business development.
5. The Bliss State Legislature, specifically assemblymen, senators, and legislative staff serving committees dealing with agricultural issues.
6. The Greatport City Council.
7. The Bliss Land Use Commission.

Customers. The large- and small-scale operators, the flower brokers in the state, are seen as the most significant and direct beneficiaries. The secondary businesses from which floral growers purchase goods and services will be indirect beneficiaries. Other commodity groups may benefit from reduced marketing costs since several kinds of marketing functions will be done jointly. Those adversely affected may be the three largest floral growers because they will have less control of prices and supply relative to demand. Also adversely affected would be flower growers not formally affiliated with the three associations because they would not have ready access to information, services, and improved cultivars produced by the proposed system. Foreign growers in certain developing countries who have expanded their exports of cut flowers to the U.S. in the past will experience reduced sales and markets in the state of Bliss as its growers fill local demand and expand shipments throughout the country. However, since land acquisition is projected to be part of the system, other agricultural commodity groups may be displaced. A remote possibility exists that residential land may be rezoned, causing fears of negative impacts on local communities.

Owners. The system will be principally owned by the three growers' associations. The nature of the membership and the relationships between members of the three groups will allow the key activities to be initiated and, as time goes by, to be monitored and revised. An advisory committee from these three groups will have authority to monitor and make changes as needed. Members will be selected on a rotating basis to ensure a balance of power. In this way, the composition of the advisory committee will be such that large- and small-scale growers' interests are represented and accommodated. The initial advisory committee will have the major responsibility for implementing the system.

The State Department of Agriculture will improve monitoring of those aspects of the system subject to regulation, particularly the setting of prices and production ceilings.

Environment. The expanded production system will operate under the following anticipated environmental constraints, which are taken as givens:

1. The majority of the commercial floral cultivars of this kind are highly seasonal during their peak production years.

2. Breeding desired characteristics into the commercial cultivars takes a long time.
3. The threat of hurricanes, droughts, or other extremes of weather is endemic.
4. Some growers will continue to import floral products to meet their market commitments.
5. Some individuals or groups will continue to sell state-developed cultivars or technology to foreign competitors.
6. Groups associated with other commodities can influence the market, reducing demand for these floral products.
7. The federal government and the European Common Market will continue to support the 1983 Caribbean Basin Initiative, which allows duty-free privileges to the nations of that region (and thus subsidizes major competitors of the Bliss floral growers).

Weltanschauungen. The world view that makes this potential expanded production system worthwhile includes the following beliefs and values:

1. Expansion is desirable because demand exists and continuous planting is required due to the growth cycle to replace plants that stop flowering.
2. The ability of Bliss to supply floral and nursery products throughout the year is necessary for the state to maintain its niche in the local, U.S., and world markets.
3. The production of these commodities is labor intensive and thus provides jobs for the people of Bliss and maintains diversified agriculture in the state economy.
4. It is possible and desirable to reduce current competition among floral growers and growers' groups in order to stimulate and expand the industry.

This example shows you what an initial definition of a human activity system looks like. Using the CATWOE mnemonic, you can create a definition like this one for each transformation statement you develop. These definitions will, in turn, serve as the preliminary description of each HAS model developed in stage 4.

Before moving on to discuss stage 4, a few words should be said about how many transformation statements to develop fully through the CATWOE procedure. Suppose that you are an extension agent working with a group of growers in a particular region. It is very likely that you will hear a large number of themes of concern as you carry out stage 1 and 2. Agricultural and natural resource situations are complex! Since an extension agent knows that the development of the CATWOE characteristics must be done by and with the growers, key commodity and industry people, as well as state and federal officials, it is very tempting to take a big reduction step and attempt to deal with only one transformation. In order to involve at least a representative sample of concerned people, an agent probably will have to make many trips to an area in need of assistance. Facilitators who have gone through this process now realize that engaging farmers, and relating them to areas of concern *they* feel are important, is critical to the thinking that goes into stage 3. It should not be avoided.

So, while you may be tempted at this stage to select only a few concerns for which to develop CATWOE features as a way of reducing discussion and effort, it is suggested that all themes of concern be developed into transformation statements and that you further develop as many as possible into full definitions. Of course, you too have to decide whether to take as *given* specific time and resources constraints. As was noted in Chapter 4, when people talk with you about the current situation during stages 1 and 2, they will tell you what key features need to change in order to improve their situation. If you made notes of these conversations, some of the CATWOE features may have already been identified. You can use them in summary form to launch stage 3's discussions with the people involved.

Precise and rigorous thinking is important during the development of HAS models and their accompanying definitions. Because words can vary in their meanings, the people working on the HAS system definitions need to work with precision and attend to detail. The analyst/facilitator plays the vital role of referee in helping the group articulate its ideas concisely and precisely (and blowing the whistle when they do not). For example, people are prone to list all the CATWOE features but not to define them precisely. One simple pointer is to develop descriptions of each feature in sentence form rather than just as a phrase. Observing grammatical rules fosters orderly and logical thinking. Ensure that your clients precisely identify the groups and people who they believe will be the actors, owners, and so on. They must be clear in attributing limits and constraints to the environment. Again, as facilitator, it is your job to conduct the discussion so as to encourage clear thinking. This sounds obvious and easy, but competence in effectively facilitating discussions takes practice.

Recommendations for Other Kinds of Inquiry

Before moving on from stage 3, recommendations for other kinds of inquiry are made to appropriate groups. What recommendations for basic, applied, and developmental research result from the transformation statements? What kind of hard systems modeling is needed? To whom does your group want to make these recommendations? What timetables must be given to those who will do this research in order for their work to be useful?

For example, in the course of stage 3, it may become clear that one of the desired improvements is to develop a plant variety that flowers or fruits all year around or that needs less light. Some basic research and technology development work is needed. Or it may be clear now that part of the transformation process calls for a reallocation of resources or inputs such as land, water, money, personnel, fertilizer, or pesticides. Yet there are questions about how much can be reallocated before things get worse rather than better. Some hard systems analysis or applied research may be needed to answer these questions. Such recommendations, made at this point, are presented to appropriate parties, for example, the state agricultural experiment station. Sometimes the group must work with those contacted in order to provide endorsements and gain funds for the required research and development work. Sometimes members of the client group will be able to do the research and developmental work themselves or with the occasional help and advice of scientists, extension agents, and service representatives of a nearby university; the state or federal departments of agriculture; or an agribusiness company.

Once several HAS models have been defined, they are developed into full-fledged models. How to formulate these models is described in the next section.

Stage 4: Conceptual Modeling

Chapter 2 described how people cycle through the phases of learning in response to their concrete experiences of a problem or situation. Modeling occupies a very important place in this cycle because it portrays a sense of an improved situation so that we can manipulate our thoughts and also externalize them for others to share. To create a model requires a *convergence* of thought in order to make an appropriate representation of an improved condition. As indicated elsewhere in this chapter, usually several models are created which address various themes of concern embedded in a given situation. In addition, depending on world view differences, several alternative models addressing the same themes of concern may be needed in order to reveal creatively how things might function and be structured in the future.

Modeling is an approach to what is referred to in the learning cycle as *abstract conceptualization*, the mental process used to make sense out of events. It is a mental construction or reconstruction of the way we organize our collective or individual views of things. In its conventional usage, however, modeling usually refers to how we make tangible what is in our minds.

Throughout this book you will find numerous diagrams and illustrations. Most represent someone's portrayal of a thought, idea, or situation in a model. The Kolb learning style grid introduced in Chapter 2 is a fine example. Diagrammatically it is simple, but it presents a wealth of ideas. It brings order and conveys understanding to subjects not normally thought about or presumed to be beyond comprehension.

One characteristic of a notable model is surprise. Does it reveal something that was not recognized before? Does it enable you to satisfy yourself or others that your thought process has been extended? Do you feel confident that you can produce and use a model as a way of introducing proposals for change to someone else or a group of people?

There are many different types of models. Many of these were described in Chapter 3. Not all models are systemic models, describing things as if they were systems using the concepts found in the systems literature. For example, a transformation statement is not a model, just a recording of an idea. This is an important distinction because the minimum work of stage 4 requires that the modeler design systems-based models. This means models are developed to include a minimum set of systemic features. There are guidelines in the literature on the minimum features to include in a model that can be said to be a systems model.

As in all models, these minimum features are modeled verbally and symbolically. They will be discussed and summarized later in this chapter (Tables 5.2 and 5.3).

In adherence to the premises underlying soft systems analysis, models are used to present a set of human activities in a novel way to illustrate what kinds of human activity would occur in an improved state. Thus, the designers' attention is not on describing *what is*, but rather on designing *what could be*. Modeling in this context becomes a vehicle for learning together and for thinking about a changed state and the ramifications of proposed changes.

While it may be useful to design several kinds of models in stage 4, one type is *always* developed. This is the human activity system (HAS) model. Why develop the human activity system model first? As noted at the beginning of this book, agricultural and natural resources management is conceptualized as a human activity system with the people involved—the farmers, growers, ranchers, agribusiness men and women, employees of natural resource industries, or families—as the subjects of our concern. There is an important role for professionals to help clients such as these to learn to manage change more effectively so that their life-styles and businesses are more sustainable no matter what happens to the environments in which they operate. Sustainability results from the interaction among the family; the business enterprise; and its physical, biological, and socioeconomic environments. Effectiveness in dealing with change is a function of the learning of family members and of employers and employees involved in the management of those environments. Peter Checkland, who developed the concept of the HAS, notes that natural, designed physical, and designed abstract systems are "inevitably linked closely to the human activity described" (Checkland, 1981). Therefore, after working out what a future state should be, the next task to work on is changes in what people will do and how they will do it in a future state. This is done by first defining each HAS model (stage 3) and then actually formulating the models (stage 4).

It is possible to develop at least one model from each definition created in stage 3. Since the development of these models takes time, there is usually serious and intense discussion among participants on which of the definitions developed during stage 3 will be modeled. Facilitators are encouraged to model at least two HAS models, not just one. This will help the people involved to understand that *the purpose of the modeling is to think deeply and creatively about how things might operate in the future* without a commitment to actually implementing any of the changes. The decision regarding what to implement comes later in the process. It often must be addressed here so that people feel free to think creatively without the fear that someone's "off-the-wall" idea is actually going to be implemented. Otherwise, creative thought is stopped. Indeed, that is what tends to happen when an outside expert or consultant delivers his or her model to clients. Creating at least two HAS models helps convey the notion that the process is still wide open to discussion and consideration of alternative possibilities.

The order of discussion in construction of a model aids in efficient use of time and in orderly thinking. The presentation of *how to construct a human activity system model*, therefore, will use an explicit order. To summarize, the most productive order is (1) clarify once again the transformation statement, (2) develop the subsystems, (3) identify inputs, (4) identify outputs, (5) locate boundaries, (6) establish measures of performance, (7) agree on the decision process, (8) clarify environmental effects, (9) use checklists, and (10) communicate the model.

The sections that follow fill in the details of this outline of how to construct a HAS model. Included is the symbolic part of one model developed by the floral growers' group. It illustrates one way of visualizing a model. The end of this discussion also includes an example of the type of narrative that accompanies the symbolic model (Table 5.4).

Clarify Transformation Statement

The development of a human activity systems model begins by reviewing the transformation statement and related CATWOE characteristics that were developed during stage 3. The transformation statement is the summary of what the entire model is about. Active language was used to construct the transformation statement, with the verb the central feature. The implicit purpose for seeming to repeat the transformation statement description process is *to sharpen* the language and, perhaps most important, *to introduce* the rest of the modeling effort.

It is very likely that the transformation statement will be clarified at a separate meeting, removed in time from earlier inquiry activities, so the idea here is to recapture the focus and momentum that characterized previous efforts.

In the floral growers' case, the modeling discussions started by *clarifying* the sense of an improved state that the participants had developed during a previous meeting: "An industry that can provide a year-round supply of floral and nursery products to be able to maintain customer loyalty." If you look carefully, you will note seemingly superficial alterations of language that were nevertheless *important* to the people involved. Next, a shortened version of the definition of the model was placed on the board for all to consider as they refined their thoughts: "Summary of the transformation we want: An expanded production system."

Develop Subsystems

The next step in formulation of the human activity system model is to discuss what major human activities would have to be in operation if the transformation identified is to happen. Facilitators lead the group in identifying all the critical activities people think might be appropriate. Use a procedure such as the *nominal group technique* (described in an addendum to this chapter), so that the identification process can proceed rapidly. The idea is to think creatively and freely without, at this point, ruling out anyone's suggestions.

To begin to identify the subsystems of the HAS model for the floral growers, the facilitator asked, "What is needed in order for this expanded production system to operate effectively?" Working in groups of four, participants generated a number of possible activities they thought had to be in such a system. The facilitators suggested that they start developing their notion of each activity by identifying a verb to act as a summarizer. All groups were asked to present their listings, which were then placed on a master list.

Several participants immediately noticed that some of the activities listed were actually subactivities pertaining to another subsystem that had been identified. Through discussion, seven key activities (subsystems) were identified as "need to have" in order for their sense of improvement to be realized. These were displayed as shown in Figure 5.3.

In order to develop the dynamic features of the human activity system further, participants were asked to identify a few subactivities under each major activity. Figure 5.4 illustrates the results of this effort.

Identifying subactivities in this way helped all participants to understand more fully what would have to be done to make each subsystem operate as intended. It also helped to weed out subsystems that were really subactivities. This also addresses the concept of hierarchy; i.e., transformation/human activity system, activity/subsystem, task/subactivity, and so on. Using systems terms in this way, participants developed a preliminary notion of the key subsystems of the model. In the usage adopted here, *subsystems are the primary human activities that make the system operate.* Furthermore, they are critical to the success of the desired transformation. As explained more fully subsequently, each subsystem should be of a similar level and order of complexity. This means you should pay attention to the notion of hierarchy. Should a given activity be considered a subsystem, or is it actually a subactivity of the subsystem you identified? In practice, this issue of hierarchical level is raised with participants in order to challenge their thinking and logic. It should not be a sticking point in discussions, however, because subsystems that are not of the same order often are deleted and redone during later discussions.

Encourage participants to examine carefully the verbs used to describe the central activities they identify, along the lines of "A system to (Verb 1)," "a system to (verb 2)," "a system to (verb 3)." . . ." Generate a listing of possible subsystems, discuss alternatives, and select the best. When this phase of the discussion is over, there should be a list of statements that clearly describe the kinds of human activities going on when the transformation is implemented.

Each subsystem will have its own CATWOE characteristics, boundaries, inputs, outputs, and measures of performance, so the discussion that follows regarding the development of the whole system also pertains to the development of each subsystem. Typically, the boundaries, actors, owners, clients, **W**, and environmental constraints for each subsystem are defined only in outline during this stage and then developed in detail after stages 5 and 6, when people actually commit themselves to implementing particular HAS models.

If the group feels that it would like to develop each subsystem in more detail, small groups are assigned the task of modeling each as a system in its own right. The subsystem models are then brought together and discussed. When this is done, the later discussions in stages 5 and 6 will usually be more detailed, since most of the nuts and bolts of the proposed changes will have been described. What typically happens, however, is that two or three models are created without this amount of detail, they are compared and debated, revised versions are created, a decision is made to implement one or more of the models, and then each subsystem is modeled as a system. This provides a finer detailed understanding of proposed changes. Much depends on the time and resources available for stage 4.

Sometimes it may be possible to identify the nature of a particular subsystem even though the key subactivities, inputs, and outputs are not known. Since a

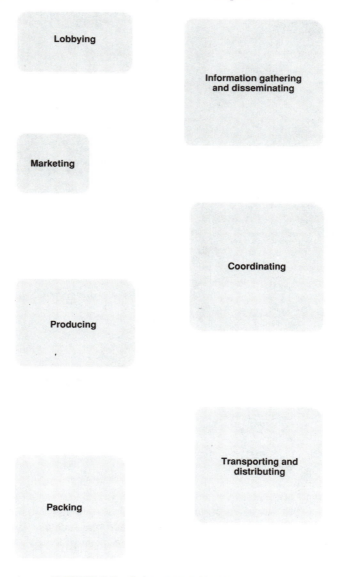

FIGURE 5.3   Subsystems identified.

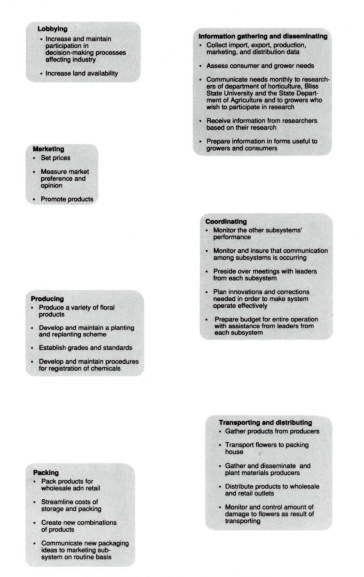

FIGURE 5.4   Subsystems refined.

hypothetical and novel activity may be involved, it is understandable that no one participating in the model construction phase knows enough about it to be able to envision all its features and dynamics. When this occurs, it means some homework has to be done before the design process can proceed. Then the group may assign a few people to call businesses, other groups, or some other appropriate organization that already has similar operations so that ideas can be gathered about how people are organized, what is done, what resources are needed, and what the results are. With information in hand, modeling can proceed, and the features of the subsystem can be developed in more detail.

For example, you will recall that in the floral growers' case, one transformation statement aimed to provide volume sufficient to support a packing house, so the packing-house operation is projected to be a part of an improved state. Yet no one in the group knew very much about how such an operation worked, what resources were needed, and so on, so they called firms that currently provided such services and visited several modern facilities to get an idea of the different ways such an operation could be organized and run.

Identify and Define Inputs

Once the subsystems are described, inputs and outputs of the system can be identified. Discussing the inputs and outputs helps participants begin to identify the relationships among subsystems and the kinds of resources each subsystem will need in order to carry out its defined functions with the expected results. Drawing and describing models can help keep people's thinking consistent and rigorous if the discussion of inputs and outputs is carried out in an orderly and detailed way.

Start with the inputs to each subsystem. An input is the resources needed in order for a subsystem to operate. Inputs typically include information, money, materials, human resources, and services. Identifying information inputs is a good place to start because every human activity system has information needs. As your discussions begin to solidify, carefully indicate in narrative form the specific information requirements of each subsystem.

Once information inputs are incorporated, move on to the next kind of input. Again, systematically think your way through the subsystems. Create a key to the kinds of inputs that are identified. Figure 5.5 is an example focusing on information inputs.

During this phase of model building, think *systematically* and *systemically*. Thinking systematically means a rigorous and orderly movement through the development process, taking careful notes as you go so as to remember the logic employed. You should also communicate your ideas about an improved condition in both symbolic and narrative forms. Thinking systemically means to be mindful of why you are identifying key properties of the system. Know the concept behind each term (*subsystem, input,* and *output,* for example) so that your ideas remain focused. Figure 5.5 illustrates the arrow symbols used by the floral growers' group to identify inputs.

Identify and Define Outputs

Once the inputs are carefully defined, identify the *outputs*. Typically, two kinds of outputs are described: (1) the results of the total system's operation—often the description of the key output of the system is a statement that closely resembles the transformation statement; and (2) the outputs of the various subsystems.

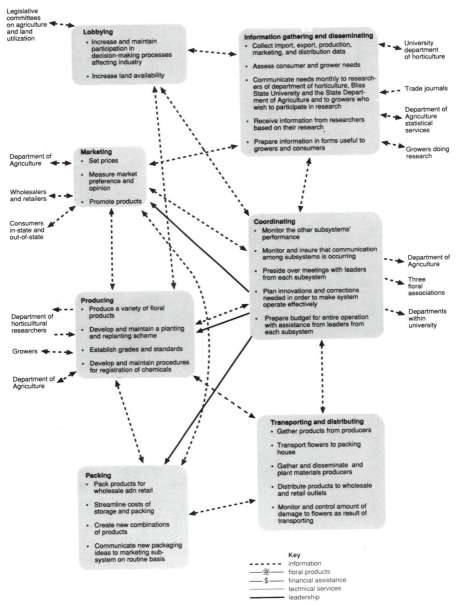

FIGURE 5.5   Information inputs identified.

Just as with inputs, create a key that identifies the kinds of expected outputs. Much of your work will be already done because many of the outputs of one subsystem are another subsystem's inputs. Thus it follows that many of the same conventions used to identify inputs will be used with the outputs. That this is so does not mean you shouldn't bother to describe outputs fully, however. If you relax and stop thinking rigorously, you will allow outputs to go unidentified. Each subsystem's outputs should be discussed systematically until all subsystems have been considered.

The degree of interdependence between subsystems will quickly emerge in your discussions, that is, the degree to which one subsystem's function is critical to another subsystem's performance. You will also be able to see what might happen if a particular subsystem does not get the amount or kind of inputs it requires. Figure 5.6 represents the initial attempt to symbolically define outputs by the group working on an expanded production transformation process for the floral industry.

Note that the arrows point to where the inputs and outputs in the visual model should flow. Arrows lost in space usually are a tipoff that there is sloppy thinking at work! Be alert also to the possibility that such arrows represent useless outputs such as waste that must be accounted for in your description of the environment.

Draw Boundaries

The issue of who is going to take responsibility for carrying out and supervising the various subsystem activities will have been discussed initially during stage 3 and again when the subsystems are developed. A particular consideration includes the identification of activities that members of the group emphatically do not want to be responsible for, services they want to hire from the outside, and those factors and processes that people view as not controllable and that must be accommodated by the system. The foregoing considerations help determine where to draw the boundary between the system and the outside or the environment. *Drawing a boundary* means determining what is under the control of the system's actors and owners and what is not. Human activity, inputs, and outputs within the boundary of a system are to varying degrees controllable and manageable. Things external to the boundary of the system are not.

Observe the following graphical standard for symbolically representing boundaries: a dotted line to represent an open system boundary and a solid line to represent a closed boundary condition. The concept of open and closed boundaries of human activity systems is largely self-explanatory. A system and its significant environment can be receptive or nonreceptive to transactions of energy, materials, information, and people in the form of inputs and outputs across their common boundary. In principle, human activity systems are said to be neither completely open nor completely closed. If a system were completely open it would be indistinguishable from its environment and if it were completely closed, it would

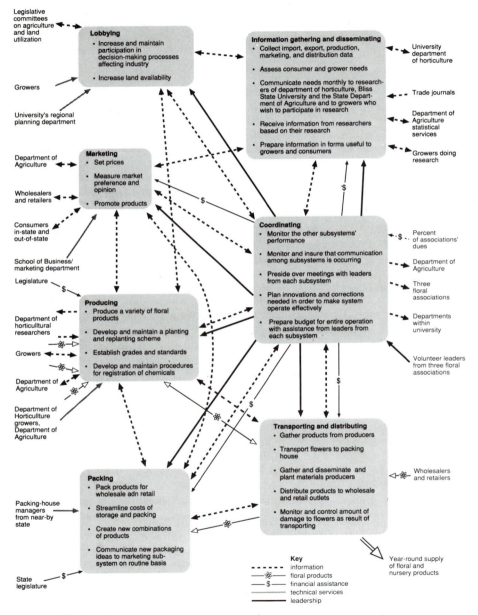

Legislative committees on agriculture and land utilization

Lobbying
- Increase and maintain participation in decision-making processes affecting industry
- Increase land availability

Growers

University's regional planning department

Information gathering and disseminating
- Collect import, export, production, marketing, and distribution data
- Assess consumer and grower needs
- Communicate needs monthly to researchers of department of horticulture, Bliss State University and the State Department of Agriculture and to growers who wish to participate in research
- Receive information from researchers based on their research
- Prepare information in forms useful to growers and consumers

University department of horticulture

Trade journals

Department of Agriculture statistical services

Growers doing research

Department of Agriculture

Marketing
- Set prices
- Measure market preference and opinion
- Promote products

Wholesalers and retailers

Consumers in-state and out-of-state

School of Business/marketing department

Legislature

Coordinating
- Monitor the other subsystems' performance
- Monitor and insure that communication among subsystems is occurring
- Preside over meetings with leaders from each subsystem
- Plan innovations and corrections needed in order to make system operate effectively
- Prepare budget for entire operation with assistance from leaders from each subsystem

Percent of associations' dues

Department of Agriculture

Three floral associations

Departments within university

Volunteer leaders from three floral associations

Producing
- Produce a variety of floral products
- Develop and maintain a planting and replanting scheme
- Establish grades and standards
- Develop and maintain procedures for registration of chemicals

Department of horticultural researchers

Growers

Department of Agriculture

Department of Horticulture growers, Department of Agriculture

Transporting and distributing
- Gather products from producers
- Transport flowers to packing house
- Gather and disseminate and plant materials producers
- Distribute products to wholesale and retail outlets
- Monitor and control amount of damage to flowers as result of transporting

Wholesalers and retailers

Packing
- Pack products for wholesale adn retail
- Streamline costs of storage and packing
- Create new combinations of products
- Communicate new packaging ideas to marketing subsystem on routine basis

Packing-house managers from near-by state

State legislature

Key
- - - - - information
—✳— floral products
— $ — financial assistance
technical services
leadership

Year-round supply of floral and nursery products

FIGURE 5.6 Development of the human activity system model's outputs.

cease to exist (Anderson and Carter, 1987). So the question is, open or closed to what? And why?

Boundary discussions are often intense because the group may need to decide how free it will be with subsystem outputs and how open it thinks groups external to it will be in supplying the inputs it needs. Based on these discussions, some groups choose to visualize the openness of subsystem boundaries by varying the distance between boundary lines. More importantly, the narrative accompanying the symbolic HAS model should note carefully the degree and kind of openness that will characterize the system as a whole and each subsystem, as well as how open and closed groups external to this system will be. For example, the people in the floral growers' case discussed at great length how open they would be in disseminating information about new cultivars to nonassociation growers and out-of-state businesses. Figure 5.7 shows how the floral growers symbolically defined boundaries. The section of the narrative on boundaries is critical, since it is easy to draw dotted lines.

Develop Measures of Performance

The next step is to develop measures of performance for the system as a whole and for the subsystems. These measures of performance resemble contract specifications and clearly define the amount, rate, and quality of output expected of the system and subsystems, as appropriate. For example, for the transportation subsystem of the floral industry, the group defined "a weekly report to the system's coordinators that states the amount of products transported to the packing house and producing subsystem." Other measures of performance may be very specific rates of performance such as "a run is made to each producer on a biweekly basis to pick up products and deliver them to the packing house," "less than one percent of all products picked up will be spoiled," or "delivery from producer to packing house shall not exceed four hours' time." Such statements specify the quality and amount of various outputs envisioned in an improved state.

Agree on Decision Processes

Finally, to complete all the minimum features of a HAS model, the decision makers and decision-making procedures must be made explicit and unambiguous. Conceptually, the idea is that systems have regulating mechanisms that control the use of inputs in order to produce outputs. Regulating mechanisms include a variety of approaches and, most importantly, people who make decisions using the measures of performance given previously. These can be a combination of the owners and actors who were identified by the group as they thought through such questions as "Who owns the system?" and "Who will be responsible for operating the system?" Who will be the decision makers of your system? What is the line and nature of authority that decision makers in various positions will have? Who will make what kinds of decisions? Where are the limits of authority

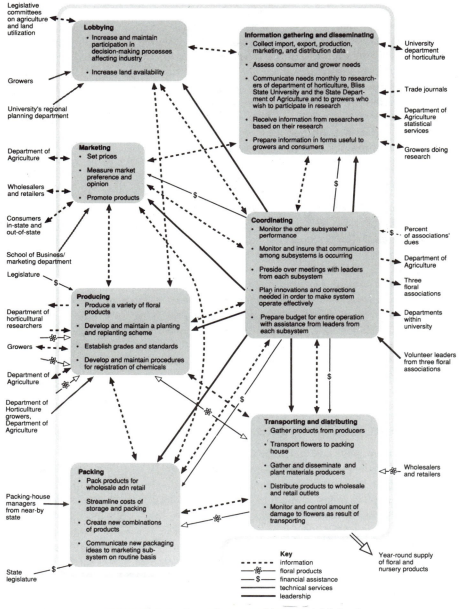

Lobbying
- Increase and maintain participation in decision-making processes affecting industry
- Increase land availability

Information gathering and disseminating
- Collect import, export, production, marketing, and distribution data
- Assess consumer and grower needs
- Communicate needs monthly to researchers of department of horticulture, Bliss State University and the State Department of Agriculture and to growers who wish to participate in research
- Receive information from researchers based on their research
- Prepare information in forms useful to growers and consumers

Marketing
- Set prices
- Measure market preference and opinion
- Promote products

Coordinating
- Monitor the other subsystems' performance
- Monitor and insure that communication among subsystems is occurring
- Preside over meetings with leaders from each subsystem
- Plan innovations and corrections needed in order to make system operate effectively
- Prepare budget for entire operation with assistance from leaders from each subsystem

Producing
- Produce a variety of floral products
- Develop and maintain a planting and replanting scheme
- Establish grades and standards
- Develop and maintain procedures for registration of chemicals

Transporting and distributing
- Gather products from producers
- Transport flowers to packing house
- Gather and disseminate and plant materials producers
- Distribute products to wholesale and retail outlets
- Monitor and control amount of damage to flowers as result of transporting

Packing
- Pack products for wholesale adn retail
- Streamline costs of storage and packing
- Create new combinations of products
- Communicate new packaging ideas to marketing subsystem on routine basis

Legislative committees on agriculture and land utilization

Growers

University's regional planning department

Department of Agriculture

Wholesalers and retailers

Consumers in-state and out-of-state

School of Business/ marketing department

Legislature

Department of horticultural researchers

Growers

Department of Agriculture

Department of Horticulture growers, Department of Agriculture

Packing-house managers from near-by state

State legislature

University department of horticulture

Trade journals

Department of Agriculture statistical services

Growers doing research

Percent of associations' dues

Department of Agriculture

Three floral associations

Departments within university

Volunteer leaders from three floral associations

Wholesalers and retailers

Year-round supply of floral and nursery products

Key
- - - - -   information
─❀─   floral products
─$─   financial assistance
───   technical services
▬▬▬   leadership

FIGURE 5.7   Boundary conditions established.

and responsibility? How will disputes be handled? Owners typically will be a part of the decision-making process in very specific ways. These should be identified now so that later, if the group decides to implement the model, the desirability and feasibility of the decision-making procedure itself can be discussed.

Clarify the Effects of the Environment on the System

All systems exist in an environment. Earlier in the developmental process, while working through the CATWOE checklist, people described the inputs and outputs that characterize transactions between the system and its environment. To clarify further the notion of environment (or E of the CATWOE), determine what additional features of the environment may affect the system. These features may be more apparent now than they were during stage 3. Lewin's force field analysis technique (presented in an addendum to this chapter) may be used to help people identify the forces in the environment that may help implement the system and allow it to operate in the intended way, as well as identify forces that may negatively affect the system's performance. When these forces are noted, some of the subsystem's activities can be modified if necessary.

A Final Check

Several checklists are available to help verify that, in developing HAS models, minimal properties of a system have been described. Tables 5.2 and 5.3 present two examples. The first checklist is based on the work of C. W. Churchman (1971), G. M. Jenkins (1969), and P. Checkland (1981) and is useful for assessing systemic thinking. The second is by the faculty of the Hawkesbury Agricultural College in Australia (Bawden et al., 1984). The Hawkesbury checklist is more extensive because the authors have added management functions to the basic agricultural systems outline. It is helpful for developing ideas about management functions in human activity systems.

Running through these checklists will signal if you missed developing one of the properties essential to a systems model or overlooked essential kinds of management detail. It is then a straightforward matter to go back and finish the job. Another useful aid is the conceptual map presented in Figure 5.8.

An example of the kind of narrative that is developed to accompany the symbolic model may be helpful. Table 5.4 presents the narrative resulting from the discussion of what the floral growers' information-gathering and dissemination subsystem would be like in the future. Remember that this was produced very quickly. The group had recorders who took notes as the discussion occurred and then, after the meeting, typed up the results. The recorders' work was aided by the facilitators who, as the discussion proceeded, wrote the key points on flip charts.

Experienced facilitators have found that the development of human activity systems models is highly motivating to everyone involved. Usually a model is developed through a whole series of discussions, rather than all at one time.

Sometimes key groups and individuals who need to be involved cannot get together at the same time. In order to get representative input from all the people and groups involved, some communication process may need to be developed to ensure that people have an opportunity to be included. The people who are left out of discussions may feel excluded, less knowledgeable about the proposed change, and finally more resistant to implementation, simply because they have not traveled the same road as the people involved in the whole process of thinking through a sense of improvement. For this reason, the communication competencies of facilitators become very important in this stage.

Since the modeling stage of the soft systems approach involves creative and detailed thought, techniques for facilitating both are needed. For stimulating creative thought, *the nominal group technique, delphi method,* and *brainstorming* are particularly useful techniques to know. These, as well as other group facilitation techniques, are explained at the end of this chapter. Facilitating detailed and precise thought involves a step-by-step procedure for leading yourself and others through the development of HAS models (and perhaps other kinds as well), so that basic features of an improved situation are described and modeled as if they were a system. Part of this step-by-step procedure is represented in the order of development for a HAS model suggested here. Other helpful aids to thinking have also been tested. These include coming to meetings with pre-cut "bubbles" with the words "A system to . . . " printed at the top to use to facilitate the development of transformation statements and subsystems. To aid in the

TABLE 5.2 Key Systems Properties (after Jenkins, Churchman, Checkland)

A system (**S**) is a system if it:
1. Has a mission, an objective, a definition of a desirable state, or an ongoing purpose, or can have a purpose sensibly attributed to it, that reflects a transforming function of one set of things (or events) to another.
2. Has a measure of performance that reflects the extent to which the purpose or transformation is being achieved.
3. Has a mechanism of regulation (a process of decision making and resource allocation) that reflects the purpose, responds to information, and governs performance.
4. Has components that themselves exhibit all the properties of **S** and can thus be labeled subsystems.
5. Has components that interact and show degrees of connectivity that permit effects and actions to be transmitted and flow through the system.
6. Exists within wider (supra) systems and/or environments (**E**) with which it interacts.
7. Can be distinguished from the suprasystem or environment (**E**) in which it exists by a boundary that represents the interface between an **S** and an **E**; and clearly distinguishes things that are under the **S**'s control from things that are not contolled and to which the **S** must be adapted or accommodated.
8. Has both physical and (often) abstract resources at its disposal by virtue of its component parts, its internal environment (**e**), and inputs that it receives from its external environment (**E**) across the boundary.
9. Has some guarantee of continuity or stability, with the capability of returning to a stable state (or qualitatively acceptable changed state) when disturbed and/or resisting forces emanating both from within the system itself (**e**), and from its environment (**E**).

development of inputs and outputs, predesigned arrows labeled "information flows," "money," "services," "materials," "flow of authority," along with some blank ones, can be used to help people to think about inputs and outputs. Preparing a listing of key features (systems properties) that will need to be designed helps people get an overview of what must be done.

Communicating the Model or Models

Finally, something needs to be said about how you present the results of your group's work to a larger audience of concerned people. The first point is that *you should never go back to a larger group with only one model representing one vision of the future.* Chapter 6 will have more to say on this subject. Another point that needs to be made has to do with the narrative that must accompany the visual representation of the model. The visual model will serve well as a memory aid to those who were involved in the discussions. While it

TABLE 5.3 Key Systems and Management Functions (following the Hawkesbury group)

1. What is the mission of the system?
2. What is its major transformation? (What is the summary verb that tells us the essence of the system?)
3. What are the features of the environment that distinguish the system from its environment?
4. What are the major subsystems?
5. What, in turn, are their purposes and major transformations?
6. What are the appropriate measures of performance of the whole system? Of each subsystem?
7. What are the hierarchical relationships among the major subsystems?
8. What are the crucial interactions influencing the performance of each subsystem and the system as a whole?
9. What flows of inputs and outputs among subsystems constitute these interactions?
10. What constitutes threats or forces to the productivity, stability, and sustainability of the system?
11. How does this proposed system relate to others, and what hierarchical arrangement can be recognized?
12. What is the major decision-making process and who are the key decision makers for day-by-day operations, for coordination among subsystem activities, for questions of overall policy, and for strategic shifts in direction?
13. How is feedback assured, monitored, evaluated, and acted upon?
14. How does the proposed system handle future innovations?
15. What mechanisms does the proposed system have for allocating resources?
16. How are the operations of the proposed system going to be monitored and adjusted relative to the effectiveness and efficiency with which it meets the stated measures of performance?
17. What major environmental constraints are anticipated?
18. What is the nature of the transactions between those who will make the system work and the environment?

will be meaningful to them, it will likely overwhelm outsiders, especially if they are just shown the visual diagram without any oral or written narrative. Too much information is presented all at one time. When it comes time to communicate the group's work to a larger audience, how you present it will be just as important as what you present.

One useful communication technique is called *progressive disclosure*. Make a series of overheads, flip charts, or drawings on a chalkboard, each having only a part of the model on it so a particular feature can be discussed without the other information being revealed. Progressively uncover and present elements of your model. Use the same order of discussion as you used originally to develop

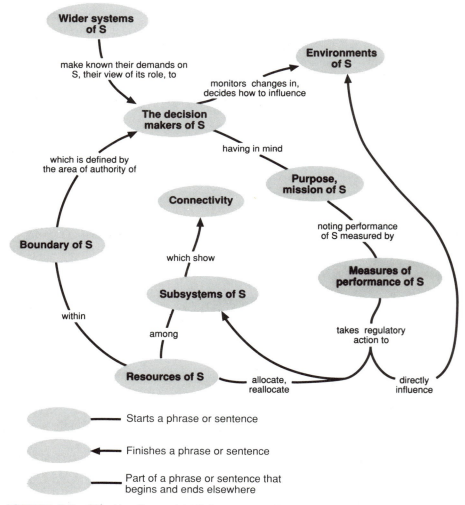

FIGURE 5.8 Checkland's model of the concept "human activity system." (Adapted from *Systems Thinking, Systems Practice*, Peter Checkland, © 1981 John Wiley & Sons, Ltd., Chichester, Sussex, England. Figure 9, p. 175. Reprinted by permission.)

TABLE 5.4 A Sample Narrative Accompanying the Floral Growers' Model of an Information-Gathering and Dissemination Subsystem

The production expansion system has seven subsystems. The first subsystem is as follows:

1. An information-gathering and dissemination subsystem that is responsible for:
   a. Collecting import, export, production, marketing, and distribution data.
   b. Assessing consumer and grower needs.
   c. Communicating these needs monthly to researchers of the department of horticulture, Bliss State University; to the State Department of Agriculture; and to growers who wish to participate in research.
   d. Receiving from researchers information based on their research.
   e. Preparing information in forms useful to growers and consumers.

Actors: The information-gathering and disseminating subsystem steering committee will have twelve volunteer members from the growers' associations, each to serve for two years. In addition, the committee will be assisted by eleven interns from the Bliss State University's college of agriculture. The internship agreements will be worked out with the appropriate college departments.

Customers: People who will directly benefit are association members and consumers who participate in the surveys and voice their opinions and needs. People who will benefit indirectly are university and department of agriculture researchers, in that their research will be seen as valuable to the growers. Indirectly, researchers will benefit from their publishing efforts; legislators will benefit from having a more detailed assessment of what our needs are.

Potential victims could be some growers whose practices are illegal or come under scrutiny because of the adverse impact their current practices are having on growers' and consumers' health and safety (such as those using unregistered pesticides or registered pesticides in inappropriate ways, or those participating in pricing practices that are illegal). They will be victims because information regarding such concerns will be made known and dealt with more quickly than is happening presently.

Since the subsystem will not try to control the dissemination of information regarding the production of new cultivars, there is a potential for state growers to become victims of out-of-state and international growers who gain access to this information and produce more or undercut the market with better flowers. However, we believe that the advantages of having better information outweigh this concern.

Owners: The people having authority to correct the operation of the information-gathering and disseminating system are (1) the 12 elected members of the steering committee elected from the three growers' associations (as described previously), (2) one member of the committee who will be assigned to act as coordinator and who will have the vested authority and responsibility to make changes in consultation with other volunteer staff and with the advisory board.

Environmental constraints include:

1. Research on some problems may take too long to be useful to growers.
2. The research may be reported in ways that are not useful or understandable to growers.
3. The university and department of agriculture may not be able to deliver information needed in a timely manner.
4. The university may not be able to place interns in a timely, ongoing fashion.
5. Growers may be unwilling to volunteer to serve in this area of activity or may not complete their assignments on time.

6. There will continue to be little financial assistance to sustain this activity (a modest amount of the association dues will go to sustaining this activity; cost-sharing arrangements will be worked out with the Department of Agriculture and the university's cooperative extension service; financial aid is anticipated in the future as a result of the lobbying subsystems activities).

World view: Major beliefs, values, and assumptions made in the design of this subsystem include:

1. Information can be obtained that will make a difference in what growers are producing and how they grow, process, market, and distribute their flowers.
2. Partnerships between the growers' associations, the Department of Agriculture, and the university are possible.
3. University researchers and extension agents from the state university are to service all growers, not just a few larger growers.
4. It will be to our long-run disadvantage to allow the present price-setting procedures to continue (seen here as an information-collection and dissemination issue). Both large- and small-scale growers will be sued by consumers, and fines may be levied by the state and federal governments. Consumer loyalty and trust are valued over gaining maximum profit.
5. Pursuing ways to make viable all growers' operations is valued over operating the associations around the interests of a few.

Inputs include:

a. Information from the Bliss State University researchers and the state's Department of Agriculture regarding progress on their research.

b. Information from the Department of Agriculture's Agricultural Statistics Service on monthly production, demand, imports, and exports.

c. Information from growers, wholesale and retail stores, and consumers regarding needs.

d. Finances to support report preparation (estimated at $600 per report using computer equipment available through the associations and the cooperative extension service).

Outputs include:

a. Specific information to University researchers, dean, and chair of the department and to the department of agriculture director and researchers regarding needs of consumers and growers.

b. Information to the lobbying subsystem before the start of the legislative session regarding specific areas needing additional appropriations.

c. Information to the marketing subsystem regarding consumer preferences, pricing constraints, predicted supply and demand concerns, potential new markets, and profiles of key competitors in other states and internationally.

Boundary characteristics:

There are some information-sharing activities that we wish to conduct in a very open way. We want to communicate as much information as we can to public- and private-sector researchers so that their work is timely and targeted to our needs. We want all legislators and appropriate government and university officials to know about our concerns and needs. We therefore will be extremely open with the information we collect on these concerns.

However, we recognize that there are several sensitive concerns that those responsible for this subsystem will need to be discreet about (i.e., less open, but not to the point of withholding information). Current grower and association concerns dealing with the use of unregistered pesticides and potential price-fixing procedures will be shared with key individuals at the university and state Departments of Agriculture, Health and Commerce. Appropriate followup will be required to determine whether the issues are being dealt with before taking them more public.

Related to the issue of what we will try to do ourselves and what we do not wish to manage and therefore get from the environment: The Agricultural Statistics Service of the State Department of Agriculture and the cooperative extension service are currently preparing certain statistics we will need routinely, so the information-sharing and dissemination system will not have to replicate these services. Rather, we envision that our system will work with these two outside services so that they will collect statistics appropriate to the growers' associations' needs, if possible. If these services are not able to collect these statistics, then those responsible for this subsystem will do the collection. Measures of performance:

The following initial standards of performance were discussed in order to correct current deficiencies and to set an initial estimate of the level and kind of performance needed:

1. Reports on market prices of flowers are delivered weekly to the marketing subsystem.
2. Six months before the start of the legislative session, key themes of concern needing legislation are forwarded and discussed with those managing the lobbying subsystem.
3. Yearly consumer needs assessments are conducted.
4. Three months after release, assessments are made of the value and use of information disseminated to members, and revisions are made before the next printing.
5. A yearly assessment is made of the appropriateness of information-dissemination media and modes (e.g., do we need written pieces, demonstrations, training sessions, radio spots, etc.?).

---

the model. If transparencies are used, then eventually all features can be laid on top of each other and the entire system seen as a whole. How detailed you get in describing each property of the system must be judged in terms of your audience and the time available.

It is also helpful to do a three- to five-minute overview based on the central transformation that the group thinks would represent an improved state of affairs and on the key human activities (subsystems) needed to accomplish this transformation. Prior to giving it, practice your presentation on someone else, so that you can discern where you need to improve it.

As indicated earlier, the Chatham River case (found in the back of this book) exemplifies the kind of situation in which several types of models would be helpful. Some of the highlights of the situation have been assembled for you, along with other representative materials. In doing so, the authors have had to organize their sense of what was happening in this community and present it to you in narrative form. From even the small amount of information you have to work with, a beginning HAS model can be created. Obviously, if you were actually in this community, your information base would be much richer, and more realistic HAS models could be created in cooperation with people there. Since some of the people involved in the Chatham River case want to change the rate of use of a natural resource, a physical model might play an important part in the investigation so that the characteristics of the present water use and potential changes could be understood better. How the resource is allocated to the various users, even the fish, which are of concern to the environmentalists, may require yet another type of symbolic model. In addition to the required HAS model, narrative, physical, and symbolic models are possible with this particular case. Some of these suggestions will be taken up in Chapter 6.

Modeling the Past to Understand the Present

Analyst/facilitators are often asked to help straighten out or evaluate change initiatives that have already been designed or even implemented, including those that are not products of the soft systems approach. Readers who plan careers in food, agriculture, and natural resources outside the lab are very likely to face this kind of situation. The soft systems approach is a useful tool for taking a retrospective look at a past situation, or at an ongoing one rooted in the past, in order to make sense out of it and provide a basis for future improvements. To illustrate this application, a case based on experience in a Third World country is introduced.

Improving a Cooperative in Papua New Guinea

While one of the authors of this book was conducting field research in Papua New Guinea in 1981, he was approached by the people of a small rural community for assistance in improving their fruit and vegetable business.

The concerns behind their request and its general context were familiar because research in this community, belonging to the Miyanmin people, began in the late 1960s. Then the situation was characterized by rapid change and very serious problems for the people. These changes included very high mortality as well as optimism about the perceived benefits of the modern world. Both aspects have been common features of the penetration of formerly isolated societies in Papua New Guinea and elsewhere in the Third World.

What seemed to distinguish this Miyanmin community from many others at a comparable stage in their histories, however, was that in the 1960s the people had assessed their own situation, developed a vision of an improved future state, and begun to implement it on their own. People were hard at work as research commenced with them in 1968. A situation summary and other basic documentation are provided in the appendix at the back of this book.

The details of the 1960s community development plan, as extracted from field data such as the materials presented in the appendix and observations of life in the community, are outlined in the transformation statement, CATWOE analysis, tentative system definition, and conceptual model (Figure 5.9) presented here. Note that because this case involves a change initiative that was not a result of using the soft systems approach, the backward look at it does not conform in all respects to the standard used with, for example, the floral growers' case. Necessary deviations will be pointed out in passing.

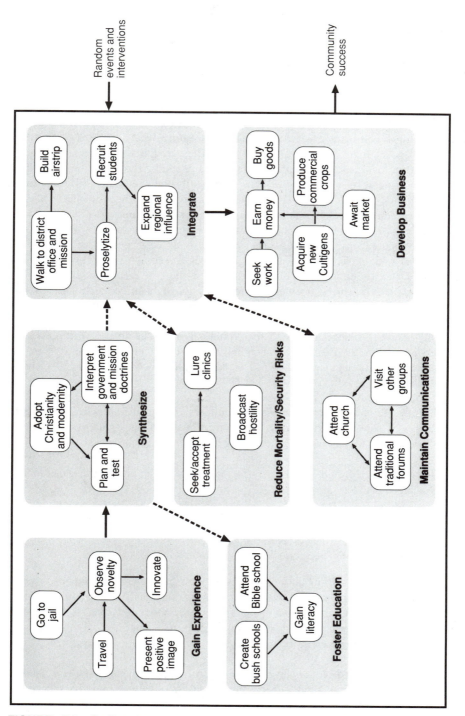

FIGURE 5.9 Outline conceptual model: retrospective analysis of the 1960s Miyanmin community modernization plan.

Transformation Statement. A system *to gain* knowledge about, *to establish* connections with, and *to adapt* to the outside world, including the modern economy.

Note that this statement is pegged at a higher level, that is, it is more abstract, than the transformation statements presented earlier for the floral growers. Multiple verbs are used in an attempt to capture what people seemed to be trying to do in their context. This seems justified by the substance of the interview with a local leader presented in the case materials. That interview describes a desire to achieve a future state that is radically and generally different from the traditional way of life of the Miyanmin. Accordingly, later in this analysis, alternative transformation statements and (sub)system definitions (for the cooperative) that are somewhat more consistent with — even similar to — the floral growers' example are presented. These can also be described as subsystems of this larger system. Thus, at a more micro-level, it seems that people adopted a vision of the future that at various times led them to accommodate external changes as they occurred; to work aggressively to establish connections with and gain experience of the outside world; to seek opportunities to earn money so that they could buy modern commodities; to look for ways of improving security, health, and well-being; to extend their political influence in the region; and to obtain education for themselves and their kids — the linchpin of their vision of the future. The remainder of the CATWOE analysis (according to the TACOEW format) follows:

Actors. Local leaders as consensus builders, young men as flexible experimenters with new ways, other men and women of the village as active participants and particularly as parents.

Customers. Most members of affected communities and future generations; some alienated elders as victims.

Owners. Men of the community as heads of families on a consensual basis. (Families, as basic social and production units, can opt out of the scheme or even leave the village.)

Environment. Within the constraints of the established life-support system, involving high mobility; extensive, shifting agriculture and hunting; limited education; managerial expertise and experience in the modern world; isolation due to restricted transportation and other infrastructure; scant medical training and technology; and no financial resources.

Weltanschauung. Since armed force is no longer a viable option for dealing with perceived external threats (including disease), old ways must be succeeded by new ones. Whites have many superior ways and technologies for dealing with recognized problems, and these should be evaluated and tested. Education and travel are sources of "revelation"; knowledge; new ideas, and novel objects, technologies, and approaches. The medium of access to these things is money earned in pursuit of a wide range of activities. Desired changes are gradual, and optimism must be balanced by patience and conservation of much of traditional culture.

A HAS Definition

"A community-owned system for gaining knowledge about, establishing connections with, and adapting to the outside world, including the modern economy, by improving security, health, and well-being; extending knowledge to other people; earning money; and obtaining education for children. This system will benefit members of the community, friendly neighboring communities, traditional allies, and future generations, possibly at the expense of certain elders and traditionalists. This system is consistent with the view that high mortality, the perceived threat of violence, and the isolation associated with the old way of life must be replaced by new ways associated with the white man. Certain environmental constraints, such as remoteness of location and deadly diseases, are taken as given only in the short run and need to be constantly evaluated. The government and missionaries must be depended on for a wide range of services and other inputs, such as medicine and formal education.

A tentative issue-based conceptual model of the Miyanmin modernization plan is presented in Figure 5.9. Since the original development of the plan involved events that were not directly observed by the researcher, the temporal order of changes is conjecture based on the people's recall to the extent possible in a society without a calendar. In any event, many of the activities have been concurrent, and they continued into the late 1960s, when the researcher could observe them directly.

Consistent with the system definition and CATWOE, the retrospective conceptual model identifies seven front-line activities, that is, subsystems, including gaining experience, synthesizing traditional and new beliefs, reducing mortality/security risks, integrating politically with other groups, developing business or money-making activities, fostering education, and maintaining transportation and communications.

As the situation summary indicates, by 1981 the scene was transformed, encompassing many changes and modifications of the plan of the 1960s, described previously. These changes, including the results of the successful activities that had been initiated over a more than ten-year period, are outlined in Figure 5.10, which is a modified conceptual model based on Figure 5.9.

This revised plan and model have several notable features associated with earlier successes. The village-level isolation, still apparent in the 1960s, has all but disappeared, along with the accompanying insecurity and hostility to traditional enemy groups. It has been replaced by openness and hospitality to these same people and common participation in a regional political and religious movement called *rabaibal* (or "revival"). Subsystem boundaries have been altered accordingly. There have also been marked improvements in access to health care and educational opportunities (see situation summary). Business opportunities still lag behind.

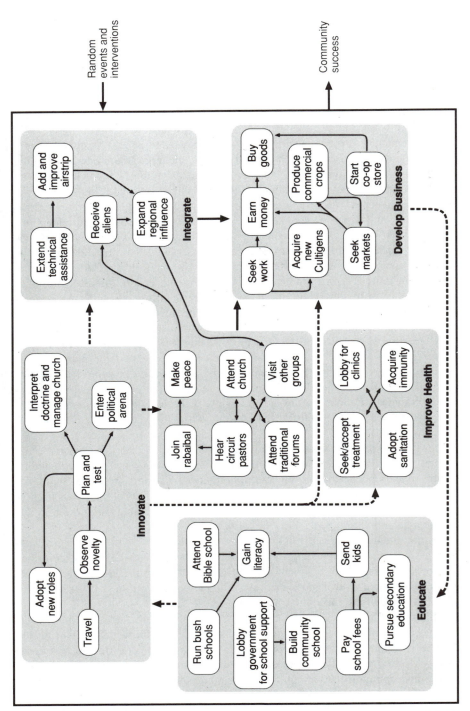

FIGURE 5.10 Outline conceptual model: retrospective analysis of the 1980s Miyanmin community modernization plan.

The updated model also sets the scene for the people's request for intervention to assist them to improve the business subsystem of their overall initiative. This involved the development of a fruit and vegetable cooperative business. The intervenor naively slipped into the expert role and quickly formed *his own* conception of "what ought to be done"—an expert designed system. This can be recast in familiar terminology. The reader should be warned in advance that this definition of a (sub)system is biased and, as described in Chapter 8, everyone involved will be forced to discard it. It stands here as a description of the intervenor's initial attempt at designing an improvement subsystem including his sense of what the people wanted and what constraints to take as givens. One of the most important immediate lessons involves the description of environmental constraints. They are real enough as factors, but they might better be viewed as concerns relating to structure rather than taken as givens. As concerns, they should give rise to transformation statements.

Chapter 8 discusses the consequences of rushing into implementation without carrying out the required steps of the approach. At that time it will be useful to compare this version with the final version, which the people themselves developed. In this way we can gain insight into what went wrong.

Transformation Statement. A subsystem *to increase* cash income from the production and sale of fruit and vegetables by introducing new crops, finding markets, and training managers.

Actors. The analyst/intervenor; selected managers with reading, writing, and computational skills previously acquired in a bush school.

Customers. Members of the community with commodities to sell; (potential) buyers; potential victims are community members who are unable to participate due to individual situational factors.

Owners. Male heads of household.

Environment. No established regional market, market center, or coordination; dependence on aircraft for transportation of commodities; low level of management experience and education; slow communications with outside world, including buyers, due to lack of radio or other modern communications; past marketing failures; lack of capital; ineffective government extension services.

Weltanschauung. Cash income is needed to assume a modern life-style and generally participate in the modern world and wider economy, particularly for purchasing consumer goods such as clothing, foodstuffs, and small electronic wares; capital goods such as tools and firearms; and services such as education. A notion of equitability means distributing proceeds of sales based on individual growers' contributions of commodities in a given sale; new crops and new techniques are desired and thought necessary to enter markets.

System Definition. An improved community-owned subsystem to produce, manage, and market fruit and vegetables in order to increase income of people who desire to participate in the modern sector of the economy. The subsystem will be operated by members of the community who have surplus produce to sell and managed by members with basic literacy skills. Ownership is consensual and participatory, with no one person or small group having the power to prevent it from proceeding. Major environmental constraints include the current lack of markets and dependence on air transport. People believe that money to purchase modern goods and services is the mark of progress.

The model of a designed system for the business subsystem based on this description is presented in Figure 5.11.

It should be emphasized again that this was *the intervenor's* conceptualization of what people wanted, needed, and were capable of doing. At the time, he was thinking in relatively mechanistic terms — focusing on *changes* rather than a desired future state. Thus the rush was to develop an implementation plan based on subsystems or even sub-subsystems along the lines of finding buyers and determining their needs; introducing, demonstrating, and encouraging production of commodities desired by customers; developing management routines and teaching local people to administer them; raising cash incomes of local people. The preceding **W**, in particular, is *his* inference rather than an accurate recording of that of the people. Indeed, he can still recall thinking at the time, "This is how I would do it."

This biased picture, along with the fact that the intervenor did not facilitate comparison and debate (Chapter 5, 6, and 7) beyond specific discussion of details such as commodity prices, possible outlets, and who the actors would be strongly influenced what followed when the plan was implemented. This is described in Chapter 8.

Summary of Stage 4

By the time you complete stage 4, you have formulated at least two human activity system models. Your focus has been on designing people's future activities in an improved state of affairs. New activities are created. Established activities are revised or abandoned. Discussing these matters does not come easily to our usual clients, the growers, farmers, ranchers, company managers, commodity groups, and on-the-ground natural resources folk. When people such as these begin to think about future actions, however, they seem to gain a new-found sense of empowerment. They now see that they are more in control of the development process and their future lives just by knowing where they want to go and why. Things they frequently had taken as givens are now reconsidered, and a new sense of self-reliance emerges.

As a result of developing the human activity system models, other kinds of modeling efforts may also be identified, as needed, and carried out. For example, as inputs and outputs are considered, questions may arise regarding the

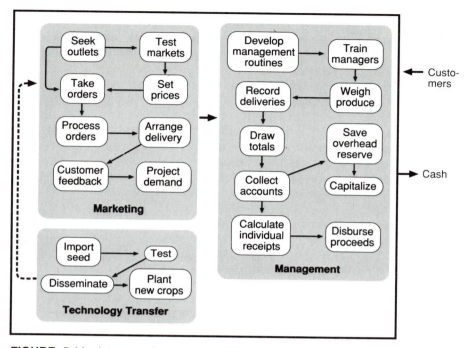

FIGURE 5.11   Intervenor's model of an improved business subsystem for the Miyanmin community, 1981.

amounts of resources required to make a particular kind of activity function efficiently and effectively. It is likely that you will find a hard systems model useful for dealing with this kind of issue.

As with the earlier stages of inquiry, some useful techniques to use in stage 4 can be suggested. As Chapter 4 argued, the techniques you use should be compatible with the basic assumptions and purposes of stages 3 and 4. Group techniques that *facilitate consensus decision-making* should be used. Voting is not necessarily the way to gain a consensus; that's a majority-rule technique for making group decisions (Johnson and Johnson, 1987:100–104). Roth (1985) reviews various group techniques that may be used during stages 3 and 4, as well as in other stages. Another good place to look is the yearly series *Annual: Developing Human Resources* (University Associates, Inc.). At the end of this chapter, several of these techniques are discussed that have been useful to facilitate group members in the design of models and accompanying definitions. The important thing to understand is that using a group technique that *forces* a choice out of a range of choices is *not* the way to go at this stage. Accordingly, such techniques as preferential voting or priority sorting, which *could* be used, for example, to weigh the order of importance of transformation statements, should be used very carefully, and some experienced workers in the field would say not to use them at all!

Why is consensus making different from "vote winning"? Let's start with voting. What happens in an election? Typically, there is a winner and one or more losers if elective office is the subject. In the case of referenda, a question is on the ballot as an up or down issue, no in-betweens, and again you end up with *people*, the advocates of one position or another, who are either winners or losers. In our society, most people *believe* in the free election process and also in such concomitants as "losing gracefully," "the loyal opposition," and so on. What that is supposed to mean in the public political arena is that the losers will not band together and organize a revolt, and decisions made in this way give every appearance of being final. The foregoing has little to do with consensus, except possibly that there is general agreement among Americans that revolts are messy, and filling vacant offices is important. This approach works mainly because it usually can count on the *loyalty* of losers to prevent them from quitting the system, dropping out, emigrating to Australia, or whatever!

Consensus means "general agreement," with everyone involved a *winner* and no *losers*. The rationale behind consensus building is that the *process* needs the people to implement changes. Moreover, the people involved in problematic situations may not belong to a single long-established and coherent group in which loyalty is high. It is thus easy for people who perceive themselves as losing in the process to simply quit. Yet if some of the people involved walk out, then the chances of improving a situation are reduced significantly. So, as a professional facilitator, you need to proceed *as if* your pay will be based on the number of people who stay on board!

In voting, *people* are excluded. In consensus building, *proposals* are excluded (or redefined, or whatever). This is expressed in the simple model that follows, which presents the situation as a tradeoff.

People ↔ Proposals

A range of procedures for consensus building and group facilitation, including brainstorming, the nominal group technique, and the TKJ technique, are described in an addendum to this chapter.

This chapter concludes with a description of the overall posture you are encouraged to adopt in order to fulfill all phases of the inquiry process.

And so with doubts, you begin to submerge yourself in a situation of concern. You make choices, but only with the full realization that refinements and greater changes are to be expected as you cycle through the inquiry process. Openness is your watchword. Yet you finally reach the point when convergence is called for. Convergence is achieved through conceptual modeling and is a fundamental step in the process of developing the potential of an improved state. Through inquiry you have learned a lot and, more importantly, created a new feeling of possibility. In this sense, modeling is another example of experiential learning. A systemic world view affects this process, whether you are pursuing a hard or a soft systems methodology, or both. And it is taken as given that an inquiry methodology is not only rational and logical, but also quintessentially social, in that each phase requires that the analyst/facilitator and the people involved authentically interact and transact with each other sequentially and interdependently.

From a learning standpoint, symbolic modeling with a visual or graphical standard of representation such as this chapter has presented engages both our abstract reasoning abilities and our intuitive and spatial patterning abilities. Our left brain only develops the linear models with their symbolic abstractions, just as it is also used for verbal communication. As Betty Edwards (1979) puts it, when we unlock our right brains, we gain access to our holistic, spatial, intuitive, and nonrational capabilities. And these clearly have important parts to play when we attempt to solve problems and improve situations. The visually rich graphical modeling approach helps us tap our right brain abilities even as our left brain works out individual linear relationships and helps us develop the written narrative.

When we shift to a systems mode of thinking, we begin to engage the right brain, or at least we move in that direction. Systems thinkers approach problem situations primarily with a sense of their wholeness (holism). They also explore their intuitive feel for the situation and examine the feelings of others who might be involved. The moment they start to conceive of an improved situation using systems thinking, they are applying mechanisms of spatial organization. And above all, when imagining things as if they were organized as systems, they are self-consciously using analogy and metaphor because systems are imaginary things existing in the minds of their creators.

Chapter 6 discusses stage 5 of the soft systems approach, which involves subjecting the models to various real-world tests.

Addendum: Techniques for Generating Ideas in Groups

This addendum presents seven techniques for group facilitation. Although each technique has its own particular strengths, they all can be employed with groups of people to generate useful ideas by building a cooperative atmosphere in which participants freely support and benefit from each other's efforts. Generally the techniques control negative communications so that threats and feelings of inadequacy are reduced, and creative thinking is encouraged. This brief guide is based on Roth's (1985) excellent review. You are encouraged to consult this source directly for additional guidance on idea generation and decision-making techniques for use in groups.

Brainstorming

Brainstorming was developed by Alex Osborn and is perhaps the simplest and best-known idea-generation technique. Its purpose is to help groups generate as many ideas as possible on a given task or problem. The greater number of idea associations participants produce, the less stereotyped and more creative their ideas tend to be. The technique also helps create a cooperative atmosphere. To guide the process, four group discussion rules are usually presented to participants:

1. Participants are requested to withhold judgment on the input of others, no matter what they suggest. Avoid judgmental comments and criticism.

2. Generate as many ideas as possible without worrying about their quality. Quantity is the goal. This helps to stimulate creativity and originality.

3. Keep up the discussion and the momentum of idea generation by exclaiming anything that comes to mind.

4. Build on the ideas others produce, manipulating them, twisting them around, and using them in any possible way.

The group discussion procedure usually includes a warm-up session using a nonthreatening theme so that people can practice. The facilitator then presents a clear statement of the task to be accomplished and the discussion rules that will be followed. There is discussion among participants to clarify the nature of the task. For example, if the task is to identify all possible subsystems, a few examples may be given, along with a very brief explanation of what a subsystem is. The group then is encouraged to begin generating possible ideas, for example, subsystem statements. Ideas are written on a chalkboard or flip chart by the facilitator as they are brought up by participants. Once the group has finished, ideas are clustered according to similarities.

Nominal Group Technique

The nominal group technique was originated by Delbeca and Van de Ven (1971). It is similiar to the brainstorming technique, in that its goal is to foster creativity and the generation of ideas. It differs in that the group process takes into account a well-known group dynamic, that a few participants can dominate a group's thinking. It therefore builds into the idea-generation process an individual reflection phase as well as a group discussion phase. Thus a nominal group is one in which individuals work in the presence of others but only interact with one another at times specified by the facilitator. In addition to a blackboard, flip chart, or other poster material, the facilitator should come with a generous supply of 3×5 index cards.

1. Usually the participants are divided into small groups, and the task is given to each individual either orally or in writing. Each individual responds to the task in writing, silently, independently, and without discussion. Those who finish early are asked not to interfere with others who are still working.

2. A volunteer from each group acts as a recorder. One at a time, each participant presents an idea that he or she generated during the individual reflection time. The item is recorded on newsprint or some other poster material but is not discussed at this time. Overlap of items is of no concern, and criticism is discouraged. "Hitchhiking" on an idea is encouraged by having members generate new ideas on their individual papers.

3. Once the ideas of all participants have been displayed, the small groups are encouraged to discuss the master list for clarification, elaboration, or addition of new items. Items are *not* condensed, collapsed into categories, or ranked. The master list is numbered item by item for identification only.

4. Each member is then encouraged to select from the master list ten items he or she thinks are most important, critical, or relevant to the task they are working on. These items are usually written on 3 × 5 index cards by name and number. If the nature of the task requires that a rank ordering eventually be achieved, each individual is asked to rank the ten items where 1 is the most important, 2 is next important, and so on, with 10 least important.

5. Each member then presents his or her selection of items, and the group recorder talleys the contributions on the master list. This results in a frequency distribution. Further discussion and clarification of the results of each small group can then be conducted.

6. Each individual is then asked to select from the master chart ten items he or she now considers most important or relevant. This is done silently and in writing on 3 × 5 cards. If the task requires this, the cards are rank-ordered.

7. The whole group reforms and then chooses ideas to deal with. They discuss them relative to the task at hand, using a *consensus* decision style rather than *voting*. Usually by the time this phase has occurred, the consensus develops fairly quickly by means of slight modifications or amplifications of some of the key ideas on which the group is focusing.

TKJ Technique

Whereas the main goals of the brainstorming and nominal group techniques are to generate useful, creative ideas, there are other techniques that help participants group, rank, or synthesize ideas. While a procedure for ranking and grouping can be used in the nominal group and brainstorming techniques, that is not their primary use or real strength.

TKJ is a technique developed by Kobayashi and Kawakita (Roth, 1985) for synthesizing the different perspectives of participants in a group into a viewpoint acceptable to all. TKJ is led by a facilitator. The materials needed are flip charts or blackboards and 5 × 8 cards. The procedures are as follows:

1. The facilitator writes an initial version of the task or problem to be addressed on a flip chart or chalkboard.

2. Participants list important and verifiable facts related to the task on 5 × 8 cards.

3. The cards are collected and redistributed so that no participant receives his or her own.

4. A randomly chosen participant reads out loud a fact from one of his or her cards. The facilitator records that fact on the flip chart.

5. Participants search through their cards for related ideas and read them out loud. These are also written on the poster.

6. When one set of related ideas is complete, another is begun with the reading of another card by another randomly chosen participant.

7. When all facts have been assigned to a set, participants choose a name for each set that reflects its essence and write the name on a card. If more than

one alternative is generated, the choice can be made by vote or by discussion and consensus.

8. The names of sets recorded on cards are then distributed, and the process is repeated until only a single all-inclusive set-name card remains. The amount of reduction necessary is of course related to the need for reduction.

Things referred to as facts in this procedure might include (1) key concerns one has regarding a current situation; (2) key groups involved in a current situation; (3) statistical data on current sales level, yield, amounts of inputs used, costs of distribution, etc.; (4) key human activities needed to alter (improve) the current situation; (5) key resources, inputs, and outputs needed; or (6) key people/groups that need to be involved.

Modified Delphi Technique

This technique can be viewed as a synthesis of the nominal group technique and the TKJ technique in that it both groups basic ideas and ranks the groupings. Its uniqueness is that participants do not meet each other until the midpoint of the process, if at all. This technique is thus very valuable for professionals operating in the field, such as sales representatives and extension agents, who often cannot get people to meetings.

The procedures are as follows:

1. A facilitator or leader presents the task or problem to be worked on to participants.
2. Participants privately list task-related facts, opinions, ideas, or answers.
3. The lists are collected by the facilitator or leader and combined into one nonrepetitive list, which is distributed to all participants.
4. Participants privately divide the items in this master list into categories.
5. The results are collected, combined, and redistributed. The combining, in this case, can be done either by the facilitator/group leader or by participants in a group.
6. Participants privately rank the categories.
7. The results are recorded and discussed in a group session.
8. A decision regarding a final listing is arrived at by consensus. The process is repeated if opinion is shifting significantly.

Dialectic

This technique was developed by C. West Churchman (1971) and is used mainly to facilitate development and debate of alternative holistic scenarios for improved futures by a group. The procedures are as follows:

1. One of the participants serves as a decision-maker and must be acceptable to everyone else involved.

2. The decision-maker forms two or more teams and provides each with a summary of a current situation and relevant background information.
3. The teams formulate improved situation scenarios. The task can be left open so that the team simply creates statements of an improved situation of their own choice, or the decision-maker can give each team instructions to create scenarios of a certain kind.
4. Each team presents its scenario.
5. The decision-maker creates a synthesis, or else a third team creates a synthesis and defends it before the other two, with the decision-maker acting as a referee.

This technique is useful for developing alternative models of an improved situation. It can also be used to debate and compare models with a current situation by having one team be for the changes and defend them and another team be against the changes and defend their ideas. The synthesis team can then take the ideas of both teams and create an improved model.

Force Field Analysis

Kurt Lewin (1951) developed a technique useful for identifying the opposing forces that press on any socio-technical system. Restraining forces are those continually pushing the current situation toward the worst possible outcome. Driving forces are those continually pushing it toward the best possible outcome.

The objective of a force field analysis exercise is to identify the factors of both sets of forces in a specific situation (present or anticipated in the future) and their relationships. Once this is done, participants determine ways to reduce the number and power of the relevant restraining forces while at the same time increasing the number and power of the relevant driving forces. Six sets of questions are usually addressed:

1. What is the current situation? What would an improved situation be like? What could be the worst possible outcome?
2. What are the restraining forces that are pushing the current situation toward the worst possible alternative?
3. What are the driving forces that are pushing the current situation toward the improved state?
4. What relationships exist between these two sets of forces?
5. Over which restraining and driving forces do we believe we have influence, and which of these are important now?
6. For each restraining and driving force subject to our influence, what specific actions or steps can be taken to eliminate or strengthen them?

For questions 5 and 6, Lewin suggests a checklist of further questions to be addressed: (1) Who will do what? (2) What exactly will be done? (3) Where will it be carried out? (4) When will it be done? and (5) How will it be implemented?

Consensus Decision-Making

Consensus decision-making procedures are neither widely known nor practiced. For this reason, a few guidelines are presented, based on the work of Johnson and Johnson (1987). These suggestions apply equally to participants and facilitators:

1. Avoid arguing blindly for your own opinions. Present your position as clearly and logically as possible, but listen to other members' reactions and consider them carefully before you press your point.
2. Avoid changing your mind only to reach agreement or to avoid conflict. Support only solutions with which you are at least somewhat able to agree. Yield only to positions that have objective and logically sound foundations.
3. Avoid conflict-reducing procedures such as majority voting, coin tossing, averaging, and bargaining.
4. Seek out differences of opinion. They are natural and expected. Try to involve everyone in the decision process. Disagreements can improve the group's decision because they represent a wide range of information and opinion, thereby creating a better chance for the group to hit upon superior solutions.
5. Do not assume that someone must win and someone must lose when the discussion reaches a stalemate. Instead, look for the next most acceptable alternative for all participants.
6. Discuss underlying assumptions, listen carefully to one another, and encourage the participation of all members.

Consensus decision-making works best when time pressures are not acute, when there is no immediate crisis, when potential participants have not yet adopted rigid positions, and when group members are able and willing to commit a high level of positive psychological energy to seeking mutual agreements satisfactory to all.

REFERENCES

Anderson, Ralph E., and Irl Carter. *Human Behavior in the Social Environment: A Social Systems Approach.* Chicago: Aldine, 1987.

Bawden, R. J., R. D. Macadam, R. G. Packham, and I. Valentine. "Systems Thinking and Practice in the Education of Agriculturalists." *Agricultural Systems* 13: 205–225, 1984.

Checkland, Peter. "A Systems Map of the Universe." *Journal of Systems Engineering* 2(2):107–114, 1971.

Checkland, Peter. *Systems Thinking, Systems Practice.* New York: John Wiley & Sons, 1981.

Churchman, C. West. *The Design of Inquiry Systems.* New York: Basic Books, 1971.

Delbeca, A., and A. Van de Ven. "A Group Process Model for Problem Identification and Program Planning." *Journal of Applied Behavioral Science* 7:466–491, 1971.

Edwards, Betty. *Drawing on the Right Side of the Brain.* Los Angeles: J.P. Tarcher, 1979.

Ford, David. "Nominal Group Technique: An Applied Group Problem-Solving Activity." In *The 1975 Annual Handbook for Group Facilitators.* J. Jones and J. W. Pfeiffer, eds. San Diego, CA: University Associates Publishers, 1975.

Jenkins, G. M. "The Systems Approach." *Journal of Systems Engineering* 1(1), 1969.

Johnson, David W., and Frank P. Johnson. *Joining Together: Group Theory and Group Skills.* Englewood Cliffs, NJ: Prentice-Hall, 1987.

Lewin, Kurt. *Field Theory in Social Science.* New York: Harper, 1951.

Naughton, John. *Soft Systems Analysis: An Introductory Guide.* Block IV. Complexity, Management and Change: Applying a Systems Approach. London: Open University Press, 1983.

Roth, William F. *Problem Solving for Managers.* New York: Praeger, 1985.

Smythe, J. C., and P. B. Checkland. "Using a Systems Approach: The Structure of Root Definitions." *Journal of Applied Systems Analysis* 5(1), 1976.

University Associates. *The Annual: Developing Human Resources.* San Diego, CA: University Associates, 1972–1988.

Wilson, Brian. *Systems: Concepts, Methodologies and Applications,* New York: John Wiley & Sons, 1984.

CHAPTER 6

Comparison: Stage 5 of the Soft Systems Approach

by D. M. Vietor and H. T. Cralle

During the comparison stage, specific proposals for change emerge to be tested by comparing one or more conceptual models to the picture of a problematical situation developed in stages 1 and 2. There are two essential goals. The first is to come down from the highly abstract model-building phase by self-consciously returning to the complex real world as originally recorded in the situation summary, composite mind maps, cartoons, or other materials developed in stage 2. The second objective is forward-looking, to get the human activity system models ready to be communicated in the next phase of the approach, when the proposals for change are debated by the people involved. The analyst continues to function as a facilitator and learner, fostering cooperation among participants, while also encouraging their involvement in the techniques for developing proposals for change. Participants are expected to gain insight about what is problematical in relation to reality and to become skilled in using the techniques of the comparison stage. Whereas assimilative learning competencies were emphasized in stages 3 and 4, stage 5 calls upon convergent skills.

Generally, it is desirable for the analyst and a selected group of participants to move fairly quickly to the comparison stage after developing several conceptual models of human activity systems. The insights derived from comparing models of human activities to the picture of reality not only yield more specific proposals for change, but enable refinements of the picture of reality and of the relevance of system definitions and models. *General discussion*, *question generation, overlaying* conceptual models on the picture of reality, and *historical reconstruction* are four documented techniques for comparing conceptual models to the picture of reality. Each technique offers an opportunity to encourage learning among the analyst/facilitator and the participants in a problematical situation. The ultimate objective of stage 5 is to prepare the conceptual models for presentation to all the people involved, including developing appropriate communications and debate techniques.

Learning Competencies

For relatively simple problems, *convergence* is the application of knowledge to solve a problem or introduce a technology (Kolb, 1984). The knowledge takes the form of a concept, theory, principle, technology, or model to enable the problem solver to deduce one or more tentative solutions. The focus is on choosing from among possible solutions by testing them and selecting whatever works. People who prefer convergence as a learning style tend to be more comfortable with technical tasks and problems rather than social or interpersonal issues (Kolb, 1983).

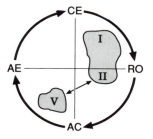

In stage 5 of the soft systems approach, convergence is practiced when one or more conceptual models of human activity systems are compared to the situation summary or composite mind map of a problematical situation. The proposals for change that arise from matching conceptual models to the picture of reality stimulate discussion among participants. The comparison stage offers the analyst/ facilitator and others involved an opportunity to determine the relevance of one or more conceptual models. At the least, comparing the fruits of conceptualization to the reality experienced in a situation can reveal new insights about what is problematical.

Expert or Facilitator

It is tempting to abandon a systemic or holistic approach — to slip back into the technology development mode — when confronted by complex situations. Instead of viewing the complexity as a whole, it seems easier to break the situation into components or parts. The parts are examined and solved one at a time. Such solutions result from comparing the knowledge and experience of the analyst to one or more parts of the complex situation. The solutions to parts of a situation can be offered to a client or those involved in the situation as expert advice. Unfortunately, the expert advice is often given without an opportunity for thorough discussions among those participants in the situation who are affected by it.

As described in previous chapters, the knowledge, perceptions, and world view of the experts will determine which parts of a situation they see as problematical. If the expert's view of what is problematical differs from that of the participants in a situation, the expert's advice may be rejected as inappropriate. If the advice is accepted and implemented, the sense of unease of the participants may worsen instead of improve. Furthermore, without discussion among experts and those affected by their advice, the experts learn very little from the experiences of the participants in a situation.

> Before reading on in this chapter, turn to the back of the book and review the Chatham River case materials located in the appendix.

That seems to have been what happened in the Chatham River case. At the point where you, the reader, enter the case, one (or more) of the main groups involved in the situation has already been "slopping at the experts' trough." While reading the description of the Chatham River case through the eyes of experts, many themes of concern are apparent:

- Reflect on what kinds of "experts" may have been engaged. Try to enumerate them. Apply an expert's knowledge and **W** to the case.
- Should the irrigators and other downstream riparians enter into litigation to capitalize on legal precedents?
- If the riparian rights of Springville are equal to those of "the town of Henderson" (cited in a court decision provided in the Chatham River case materials), does the town have the right to sell river water to nonriparians?
- Conversely, is the town council's proposal to pay compensation to friends and irrigators feasible?
- Would you, an expert, support the argument of the townspeople that industry will generate more income per unit of water than is possible from irrigation of crops in an already declining agricultural economy? (Note that the town council and commercial interests are encouraged by the previous court ruling that Nantahala Power and Light need pay other riparians no more for river water than the revenue those riparians realize from their current use of the water.)
- Conversely, if you give expert advice favoring the downstream riparians over the townspeople, will the townspeople be receptive to your intervention?

If you have already practiced divergence through reading, reflecting upon, and recording a pictorial description or mind map of the Chatham River case, the appropriate response to each of the previous questions may be obvious. The disadvantages of advice offered from only one **W**, and specific to only part of the Chatham River case, may also be obvious. How can you use one or more definitions and conceptual models of a relevant system to systemically validate expert advice or solutions for the Chatham River case?

The role of the analyst in the soft systems approach is that of facilitator and active learner rather than of expert. Although an analyst may be hired to serve a particular client, the goal is to develop mutual awareness and trust among all those involved in a problematical situation. The analyst needs to recognize and record each of the diverse **W**s represented, including his or her own, during the initial stages of observing and assimilating a problematical situation (see especially Chapters 2 and 4).

In convergence and in the comparison stage, the analyst needs to involve representatives of the diverse **W**s in discussions concerning proposed improvements in a situation. One goal of the comparison stage is to encourage the client and other participants to learn and develop by self-direction and self-discovery. The hope is that the participants in the discussions about improvements will perceive themselves as pursuing activities leading to cooperation and negotiation with others in the situation.

Achieving such mutuality is no easy matter among people with diverse **W**s such as the downstream riparians (irrigators), townspeople, and environmentalists in the Chatham River case. Representatives of each world view are likely to have a definite sense of their own identity, interests, and autonomy. Each is likely to be argumentative and to blame others in the problematical situation for the disorganized mess that has evolved.

This is especially true in the Chatham River situation, since it is apparent that some parties have already engaged in considerable planning and have, through ignorance or design, excluded other parties from the process. In other words, the situation is well beyond the formative stage, and the positions of the different parties are strikingly polarized. With such divergent **W**s involved, it will be difficult, but not impossible, to engage all parties in a constructive dialog. There are several directions such an effort might take, but for the time being, readers themselves should consider the alternatives, including scenarios in which the soft systems approach is not useful at all!

How can an analyst facilitate continued involvement and cooperation among participants during the comparison stage? It helps if the work of previous stages has been sound, for example, if the conceptualization process yields one or more definitions and models of systems that are relevant to the concerns of all participants in a problematical situation. It may be necessary to conduct several iterations of the cycle of gathering information, defining and modeling human activity systems, comparing the models to a synthesis report or mind map, and debating proposed improvements and insights. It is important to encourage expression of individual feelings and opinions during each stage of the methodology, including the comparison stage. Cooperation results only when participants are confident that they have been heard and are full participants in improvement activities. The analyst learns and develops skills to encourage cooperation among participants during each stage and iteration of the cycle. In addition, each iteration offers lessons enabling the analyst/facilitator to improve his or her techniques.

Ideally, as clients and other participants agree upon improvements as a result of discussions during the comparison and earlier stages, they will have integrated the perspectives of others with their own. Each participant will have learned to balance group and individual interests, while also maintaining constructive interpersonal relationships with others.

Both the individual and group activities of participants must be valued. Participants should be encouraged to develop their own standards for learning during the comparison and other stages of the approach. Yet they must accept the multiple standards of performance, interpretations of experience, and sources of information of the other participants. The result can be very positive mutual and interdependent relationships among participants and the analyst/facilitator. Again, in addition to making decisions and improvements concerning the current situation, the participants in the problem situation become more capable of practicing a systems methodology for tackling their own problematical situations in the future.

Advice, new information, or challenges to the beliefs of participants may be appropriate during the comparison stage but should be posed as questions rather than as declarative or judgmental statements. For example, advice is most effective when posed as a question: "May I offer you a suggestion?" or "May I share some information I have on this?" As a result, the participants in a problematical situation can reject an unwanted intervention before it occurs. In addition, mutual trust and cooperation among participants and the analyst are sustained.

Timing

When should the analyst stop modeling and begin comparing one or more conceptual models to the situation summary or mind map of the problematical situation? Deciding when to begin the comparison stage is a judgment call, but usually you begin after checking each model of a human activity system against a formal systems model using the checklists presented in Chapter 5. You will recall that the checklists are a time- and trial-tested set of guidelines of the essential features of various kinds of systems, including human activity systems.

To recapitulate, a conceptual model should have (1) a clear mission, goal, or direction, which is continuous; (2) a means of measuring relative success or failure in trying to achieve the continuous purpose or objectives; (3) a decision-making process or role; (4) components or subsystems that are themselves systems; (5) a degree of interaction among the subsystems; (6) an environment with which the system interacts; (7) a boundary separating the system (in which the decision-maker has jurisdiction) from the environment; (8) resources that are at the disposal of the decision-making process; and (9) long-term stability or potential to recover from some degree of disturbance (Naughton, 1984).

After the conceptual model has been checked against the guidelines of the formal system, Checkland (1981) advises moving fairly quickly to the comparison stage. He argues that allocation of large amounts of time, money, and resources to modeling without comparing the models to reality can be wasteful. For example, an analyst may expend great effort modeling a system in response to a farmer's expressed desire to enhance the productivity of corn. In support of a subsystem for deciding about allocation of resources for production, the analyst may welcome the chance to apply his or her expertise to developing mathematical models of relationships among physical and biological components of a corn-production system. The model could then yield recommendations for the optimum combination of inputs to maximize yields per acre for a particular environment. *When the system to enhance productivity is compared to reality, however, it quickly becomes apparent that current surpluses of corn and low prices per unit of corn produced make it uneconomical to enhance the productivity of corn.* The farmer may also shun the use of herbicides and insecticides that has been factored into the model. The recommendations suggested by calculations from the quantitative model are technologically sophisticated and exact, yet irrelevant when compared to broader concerns of the farmer in the current (to say nothing of the future) economy and environment.

At the other extreme, there is the risk that a hastily prepared conceptual model will similarly yield irrelevant proposals for change when compared to a picture of reality. A conceptual model can conform to the guidelines of a formal systems model and still lead to proposals for change that are rejected by participants in a problematical situation.

Deciding to begin the comparison stage also depends on the emotional state or climate of the relationships among people with differing **W**s who are involved in the situation. A hastily prepared model may represent only one among those diverse **W**s. Bias toward one world view can fuel hostility among participants during the discussion of proposed changes. Proposals for improvement should facilitate cooperation rather than destructive conflict among the diverse viewpoints of the people involved. If the analyst/facilitator is rejected or fired by the owner of the system, there may be no further opportunity for another iteration of the approach or for a contribution to effective improvement.

Yet, in support of Checkland's (1981) argument, remember that the soft systems inquiry process can be repeated. Early comparison of even relatively simple models to pictures of reality can yield valuable insights to the analyst and participants about what is problematical. Learning by the analyst and participants is facilitated by the process of preparing and discussing proposals for change during and after the comparison stage. The analyst gains insights while translating from the formal language of systems models to terms understood by participants in a situation. Once translated into proposals for change, the conceptual model serves as a source of new information for discussion and learning among the participants. As participants in a problematical situation assess and discuss proposals for change and conceptual models in terms of their knowledge and experiences, they contribute valuable feedback to the analyst. The feedback can stimulate the analyst to rethink the comparison stage and earlier stages of the soft systems approach.

As the participants in the problem situation become better informed, their viewpoints are clarified through discussions, and they can be encouraged to integrate the ideas of others into their own thinking. Furthermore, as they participate in each cycle or iteration of the soft systems approach, they grow not only in knowledge, but in their ability to use the methodology to tackle their own problems.

If participants are familiar with systems concepts and language, they can be actively involved in the comparison stage, and the analyst doesn't need to translate the definitions and models of relevant systems for them. The analyst and participants may already have developed a common vocabulary of terms if they mutually contributed to conceptual modeling of one or more relevant systems.

To summarize, this stage is when you take the time to reflect on whether or not the conceptual models seem to do what is intended. You say to yourself, and to others involved, "Time out. Let's go back a few notches and recall what the situation looked like when we began this process." In many respects, and as will be made clear in the following section, it can be seen as an explicit return to stage 1 and 2 inquiry activities. Here are the kinds of questions you ask:

- Do the proposals for change embodied in the models address real problems and issues previously identified by people?
- Will the models at least contribute to future debates about change? (For example, do they provide the people involved with a novel learning situation that will prepare them to participate better in a debate on changes, improvements, and their overall future?)
- As a result of practicing comparison, do the people involved now see new aspects of their situations that may justify another round of system definition and modeling?
- Do the models and associated proposals fit into the existing structure in such a way that established primary tasks can be sustained? Or do the models and proposals explicitly call for the abolition of the primary tasks, in whole or in part, in a future state?
- Do they call for improvements that might be accomplished in less costly, disruptive, or divisive ways?
- Does the model provide standards of performance that *make sense* in relation to our understanding of the situation?
- Do the models present alternative scenarios, contingencies, and probable outcomes so that the people involved know what they could be getting into (and hence what it is they are debating)?

Using Comparison Techniques for Gathering Information

Wilson (1984) applied techniques from the comparison stage to the task of gathering and organizing information during the initial stages of finding out about a problematical situation (stages 1 and 2). He advocates that the analyst define and model a system as soon as enough information is collected to discern a purpose or transformation. Questions are generated while comparing the model of a relevant system to the relatively unstructured problem situation. These questions are used to plan a structured and coherent set of interviews, rather than to propose improvements in a problem situation.

Note that this is a seemingly radical departure from the approach presented so far. It is, however, consistent with the iterative character of the soft systems approach and is potentially useful, as discussed subsequently. For example, this technique may be particularly appropriate to territorially extensive issues involving many interests and to potentially concerned groups in which it is unclear where to start collecting information. In such cases, conceptual models and ultimate proposals for change also are likely to be very complex.

Because the system definition and preliminary model are constructed by the analyst to aid stage 1 inquiry, participants need not be included in the question-generation process. Indeed, *the comparison stage in general* requires that the analyst/facilitator be very calculated regarding who is to be involved in discussions. It may be desirable to exclude participants altogether if the model even superficially seems to represent a **W** that may aggravate distrust among diverse groups and viewpoints. Conversely, one or more of the participants may provide the inputs necessary to construct the initial definition and model of a system relevant to their situation. And, in general, people left out or groups left unrepresented can pose serious difficulties when debate and implementation are on the agenda.

When in doubt, err on the side of *inclusion*. Participants can begin to visualize human activities as they contribute to conceptualization of a preliminary model of a system. They can also think of their own role in the activities called for in a preliminary model.

Wilson (1984) believes the novice analyst spends far too much time gathering information from randomly generated questions in an environment where the people involved in a situation, such as managers, have little time to spare. The novice analyst benefits from the more organized inquiry that comes from application of the comparison techniques early in the process.

This alternative approach of using comparison techniques to aid stage 1 and 2 inquiry can be illustrated using the Chatham River case. An allocation system for river water appears relevant to the concerns of many of those involved in the situation. In this example, the definition and model of a water allocation system can be compared to the unstructured situation described in the case materials. A realistic possibility, only hinted at in the case materials, is that an appropriate state agency might attempt to intervene in such a complex dispute. Note that, inevitably, the **W** of the analyst or formulator of the model is strongly represented in the definition of the system:

> A riparian-owned system for efficiently allocating river water among riparians according to established legal precedents of North Carolina and to constraints created by seasonally variable rates of river flow and of water usage by riparians.

Each of the CATWOE criteria (Checkland, 1981) can be identified in the definition:

C = Downstream riparians (irrigators)
A = Downstream riparians, Friends of Chatham, members of town council, townspeople
T = To allocate river water among riparians
W = River water should continue to be allocated among the needs of riparians according to the laws of North Carolina
O = Riparians
E = Variation of rates of river flow and water usage by riparians

The major human activities in the conceptual model presented in Figure 6.1

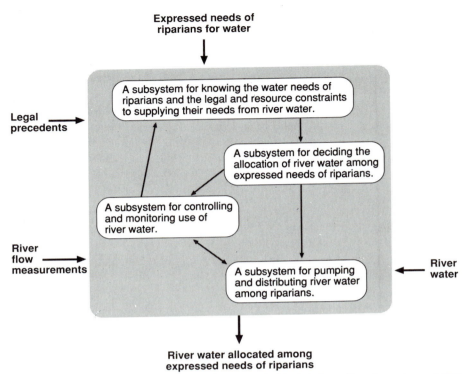

**Expressed needs of
riparians for water**

A subsystem for knowing the water needs of
riparians and the legal and resource constraints
to supplying their needs from river water.

**Legal
precedents**

A subsystem for deciding the
allocation of river water among
expressed needs of riparians.

A subsystem for controlling
and monitoring use of
river water.

**River
flow
measurements**

**River
water**

A subsystem for pumping
and distributing river water
among riparians.

**River water allocated among
expressed needs of riparians**

FIGURE 6.1 A preliminary conceptual HAS model for allocating river water among expressed needs of riparians. Unlabeled arrows denote flow of information.

suggest questions for gathering information and making sense of the Chatham River situation:

1. Does each subsystem or human activity exist?
2. How is the subsystem or human activity presently carried out?
3. Who is responsible for the activities within a subsystem?
4. Is the activity done well or badly?
5. Do the interactions pictured in the model exist?
6. What are the forms of interactions?

The comparison of the preliminary model and current perceptions of a situation can also help determine who could best answer the questions. Would you address the questions to downstream riparians only? Who are *all* the possibly numerous concerned parties, communities, and interests that need to be contacted if a truly open and constructive process is to be initiated?

Notice that the model-generated questions focus on subsystems and human activities (verbs). The focus on human activities complements information relating to structure (institutions, things, etc.) and people (roles, attitudes, etc.) that become apparent as information is gathered and interviews are conducted. The composite mind map presented with the case materials in Figure A.1 is an example of an effort to map themes of concern and human activities related to water allocation for local and state governments and interest groups. It also includes people's feelings and other factors gleaned from the Chatham River case. Look upon this composite (or second edition) mind map as composed of the original individual mind maps together with descriptions of human activities and answers to questions that resulted from formulating a preliminary model in accordance with Wilson's procedure.

Questions generated by the preliminary model (Figure 6.1) cause both the analyst and participants to think in terms of human activities, that is, how various functions are or are not done. The more participants can come to terms with their own activities, the better they are prepared to contribute to the comparison stage. Skillfully worded questions can solicit answers in terms of human activities and prepare participants to contribute to and evaluate models of relevant systems and proposals for change during later stages of the soft systems approach.

The novice analyst must guard against becoming too attached to the definition and model used to generate questions for finding out about a situation. This is the possible danger of using Wilson's technique. A preliminary definition and model should not discourage development of new and different definitions and models during stages 3 and 4 of the soft systems methodology.

Using Comparison Techniques to Propose Change

The task of proposing changes during the comparison stage can be clarified by a brief description of the criteria of desirability and feasibility for evaluating such proposals (see Chapter 7 for expanded discussion). Typically, the proposals are evaluated in discussions with the participants who are concerned to make changes in a problematical situation. The discussions may follow the comparison stage as a separate stage of the soft systems methodology. Often, the discussions of that debate stage overlap with the comparison stage.

During the discussions, the desirability of proposals for change or improvement in a problematical situation are assessed in terms of the **W**s of the participants. Proposals for change will be judged desirable if they are consistent with the beliefs, knowledge, and desires of participants. Here the question is "Is this what we want?" Changes are feasible if they can be implemented within the environmental and organizational constraints present in the collective reality experienced by participants in the problematical situation. Here the questions are "Will it work?" and "Can we put it into effect?" In short, proposals for change are desirable and feasible if (1) they fit into established ways of doing and viewing and (2) implementation is possible within the constraints of physical, economic, and human resources and of organizational rules and structure.

Ideally, changes that are both desirable and feasible will emerge during the comparison stage if the **W**s and constraints represented in the definition and model of a human activity system match the reality faced by the participants. More often, a number of iterations of the soft systems methodology are needed, including the definition and formulation of several models and their comparison to mind maps or situation summaries, before participants agree on proposals for change. The discussion of what is desirable and feasible following or during the comparison stage is essential to each iteration of the methodology if proposed changes are to be accepted and implemented. You should realize that the participants' view of what is desirable and feasible can also evolve during each repetition of the approach.

Techniques

Four general techniques for comparing a conceptual model to a mind map, other pictorial representation, or situation summary have arisen out of the experience of soft systems practitioners (Checkland, 1981; Naughton, 1984; Wilson, 1984): (1) general discussion, (2) question generation, (3) overlay, and (4) historical reconstruction. In order to provide a more detailed illustration of the comparison stage, each technique is displayed using the Chatham River or Mucho Sacata cases found in the back of this book.

General Discussion Technique

The general discussion technique is used to raise questions about those human activities of the conceptual model that are especially different from present reality. Attention is given to fundamental issues arising out of the comparison rather than more detailed questions concerning specific ways of doing things. Major differences between the conceptual model and reality suggest points to be revised. Often strategic issues regarding overall objectives rather than specific operations or activities are involved. The utility of the discussion method is greatest when all participants are familiar with the systems language (Wilson, 1984). It is also common, however, for the analyst to translate the models of human activity systems into terms understood by participants.

Assume that the analyst/facilitator has been called into the Chatham River situation by a farmer/riparian advocacy group. The general discussion technique can be used in collaboration with representatives of downstream riparians or irrigators to compare the conceptual model (described earlier in Figure 6.1) with the situation summary and composite mind map (in Figure A.1). It quickly becomes apparent that the subsystem for knowing the water needs of all riparians in Figure 6.1 does not correspond to a theme of concern in the mind map describing reality in Figure A.1. The greatest concern of the irrigators is that there is not an adequate procedure to determine and express their river-water needs. Communicating the conceptual model (Figure 6.1) in the irrigators' language may enlist their cooperation and continued participation in a subsystem for the greater good of knowing the needs of *all* riparians.

At the least, the knowledge subsystem strikes participants as a better alternative than the relatively uninformed and oppressive role that the town council of

Springville is currently playing. A subsystem to acquire knowledge of water needs, including knowing legal constraints, may be particularly appealing if the irrigator is aware of the previous legal precedent (*Pernell* v. *City of Henderson*) that precludes a town such as Springville from selling river water to nonriparians.

Through the **W** of the irrigators, the subsystem for deciding the allocation of water is also unrepresented in the mind map. The function of deciding the allocation of water is implicit in the current managing activities and initiatives of the town council, but the irrigator would likely feel alienated if the town council were deciding for all riparians.

So irrigators and other participants focus on major differences arising from the comparison of each subsystem of the conceptual model to reality. Proposals for changing the current reality experienced by participants may grow out of discussions of the differences between the model and reality. Or new insights into the reality of the problematical situation may be revealed, discussed, and incorporated into the mind map to begin another iteration of the methodology, leading to alternate definitions and formulations of conceptual models.

Discussions of differences between the model and reality may also reveal a need for a detailed model of the activities comprising a particular subsystem in the original model. For example, what activities comprise a subsystem for controlling and monitoring use of river water (Figure 6.1)? The definitions and models of one or more subsystems within a model will provide more detail for subsequent comparisons to reality. Specific proposals for change may be apparent as discussion focuses on the detailed model of a subsystem compared to reality. In the example in which the irrigators are viewed as your clients, it is desirable to involve them in as many of the stages of each iteration as their interest will bear. Remember, the goal is to equip the participants to grapple with their own sense of what is problematic.

- Will the conceptual model used in this example satisfy all participants in the Chatham River situation?
- How would the Friends of the Chatham (particularly since many are not riparians) respond if they participated in the comparison stage using the same conceptual model?
- Would the friends be satisfied with a definition and conceptual model that made no provision for sustaining the natural beauty and recreational resources of the river?
- Will the members of the town council surrender their anticipated decision-making power and accept the subsystem for deciding the allocation of river water in the conceptual model? For example, they may argue that new industry could return far more dollars per unit of water to their town than any of the other viewpoints represented in the Chatham River situation.

Imagine using the general discussion method to complete the comparison stage using the model and mind map with participants representing all of the diverse **W**s in the situation, including their lawyers and other hired experts such as hydrologists and economists. The abilities of an analyst/facilitator would be severely tested under these conditions. Indeed, an alternate technique is *mediation*, an approach to debating change discussed in Chapter 7.

After answering the questions in the previous paragraph, another iteration of the soft systems methodology was carried out. This resulted in a revision to the original definition of a human activity system. The new transformation, constraints, and **W** of the revised definition are an attempt to encompass the range of diverse concerns and viewpoints represented in the Chatham River case:

> A riparian-owned system for improving and sustaining the quality of life of riparians, townspeople, and sportsmen through development, management, and conservation of limited water resources according to state law, and constrained by a slumping economy and large seasonal fluctuations of the availability of and demand for water resources.

An itemized CATWOE analysis clearly reveals the broader scope of the definition:

C = Consumers of water resources
A = Townspeople, irrigators, recreation seekers, officials of town council
T = To sustain and improve the quality of life
W = A negotiated and managed use of water resources will improve the quality of life of both riparians and nonriparians
O = Riparians
E = Limited water resources, declining agricultural economy, potential population growth, and state law

The revised conceptual model presented in Figure 6.2 will be used to complete the comparison stage with representatives of the diverse **W**s present in the Chatham River case. As suggested earlier, the greater complexity of the model and the large number and diversity of viewpoints of participants would make the general discussion technique difficult to implement and might favor the use of alternative approaches to reach agreement. Yet another option would be further iteration, leading to a new system definition; for example, ownership might be vested in the state department of water resources along with a public advisory board appointed by the governor.

Question-Generation Technique

The question-generation technique is one alternative to the general discussion technique for comparing a complex human activity system model to reality. As the name of the technique implies, ordered questions about the reality of a problematical situation are generated from knowledge of the activities in a conceptual model (Checkland, 1981). Once the questions are written down, they can be answered in a systematic way by the analyst and other participants in the comparison stage.

Wilson (1984) suggests using a tabular format to systematically list and pres-

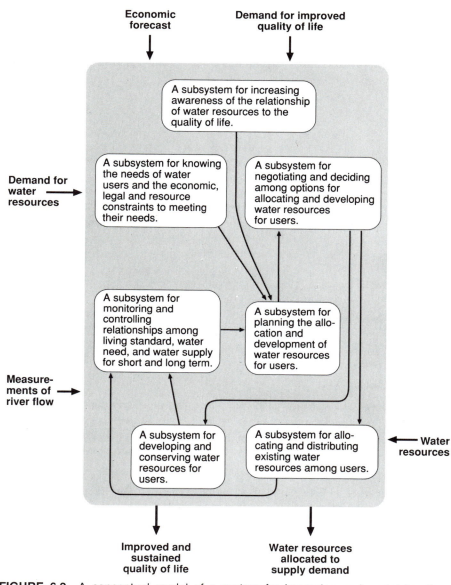

Economic forecast

Demand for improved quality of life

A subsystem for increasing awareness of the relationship of water resources to the quality of life.

A subsystem for knowing the needs of water users and the economic, legal and resource constraints to meeting their needs.

A subsystem for negotiating and deciding among options for allocating and developing water resources for users.

Demand for water resources

A subsystem for monitoring and controlling relationships among living standard, water need, and water supply for short and long term.

A subsystem for planning the allo-cation and development of water resources for users.

Measure-ments of river flow

A subsystem for developing and conserving water resources for users.

A subsystem for allo-cating and distributing existing water resources among users.

Water resources

Improved and sustained quality of life

Water resources allocated to supply demand

FIGURE 6.2 A conceptual model of a system for improving and sustaining the quality of life of riparians, townspeople, and sportspeople.

ent the questions. This is reproduced in Figure 6.3. Columns 1, 2, and 3 in the table are questions asked and answered in systems language, presumably by the analyst, with help from participants in the problematical situation. Columns 4 through 6, written in the language of the existing situation, present insights into what is problematical in the situation. The insights, including proposals for

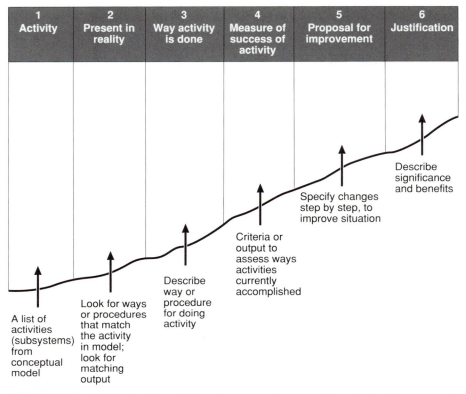

1 Activity	2 Present in reality	3 Way activity is done	4 Measure of success of activity	5 Proposal for improvement	6 Justification

A list of activities (subsystems) from conceptual model

Look for ways or procedures that match the activity in model; look for matching output

Describe way or procedure for doing activity

Criteria or output to assess ways activities currently accomplished

Specify changes step by step, to improve situation

Describe significance and benefits

FIGURE 6.3 A tabular Presentation for generating questions during the comparison stage. (Adapted from *Systems: Concepts, Methodologies, and Applications,* Brian Wilson, © 1984 by John Wiley & Sons, Ltd., Chichester, Sussex, England. Reprinted by permission.)

improvement, are revealed through an assessment of the relevance of each subsystem or activity in relation to the picture of reality. This table, or a similar instrument developed to meet a specific need, is a useful tool for translating the conceptual model into the language of the existing reality while answering the questions. In the debate stage (Chapter 7), it is particularly effective in catalyzing discussion because it illuminates features of the model by building a bridge to the current situation in a very clear-cut manner.

The planning subsystem of the revised model of a human activity system (Figure 6.2) provides material for demonstrating the question-generation technique. Planning is entered in column 1 (Figure 6.3). The task is to find a theme of concern or activity in the mind map (Figure A.2) that coincides with the subsystem for planning in the model. This composite mind map of the Chatham River case shows planning, punctuated by a question mark, linked to the managing activities of the town council. The notations linked to planning suggest it was, at the least, incomplete at the time the mind map was revised. Presuming that the participants agreed that the second edition of the mind map accurately recorded their perceptions of the problem situation, there is already some awareness of a lack of comprehensive planning of water allocation among the various demand sectors of the town and the surrounding region.

To evaluate the presence of the planning activity in the reality described by the mind map, it would be helpful to look for some output of a planning activity. The recorded testimony of Mayor Hall and Mr. Priest in the public hearing (see Chatham River case materials) suggests the output relevant to planning by the current town officials and town council is little more than speculation. This speculation is accompanied by uncoordinated initiatives to attract industry. Furthermore, the mind map records separation of activities of the town council from the concerns and activities of conserving and irrigating. The council's view that the petition from conservationists is petty carping, and the mismatch between the council's idea of fair compensation compared to riparian irrigators' perceptions, indicate that the interests of other participants have been unrepresented in any planning output.

The mind map suggests a "yes" should be recorded in column 2 of the table, and the uncoordinated initiatives by city officials could be recorded in column 3. The way an activity is conducted as described by the mind map corresponds to the most detailed level of resolution that would be conceptually modeled. At this detailed level, there are several ways to accomplish a general activity, such as planning.

In this example, what is the measure of success of the town council's uncoordinated initiatives toward allocation of water for economic development (Figure 6.3, column 4)? The mind map suggests the council will measure success in terms of economic prosperity and, more specifically, tax dollars returned per unit of water. A general proposal for debate, in column 5, would be to institute and involve all interested participants in deliberate procedures for planning the allocation and development of water resources. The proposed improvement can be justified based on the benefits of developing cooperation among riparians and nonriparians such that the quality of life, measured in terms agreeable to all of the diverse viewpoints in the problem situation, will be improved.

Discussions among participants while answering the questions in the table will facilitate development and evaluation of proposals for change, including the proposal to institute planning procedures. Where participants agree to a proposal, the viewpoints and ideas they express during discussion can be used to begin another iteration of the soft systems methodology aimed at specifying what activities comprise a subsystem for planning (i.e., you can model a subsystem in more detail). The model of a human activity system may eventually be detailed enough to specify several different ways in which participants could implement improvements. In some cases, participants may take their own initiatives to implement specific activities or changes after they are convinced of some generic need, such as planning.

Notice that the question-generation technique enables the analyst to develop a specific agenda of proposals for improvement in the language of the participants.

Although the analyst may choose to involve the participants in assembling and debating information in columns 1 through 4, using column 5 to focus debate among the diverse viewpoints represented in the Chatham River case has the greatest intuitive appeal.

Direct Overlay Technique

For models concerned with organizational structure, Wilson (1984) particularly recommends configuring the conceptual model to resemble a picture of the problematical situation. For this comparison technique, the similar configuration of both models permits one model to be overlaid on the other. For example, differences between the decision-making boundaries in a conceptual model and the areas of authority in the actual organization can be revealed and debated.

If you have not already done so, review the Mucho Sacata Ranch case materials in the appendix.

The Mucho Sacata Ranch is a case where the overlay technique is appropriate for comparing a conceptual model to a picture of reality. The situation summary description and initial pictorial representation of the case in Figure A.3 reveal the anxiety of Andy, the manager of the Mucho Sacata Ranch. One issue is Andy's confusion about lines of authority in relation to the general manager and to the owner of the overall enterprise that includes Mucho Sacata. Following the owner's employment of the general manager, Mr. Book, Andy attempted to report his initiatives and progress to both the owner and the general manager. The owner sent a terse note directing Andy to submit his reports through "proper channels."

The general manager gave Andy no verbal feedback during an on-site evaluation of Mucho Sacata but later criticized Andy for deciding on culling and replacements for the beef herd. Andy was criticized for exceeding the bounds of the general manager's perception of a ranch manager's authority.

The pictorial representation of Mucho Sacata is revised to present the activities of the owner, general manager, ranch manager, and laborers in relation to a hierarchy of authority perceived by the analyst and general manager. This is shown in Figure 6.4. This revised picture of reality reveals the importance of coordination among the levels of authority to the development of the owner's goals. A system relevant to coordination of lines of authority can be defined and modeled to illustrate the overlay technique:

A system for developing and achieving the goals of the owner through coordination among the owner, managers, and laborers for production and marketing of products from several agricultural enterprises, according to society's notion of land stewardship and constrained by a political and economic climate favoring cheap food for the consumer.

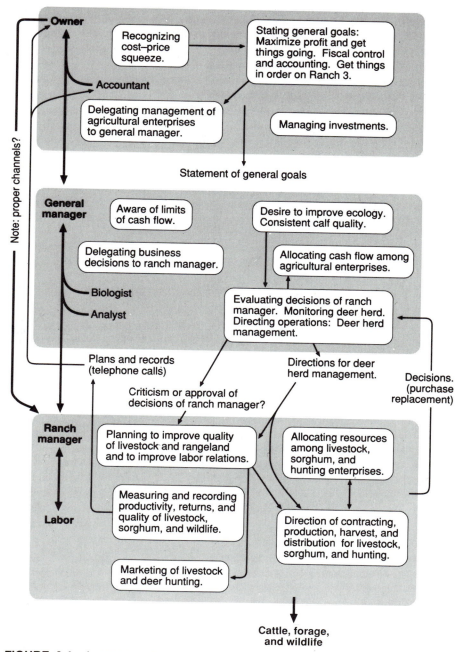

FIGURE 6.4 A second-edition pictorial representation of Mucho Sacata Ranch, emphasizing structure and relationships among main actors in the situation.

The CATWOE criteria for the definition:

C = Mr. Bell (owner)
A = Owner, general manager, ranch manager, buyers, hunters, consultants, laborers, suppliers, and families
T = Developing and achieving goals
W = Goal-setting and goal achievement can best contribute to long-term sustainability of the system when practiced consistent with social, economic, and political constraints
O = Mr. Bell
E = Land stewardship important, cheap food policy

The initial conceptualization of a model of the system as described in the definition in Figure 6.5 is not useful for the overlay technique. As suggested by Wilson (1984), the conceptual model is revised and configured to portray human activities in relation to the lines of authority and decision-making boundaries described in the picture of the reality of Mucho Sacata. This reformed model is presented in Figure 6.6. Now this conceptual model can be superimposed on the picture of Mucho Sacata (Figure 6.4) to complete the comparison stage.

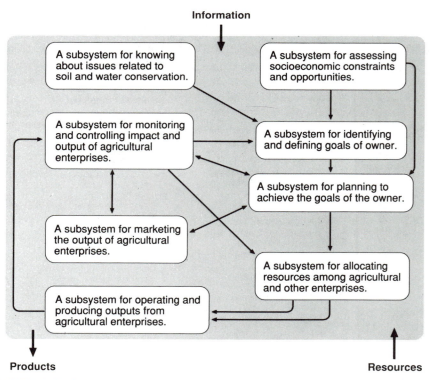

FIGURE 6.5 A first-edition conceptual model of a human activity system for developing and achieving the goals of the owner of Mucho Sacata.

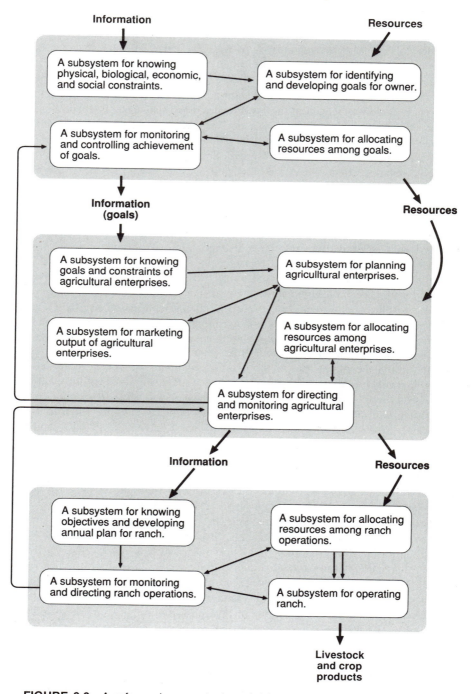

FIGURE 6.6 A reformed conceptual model for use with overlay technique.

For example, included in the HAS model represented by Figure 6.6 is a sub-system for planning agricultural enterprises (the profit centers described in the case materials), one of several subsystems relevant at the general manager's level. The picture of reality (Figure 6.4) illustrates an expressed desire at the general manager's level to improve the ecology of the ranches and the consistency of calf quality (an output of planning). Planning of specific agricultural enterprises, however, is largely practiced at the level of the ranch manager of Mucho Sacata. The general manager periodically evaluates the decisions resulting from plans of the ranch manager concerning cattle operations, but he does not give detailed instructions, plans, or objectives. Notice, however, that the general manager does give detailed directions to the ranch manager of Mucho Sacata concerning management of the deer herd, but he does this without considering the conse-quences for the whole of Mucho Sacata. This and other mismatches, revealed when the conceptual model is overlaid on the picture of reality, provide the basis for discussion among the participants. Will the owner, general manager, ranch manager, and other participants agree that planning of agricultural enterprises should shift to the level of general manager?

At the owner level, the conceptual model (Figure 6.6) suggests a subsystem for monitoring and controlling achievement of goals. Such activities are currently delegated to the general manager, at least for the agricultural enterprises (Figure 6.4). Will an owner who is uncertain about his own ability to manage agricultural enterprises agree to the change suggested by the mismatch between the two figures?

As was true with the question-generation technique, the discussion among par-ticipants during the initial trial of the overlay technique can reveal more details about the reality of Mucho Sacata. For example, feedback concerning planning activities currently conducted by the general manager will provide insights useful for defining and modeling the subsystem for planning agricultural enterprises.

The overlay technique is less applicable to the Chatham River case than to Mucho Sacata. The composite mind map (Figure A.1) suggests that the Chatham River situation is not perceived as an organizational–structural problem. For example, the conceptual model of a system for improving and sustaining the quality of life of participants (Figure 6.2) is difficult to configure in a way that will produce much insight if overlaid on the mind map (Figure A.1).

Alternatively, it is possible to diagram the picture of the reality of Chatham River to resemble the conceptual models developed during stage 4 of the soft systems approach. For example, in Figure 6.7, reality is portrayed in a form similar to the earlier example of a human activity system in Figure 6.2. The con-ceptual model can then be superimposed on the picture of reality. Unfortunately, picturing reality to match a conceptual model can lead to a perverse application of the overlay method. This occurs if perceptions of reality are narrowed or biased to match the conceptual model. For example, an analyst may diagram those current activities that match his or her ideas about a relevant system while overlooking issues and concerns of comparable or greater importance in the minds of participants in a problematical situation. Remember, the goal of the ap-proach is to define and model systems relevant to the concerns of participants in a situation that envisage an improved future state of affairs.

Legal precedent

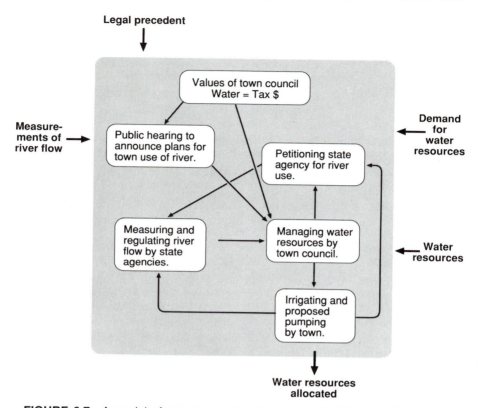

FIGURE 6.7 A model of one perception of reality in the Chatham River case.

It is possible to avoid this perversity by involving participants in any new attempt to (re)describe reality, aimed at matching the form of one or more conceptual models. As described earlier, the picture of reality should represent the diverse **W**s and concerns of all participants if proposals for change arising from the comparison stage are to be accepted and implemented.

Compare Figures 6.2 and 6.7 and note the differences. Can the regulating activities of the state agency (Figure 6.7) adequately achieve the functions of the monitoring and controlling subsystem in the conceptual model (Figure 6.2)? Does the public hearing documented in the case materials enable the participants in the reality of Chatham River to know the needs, as specified in the conceptual model?

Historical Reconstruction Technique

The historical reconstruction comparison technique is appropriate when the client and other participants are interested in gaining insight about improvements through examining how the current problematical situation originally developed. The current structure of a situation is often the result of a process of "muddling through"; in the past, small, incremental, and unplanned changes in organization or operations were made in response to other internal and environmental changes. Nobody intended or planned the overall result. Perhaps they took for granted established ways of doing things. Here it is anyway, and people now begin to perceive a need to do something about it. This technique requires that participants *reconstruct* the sequence of activities or events. The reconstruction can then be compared to a sequence of activities and events specified in the conceptual model.

A historical reconstruction could be used to compare conceptual models with reality for either Chatham River or Mucho Sacata. For example, in the case of Chatham River, if the analyst/facilitator had been called in by the mayor or some town official, the history of official efforts to achieve a goal could be compared to one or more conceptual models of systems relevant to achieving the goal over the same time frame. Presume that the mayor provided the leadership for planning and managing the water resources of Springville during the six years preceding the current description of the "mess." A historical reconstruction of the situation, compared to a conceptual model of a system for planning and managing water resources over the same period, would enable the mayor to learn from his mistakes. Yet any comparison between the conceptual model and the developing situation must be done skillfully to avoid any judgmental overtones that could alienate the client or other participants in the comparison.

A human activity system to account for historical changes that might shed light on features of the model presented in Figure 6.2 has not been modeled. Nevertheless, it would be useful to speculate about the probable results of a historical reconstruction for Chatham River. Judging from the public hearing documented in the case materials, did the mayor and other town officials adequately research legal precedents, the **W**s of downstream riparians, and the Friends of the Chatham? Did they make provisions to know the needs of water users and the constraints to meeting those needs?

What seems to have happened is that Springville attempted to pursue its time-tested policy of enhancing its water supply by simply digging more wells. Suddenly this approach no longer worked. So the town decided to take an utterly new direction without really looking where it was going, perhaps envisaging the Chatham as just another water source like a well to be tapped.

Historical reconstruction also appears applicable to the comparison stage of Mucho Sacata. The owner could benefit from examining the brief period from the time the ranches and other profit centers were inherited to the present. Although the system for developing and achieving the goals of the owner (Figure 6.6) is not modeled over time, it can provide material to illustrate the use of historical reconstruction. Were the activities modeled in relation to the owner's practices since he inherited the ranches? Were the activities modeled in relation to detailed job descriptions for positions, such as general manager and ranch manager, that would have been considered or negotiated between the owner and new management-level employees when they were hired? Would the practice of the activities modeled in relation to the owner, general manager, and ranch manager avoid the development of the current confusion about proper channels and lines of authority?

In both cases, the proposals for change arising from the comparison between historical events and a time-dependent conceptual model will enable participants not only to avoid repetition of past mistakes, but also to understand the need for fundamental processual and structural changes in their model of an improved future situation. In other words, one key to understanding where you are going is to examine where you have been.

Communicating the Results

The aim of the comparison stage is to test conceptual models developed in stage 4 *and* to prepare to communicate the proposals for change embodied in those models, to be debated by the larger body of people involved in the situation. So far, this chapter has presented a variety of comparison techniques appropriate to differing circumstances. The learning competencies, essentially those associated with convergence, that are necessary for all involved at this stage of the process have also been discussed. The final section of this chapter reemphasizes the importance of communications competencies in relation to the task of presenting proposals for change.

Where the comparison has been conducted with *full* participation of those involved, communications and the form of proposals obviously present few problems. Indeed, experienced practitioners point out that the comparison and debate phases in such circumstances are often hard to distinguish. In other cases, however, the comparison will have been conducted by the analyst/facilitator alone or with a highly selected group of participants and, occasionally, even those participants will be isolated from each other. In those cases, communications may not present great difficulties but nevertheless must be carefully prepared. This is because the people who were not involved in the discussions will have a certain amount of catching up to do to understand the proposals and their implications and to adjust their level of commitment.

Figure 6.3., Brian Wilson's (1984) tabular outline for generating questions during the comparison stage, which was introduced earlier, provides an excellent format for presenting proposals (column 5) and their justifications (column 6). Used in concert with visual mockups or projectable transparencies of the conceptual models, the format can be used with fairly large groups of participants.

In situations that continue to be characterized by wide differences in the distribution of perceived benefits and costs, or conflicting interests and **W**s, the actual packaging of proposals may be necessary. Note that factions are often distinguishable in terms of their **W**s. At this stage the concern is that factions "value the same things differently" (Susskind and Cruikshank, 1987:120) and therefore will respond differently to the elements of proposals. Accordingly, the form in which proposals are presented to those involved can be critical.

Packaging means casting proposals in "all-gain" terms rather than "win–lose" or "yes–no" terms, carefully integrating alternatives. You might review the discussion of consensus approaches at the end of Chapter 5. In the case of the floral growers, discussed in Chapters 4 and 5, potential factions exist in the division between large- and small-scale growers. The factions or interest groups in the overall Chatham River situation seem fairly well defined, but it might be useful to speculate about the possible existence of factions within particular interest groups, for example, townspeople who are pro- and anti-growth. The more complex and contentious the situation, the greater the analyst/facilitator's burden to identify all the interested parties, involve them or their representatives in the process of inquiry, and then bring final proposals to the floor that tend to foster general commitment.

Mediation experts, who specialize in facilitation in the most difficult situations, have some tools that are potentially useful at the end of the comparison stage (Susskind and Cruikshank, 1987:120–124). When participants representing different **W**s in the comparison stage complete their work, all should possess a common understanding of the proposals agreed upon. It should be recorded rather than left to individual memories. There are "good" and "bad" ways to achieve this. The least helpful is to ask each group identified with a **W** to prepare its own written version. A better approach is called the *single-text procedure*. Here, one person or a small group produces a draft, which is then circulated until closure is achieved. Participants are asked *not* to produce written comments, but to improve the draft by adding more acceptable language.

Another suggestion on packaging proposals is that they should emphasize *contingencies* by using "what-if" and "if-then" statements capable of integrating varied assumptions about the future. According to this form, single proposals may then be acceptable to pessimists as well as optimists. Again, the overall strategy is the politics of inclusion, preparing for the debate stage of the soft systems approach in ways that will keep everyone on board and moving in the same general direction toward agreed-upon goals.

REFERENCES

Buzan, Tony. *Use Both Sides of Your Brain*. New York: E.P. Dutton, 1983.

Checkland, Peter. *Systems Thinking, Systems Practice*. New York: John Wiley & Sons, 1981.

Kolb, D. A. *Experiential Learning: Experience as the Source of Learning and Development*. Englewood Cliffs, NJ: Prentice-Hall, 1984.

Kolb, D. A. *Organizational Psychology: An Experiential Approach*. Englewood Cliffs, NJ: Prentice-Hall, 1983.

Naughton, John. *Soft Systems Analysis: An Introductory Guide*. Block IV. Portsmouth: Open University Press, 1984.

Susskind, Lawrence, and Jeffrey Cruikshank. *Breaking the Impasse: Consensual Approaches to Resolving Public Disputes*. New York: Basic Books, 1987.

Wilson, Brian. *Systems: Concepts, Methodologies, and Applications*. New York: John Wiley & Sons, 1984.

CHAPTER 7

Debating Desirable and Feasible Change: Soft Systems Approach

The soft systems approach addresses both sides of the problem of how to bring about desirable and feasible change in a situation. First, it helps the people concerned to understand and analyze what is already going on and how to conceive ideas about problems and improvements. Second, it provides a framework to test and introduce proposals for change with the participants in the situation. This testing occurs in Checkland's stage 5, *comparison*, while the procedure for introducing proposals for change is embodied in stage 6, *debating desirable and feasible change*. Carrying out previous phases should have left participants believing that their situation can indeed be improved and that they are able to work cooperatively toward this goal. Now the task of the analyst/facilitator is to assist the people involved to discuss specific changes using conceptual models as guides.

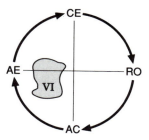

The comparison and debate stages of the soft systems approach have much in common. Both aim to test the conceptual models developed in stage 4. They also represent departures from the conventional expert advise-giving–client recommendation-receiving approach. Each favors the use of convergent learning. The aims of both stages are to assure that models and their output are anchored in the reality of a situation as seen by the people involved. Certain specific techniques can be employed in the tasks of both stages.

Nevertheless, there are many important differences between the two stages. The comparison stage was a pause, requiring that those who formulated the conceptual models look back to their original description of issues and themes of concern characterizing the problematic situation. The debate phase looks forward, taking the recommended changes of the model and discussing if they are needed and workable. The comparison task was to range the models against the situation summary and other supporting materials to ask the question "Does this make sense in relation to what we know about the situation?" The debate task is to take the recommendations for action and their costs, benefits, and the overall picture of a future state they represent and ask, "Is this what we actually want to be?" And "Can we really implement the specific changes necessary to get there?"

This chapter begins with a discussion of the meaning of *desirable* and *feasible*, particularly in relation to features of the original CATWOE definitions of human activity systems. It then focuses on the relationship of the analyst/facilitator to participants and their needs and goals. Some of the learning and action competencies necessary to fulfill this role are outlined also. The topic of intervention, originally introduced in Chapter 2, is reviewed briefly, particularly emphasizing that the purpose of facilitation is to assist the people involved to gain self-reliance. This is followed by a review of specific debate techniques, including some already introduced in Chapter 6 and others that are new. The rest of the chapter discusses *debate arenas*, the contexts in which debates about improved future states take place. These arenas are distinguished in terms of the nature and scale of the groups involved, ranging from families to whole nations, and the intensity of prospective debates, from calm discussion to acrimonious disputes. How facilitators might fit into these contexts is an important theme.

The importance of carrying out this phase of the soft systems approach cannot be overemphasized. Experience shows that it is quite possible to formulate plans carefully and faithfully follow them into total confusion, even disaster. The case of the Papua New Guinea cooperative, introduced in Chapter 6, well illustrates this point. The chance of this happening is enhanced when the potential beneficiaries (or victims) of the plan are alienated from the process of formulating it. Even their involvement, however, does not guarantee success.

Theory and *practice* are intertwined, even continuous. This is because all stages of the soft systems approach, including that of debating feasible and desirable change, engage the analyst/facilitator and the people experiencing a problematical situation in a mutual learning process. In this stage of the process, the analyst/facilitator and the people involved share perceptions and ultimately debate the reality, focus, appropriateness, usefulness, and potential effectiveness of proposals for change. The people should not be detached from the analysis, nor should the analyst/facilitator be alienated from the problematic situation.

In the language of soft systems, specifically the CATWOE, *desirability* refers to the **W**, embracing the overall aspirations, expectations, established styles of action, and perspectives of the people who will participate in the debate. The concern is with how they see the world and what they are trying to achieve in the future as individuals as well as collectively, with *or* without change. To be desirable, the changes must be desired by someone in relation to features of his or her **W**.

Feasibility refers to **E**, the environmental and internal constraints to change, which are also described in the CATWOE. *Feasible* means two specific things:

1. A change can only be implemented with the resources, staff, accessible technology, structures, capabilities, and so on *at hand*.

2. A change is feasible only when it is environmentally appropriate, in terms of avoiding unacceptable, especially irreversible, costs and in terms of involving factors over which people have control. The watchword for feasibility is "You cannot get blood out of a stone."

If participants (especially the analyst) fail to get the **W** and **E** right in their CATWOE, then, during the debate or implementation stages, gaps will emerge between the conceptual model of the improved situation and the people's vision, desires, and needs (see Chapter 8). Several generic pitfalls in identifying **W**s suggest themselves. The first is that *analysts must examine their own* **W**s *very carefully and compare them to the* **W**s *attributed to participants*. The second is that, in the course of debate, a proposed change may sound right to participants on a priori grounds, such as the sentiment that "that is how real business people operate" or an exaggerated sense of the analyst's expertise and authority. It goes without saying that it is the duty of the analyst/facilitator to guard against both pitfalls.

You may run across other usages of this terminology. Some applied scientists have in good faith gone out of their way to sensitize themselves to the perspectives of the social sciences. They may use the term *desirable* to refer to the cultural dimension, as in "culturally desirable," and reserve the term *feasible* for technical and environmental factors. Here, both concepts are presented more holistically. Yet another variant is Checkland's usage of "systemically desirable" and "culturally feasible" (1981:181). According to the usage in this book, *both* desirability and feasibility have sociocultural dimensions. The actual determination is made by the individuals who are called upon to debate the substance of proposals. For them, the issue of the fine distinctions between *cultural, technological, systemic,* or *environmental* are not very relevant. Hence the need for a carefully selected and representative group of participants so that all **W**s are expressed.

The term *debate* is used in this chapter advisedly. Some readers may have a mental image of a debate as a combative exchange of arguments around an issue that produces winners and losers, as in a debate between candidates for public office. The dictionary presents a less combative picture, "a discussion or examination of a question that presents and considers arguments on all sides." Sometimes *discussion* will be a more apt description. Nevertheless, Checkland (1981:251) states explicitly that the debate is "intrinsically concerned with conflict and change," with conflict developing particularly around the various **W**s that participants use as bases to judge current questions. He goes on to say that the debate phase of the approach is "a device for helping actors to explore aspects of the problem situation rather than a description of part of the real world" (252).

The Facilitator's Role

The role of the facilitator is to ensure that a debate occurs among the participants who may choose to pursue a change and ultimately must live with the consequences. From the standpoint of the analyst/facilitator, the purpose of the debate is to *promote* learning on the part of participants, *test* the desirability and feasibility of conceptual models against the client's reality, and *complete* his or her own learning cycle.

The stance of the analyst/facilitator in the debate process boils down to two points. First, *the soft systems* analyst *must be neutral with respect to both means and ends.* Second, the *facilitator* must identify with the people involved in the situation.

The first point came up earlier in the book in a discussion of the need to approach each situation without preconceptions (Chapter 2). It was a tall order then, and it is even more important now, in the debate phase, that the *analyst* not have a stake in particular conceptual models and proposals. As stated earlier in this chapter, examine your **W**. If you cannot follow through, then the soft systems approach is not for you.

This goes against the grain of people who have been trained in an applied science or technology. By definition, the applied sciences take for granted the application of their established knowledge. Hence, it is argued that the applicability of bodies of knowledge is an open question until and unless the people involved achieve a sufficient level of understanding and agree (or disagree). *Analyst/facilitators must be prepared at all times to recognize and adjust to situations that do not call for their scientific or technological expertise.* Analyst/facilitators should not only be sufficiently open-minded but well and broadly educated to seek alternative knowledge when it is called for.

While the *analyst* must in some sense remain neutral, the *facilitator* is, by definition, *on the people's side.* Therefore, as the process of debate unfolds, facilitators cannot be (potential) winners or losers. Only the people can. In other words, the facilitator is *not* the proponent of a conceptual model, but rather is the advocate of the people's success.

Some of you may one day find yourself called upon to use this approach as an insider and employee of the group to be assisted. This raises some thorny issues, not the least of which is that you cannot withdraw without losing your job! Insider analysts are first faced with the problem of establishing or maintaining their objectivity. If this can be done self-consciously and satisfactorily, then there is an immense advantage in terms of the activities required in the finding-out stage. Insider facilitators may, however, be faced with questions regarding their neutrality from the various interests and departments in the organization. These are difficult to resolve, especially if the analyst/facilitator is normally assigned to one of those departments. Otherwise, knowledge and familiarity with the climate and customs of an organization may advance the debate process.

An emphasis on the role of analyst/facilitator should not be interpreted as downgrading the role of the *expert*, but rather to distinguish the two roles. Many problematical situations faced by people are made worse when outside authorities, backed by a staff of experts, propose remedies. Because they lack experts of their own, the people seem to have few options. Conversely, merely having their own expert, someone who can articulate their concerns, answer questions that the people's own experience suggests, and speak the same language as the other side's experts, accords people the ability to deal with authorities on a much more equal basis.

This analyst/facilitator role and the soft systems approach generally contrast with established approaches to rural development, agricultural extension, and environmental management, in which the relationship between expert change agent and people is strongly imbalanced. Typically, the change agent approaches clients and their problematic situation with a preconceived and often conventional understanding of what has to be done and a considerable amount of power and authority to impose solutions. In terms of self-images, conventional change agents are not only de facto members of an elite, but they see themselves as such: educated, well paid, superior, condescending, and with no particular desire to reveal their own values or to enter the lives of clients in a meaningful way. In the extreme, subjects of this style of intervention have but two general options: to bow to authority and comply more or less in the dark or to reject solutions, refuse to cooperate, or actively resist.

The soft systems analyst/facilitator attempts to break new ground by applying different competencies and operational principles and by adopting a new role in relation to his or her clientele. This becomes most critical as the practitioner moves beyond the role of analyst to test the desirability and feasibility of conceptual models against the reality of the people involved.

Key Learning and Action Competencies

At this point in the approach, the analyst/facilitator and participants are oriented toward finding out through *abstract conceptualization* and *active experimentation*, with the possible consequences of implementing particular changes. Note that the facilitator particularly must keep mobilized the competencies of a well-rounded learner. Thus, some of the main competencies employed in the debate phase are

- Sensitivity to people's feelings regarding proposed changes.
- Ability to listen with an open mind.
- Skill to gather information about the present situation.
- Ability to absorb information from people as they talk.
- Capability of imagining the implications of ambiguous situations.
- Willingness to engage in discussions of change fully, openly, and without bias.
- Ability to assess the implications of change and facilitate others to do the same.
- Flexibility to rethink and revise conceptual models.

Although many action competencies suggested for the role of analyst/ facilitator are carried over from earlier stages, the emphasis is now on the *active experimentation* dimension of learning. The two aspects are merged together because people with strong accommodative learning styles need to be reminded constantly of the need for reflection, and people biased toward the diverger style must be goaded to action. Ultimately, facilitators attempt to tap the active experimentation skills of participants as well as their own. The facilitator's effectiveness will be enhanced by some or all of the following commitments:

- Long-term familiarity with and to the people involved.
- An active learning style in which the objective is not only to help others but to gain personal understanding from the experience.
- An understanding of the language, customs, social patterns, and other established cultural features of the participants.
- Willingness to help people do what they themselves determine, including helping them to learn to learn and validating their self-determined conclusions.
- Favoring facilitative over authoritative roles and behaviors.
- Ability to foster experimentation and speculation through assessment of alternatives and consequences in ways that reduce inhibitions and overcome biases.

Note the *practical meaning* of active experimentation in the debate phase. Clearly, it is not intended to *actually* implement changes in order to see what happens. Rather, the participants must be encouraged to imagine implementation, to use their own experience, training, and expertise to run a proposal inside their heads and then share observations. Here are some key phrases, hopefully the words of the participants, that exemplify the meaning of proposals and observations in the context of a debate:

"Try this: _____."
"What would happen if _____?"
"How can we avoid _____?"
"Is there any way we can bring you (or so and so) on board?"
"How do you think X will respond?"
"Why do we have to worry about Y?"
"Let's cost it out and look at it again."
"How does A compare to B?"
"Do we really want _____?"
"I object to _____."
"Do you mean to tell me that _____?"
"That is totally unacceptable."
"I have decided _____."

These phrases represent not a complete scenario, but the kind of language — declarations and questions, some confrontational — that might be used by participants in a constructive debate about desirable and feasible change as they progress toward conclusions.

As was indicated earlier in this chapter, the facilitator must bring to his or her interventions the consciousness and expectations of the people involved. To discover this is an aspect of the grounded research that initiated the soft systems

analysis in the first place. This is especially important *and* difficult when working in a cultural context other than your own.

One of the hardest lessons for an outsider bent on intervention to learn when
working in rural communities around the world is the meaning of *an established
tradition of autonomy.* This tradition finds expression both in individuals and at
the community level. People are used to assessing their own situations, ordering
their options, making plans, trying them out, and observing the results, including
the successes and failures of others as well as their own. By and large, they are
not used to taking orders or accepting direction except in the restricted circumstances of authority that they already perceive to be or accept as legitimate.
Arguably, the action of seemingly acceding to the superior power of outsiders is
a special case. It is comparable to absorbing or otherwise coping with capricious
and unpredictable natural events such as earthquakes or droughts. People respond
until the event passes, the emergency ceases, and then they move to reestablish
a semblance of normalcy. They may do the same when intervenors also go
away, as they likely will.

People inevitably set limits on intervention; have the power to resist any process; and attempt to influence its course in relation to their own problems, needs,
and capabilities. People are also capable of manipulating intervenors. Following
Paulo Freire (Shor and Freire, 1987:180ff.), you can improve the quality of your
intervention and facilitation if you stick to the following rules:

1. Understand the limits of your task. You cannot change the world or the
   Ws of the people.
2. Improve your humility vis-à-vis the people by working *with them*, "not as
   a tactic but as a necessity." The people involved already knew many things
   before they entered the relationship, and you have an opportunity to learn
   these things as well as share your own knowledge.
3. Be prepared to reknow what you think you already know.
4. Apply a critical understanding of the role of your current activity in the
   larger system.
5. Attend to language; transform your scientific idiom into a vernacular that
   the people involved can share.
6. Diminish the distance between expressions of reality conveyed in books
   and expressions of reality — a statement of experiential learning.
7. Be full of respect for what people say; there is no such thing as a stupid
   question from a participant or a right answer from a facilitator. In other
   words, avoid dogmatism and disrespect.

The emphasis on helping the people involved to gain self-sufficiency in and
through learning in no way detracts from the practitioner's role as analyst.
Rather, it frees the analyst to learn alongside the client. From the perspective of
the soft systems approach, the function of the practitioner is to facilitate change,
in particular to catalyze debate regarding the desirability and feasibility of proposed changes and improvements.

Debate Participants

As Chapter 6 indicated, the facilitator needs to be very calculated in deciding who is to be involved in comparison and debate as well as what debate techniques to employ (discussed subsequently). In stage 1, one of your first tasks was to identify the key people in the situation whom it was necessary for you to interview. The analyst/facilitator should go back and review those carefully crafted notes and check the major categories of involved people identified in the CATWOE definitions.

To review them briefly: *actors* are people having front-line responsibilities for carrying out current and future human activities specified in the models; *customers* are people who will either benefit from or be victims of proposed changes and *may be from groups outside* the original circle of participants; *owners* are people who have sufficient authority to stop the whole thing if they wish. Leaving any one of these groups or types of actors unrepresented will create real problems, which may not be apparent until implementation is attempted. To cite just a few of these problems, the facilitator and some participants will have to spend additional time trying to communicate with them about what happened in the discussions they missed. The group of participants and the overall process will lose their potential input. The people omitted will feel that they were excluded and will be less committed to the proposals accepted by participants or even hostile to them. Existing divisions in the group can be widened; for example, meetings can deteriorate into mud-slinging fests directed at unrepresented parties.

Debate Techniques

Checkland (1981:178–179) has described four techniques for carrying out the debate phase of the analysis to ensure that it will be "conscious, coherent and defensible" (179). Note that these were also introduced and extensively discussed in Chapter 6 as comparison techniques. That reflects the possibility that the comparison and debate stages may be carried out together if circumstances warrant and that, in any case, one of the objectives of the comparison stage is to prepare to communicate conceptual models to a larger body of participants. To Checkland's original four are added three other debate techniques that may be appropriate in various settings: (1) general discussion, (2) question generation, (3) direct overlay, (4) historical reconstruction, (5) alternative futures, (6) game simulation, and (7) mediation.

General Discussion Technique

Occasionally the analyst/facilitator and the people involved complete the conceptual modeling stage with fundamental questions about objectives rather than operations. This calls for a general approach. Rather than questions about current procedures, the facilitator raises strategic questions, possibly with the manager who commissioned the study, about the overall objectives of present activities. In other words, "Why do you want to do this in the first place?"

Question-Generation Technique

As a novel synthesis of human activities relevant to future improvement, the conceptual model developed by means of the earlier stages of the soft systems approach and preliminarily tested in the comparison stage provides a basis for *posing carefully ordered questions.* The analyst/facilitator uses the conceptual model to generate focused questions about the situation. The questions are written down and answered systematically by the participants. As discussed in Chapter 6, Wilson's tabular format (Figure 6.3) is a useful instrument. These questions serve to illuminate the problem for clients. As an application of the catalytic mode of facilitation, it may help them to come up with key changes themselves, rather than merely providing for the implementation of the changes suggested by the conceptual model.

Direct Overlay Technique

This approach appears to be particularly appropriate to situations in which a firm or other organization seeks to rationalize its organizational structure. With this procedure, the conceptual model is superimposed on a model of the established activity system. In this way, mismatches may be discovered, and recommendations to shift or modify activities may emerge. The reader might go back to Chapter 6 to examine the application of the direct overlay method in the Mucho Sacata Ranch case.

Historical Reconstruction Technique

This approach, which might also be called the Monday-Morning Quarterback, calls for the reconstruction of a sequence of real events in the past that led to a flawed outcome related to the problem situation. It seems to be most appropriate for primary task concerns, that is, systems concerned with processes that define an organization's reason for being. The reconstructed sequence is compared to a sequence of events that would occur if the changes of the conceptual model had been implemented.

Checkland points out that this approach must be used delicately because participants may see it as offensive recrimination regarding their past performance and competence. In other words, it may be interpreted as an expression of authoritative-confronting intervention rather than as *facilitative-catalytic.* Nevertheless, it is interesting to note that a variant of this technique is common in such sports as football, where experiential learning reigns supreme and both game plans and individual performances are scrutinized through the use of real-time film or video records.

Alternative Futures Technique

This and the other debate techniques that follow were developed independently of the soft systems approach but are included here because they seem appropriate to the aim of debating change. The alternative futures approach appears to be applicable to situations in which specific goals or program guidelines have already been established, perhaps by some central authority, but local implementation allows scope for the consideration of alternatives. It is seen by proponents to offer a significant corrective to conventional centralized planning because it tends to affirm the point made by planning critics that there is no single certain "future" of the sort that is often assumed. Such critics have claimed that most planning prepares for a future that will never occur! According to the alternative futures approach, experts develop a limited set of scenarios to describe alternative futures that otherwise reflect program guidelines and other parameters, such as the planning horizon. These scenarios are then offered to a broadly representative group in a workshop format for detailed discussion, debate, revision, and consensus on both likelihood and desirability.

As is pointed out later in this chapter, *an analyst/facilitator may find himself or herself injected into a situation late in the process but must nevertheless start using the soft systems approach.* This technique and those that follow may help to establish a context in which this is feasible. In the case of centrally designed programs, the soft systems strategy implied by the alternative futures technique is to treat the program guidelines as *environmental constraints* in the CATWOE.

The alternative futures technique was recently used in Vermont to produce a 50-year plan for the management of the Green Mountain National Forest in response to a U.S. Forest Service plan of national scope. Participants were drawn from advocacy groups, commercial timber interests, recreational users, the general public, and public-sector agencies (Institute for Alternative Futures, 1982; Brighton et al., 1987).

Game Simulation

Game simulations have been used in a number of organizational settings to stimulate debate about change and to catalyze learning among owners and actors. In some instances, a preexisting simulation has been used with international banking firms, economic planning ministries of developing nations, and other groups (e.g., Greenblatt and Duke, 1979). In other instances, an organization itself has commissioned the creation of a game simulation, based on a conceptualization of its mission or organization, and then used the game as a planning tool to debate high-level change (Gagnon and Greenblatt, 1975) or specific proposals for change (Rundle, 1986).

Mediation

Modeled after historic experience in international conflict resolution and labor–management negotiation, this is an approach to public dispute management that is regarded as particularly appropriate to aggravated situations that are on the brink of, or have already achieved, high-level conflict. According to this approach, parties to a serious dispute agree to the appointment of a facilitator, technically known as a mediator, who then guides participants through a process of gathering information and debating possible improvements to the issues characterizing the original situation.

Mediation has been popularized and has taken hold in a variety of new areas (Fisher and Ury, 1981), including the environmental field, to deal with conflicts on issues ranging from the uses of natural resources to the siting of hazardous facilities. One accounting (Talbot, 1983) cited more than 40 site-specific cases of environmental mediation. Institutions devoted to environmental mediation/public dispute management exist, and an extensive literature has emerged. This is reflected in the range of situations to which it has been applied (e.g., Amy, 1982; Bingham, 1981; Crowfoot, 1980; Harter, 1982; Lake, 1980; Schuck, 1979; Susskind and Weinstein, 1982; Susskind and Cruikshank, 1987; Talbot, 1983). From a public-agency perspective, it may be the approach of last resort before the abandonment of an initiative that has previously failed to authentically involve concerned members of the public and has become mired in the courts. This debate technique is discussed more extensively in connection with the Hudson River case, presented later in this chapter.

Debate Arenas

These general statements about the behavior and role of analyst/facilitators can be made more concrete by looking at the contexts in which intervention and debate might occur. This is necessary because Checkland's (1981) work, for the most part, focuses on single organizations, particularly what are referred to subsequently as formal organizations. Up until this point, this book also has presented the approach as if it were applicable only to very limited kinds of organizational contexts. As Kenneth Boulding (1982) has argued, the soft systems approach can be used in other social–organizational contexts: communities, family enterprises, joint ventures of various kinds, as well as political organizations and processes, including those involving conflict. Although the following discussion focuses primarily on the conduct of *debate* in these varying contexts, other stages of the approach may be affected as well. For example, in situations involving a great deal of conflict, stages 1 and 2, which require the discovery of basic facts and concerns, may entail just as much contention as the discussion of proposals for improvement in stage 6.

Accordingly, this section provides some facts and insights on the varied contexts in which you might attempt to use the soft systems approach. These contexts are referred to as *debate arenas,* although, as indicated, more than the tasks of stage 6 are involved. In addition to presenting an outline of the kinds of group contexts in which you, as an analyst/facilitator, might be called upon to work, there is also discussion of the *nature of debate* in small face-to-face communities, businesses, multiparty conflicts, the courts, bureaucracies, and government in unconventional terms. This is not a lecture in abstract sociology or political science. From the standpoint of *critical understanding*, an analyst/facilitator needs to know how our social institutions really work, not how they are supposed to work.

The identity and organization of the people involved in possible situations covered in this book are diverse. They include agribusiness firms, towns, family and corporate farmers, co-op and commodity groups, governmental bodies, and international agencies; and these may be located in the U.S. or overseas. In order to expose these possibilities and broaden the applicability of the soft systems approach, five arenas have been tentatively identified. They are distinguished by their scope, hierarchical level and inclusiveness, mode of organization, complexity, intensity of conflict, and the intervention and debate techniques particularly appropriate to them, when these can be specified.

Timing, particularly the point at which the analyst/facilitator enters an ongoing situation, is very important. Although success is never assured, the best time to initiate the soft systems approach is early in a situation, when the people involved are uneasy enough to ask for assistance but have not otherwise entered into or been caught up in an uncontrolled sequence of actions, consequences, and conflicts. As is the case with a fire, early involvement of an analyst/facilitator, as of a firefighter, is preferred to later and more costly intervention. But even conflagrations can be successfully controlled if appropriate techniques are used.

The outline of debate arenas follows from this. It reflects the view that responses to a problematic situation may follow a pattern of escalation from level to level, as from individuals to groups of various kinds, and involving greater potential for conflict, until the situation emerges as an issue for public policy-making. Escalation, however, is by no means inevitable and may not be relevant in many situations of concern.

Some of you will find yourselves in situations where a process of action has already been initiated, and debate is well advanced. It is likely that the unassisted process will have leaped over many of the prior tasks you have learned about in this book. Nevertheless, it is feasible to begin using the soft systems approach right away, or at least after applying certain preparatory techniques, since the basic objective of improving an ongoing situation, even one characterized by complexity and conflict, is the same. *Preparatory techniques* means facilitation aimed at helping people to enter into the full cycle of mutual learning involved in the approach.

In any event, the following breakdown of arenas, intervention techniques, and modes of facilitating debate should not be interpreted rigidly. The original injunction to enter into situations without preconceptions applies as well to the identification of relevent groups and contexts. Moreover, techniques described in connection with one arena are likely to prove useful in others. In all arenas, the facilitator must establish and maintain a relationship of trust and respect. The people involved must relate well to the facilitator as a person as well as to the process of facilitation. They must be brought to the point where they are hearing and articulating new things with utter credibility.

Individual and Small Group

Here, debate participants are individuals, families, small enterprises, cooperatives, and members of small face-to-face communities such as rural towns and villages, small municipalities, and some urban neighborhoods. *Face to face* means everyone knows everyone else. These are the basic units of social action, and their social organization may be loose, personal, and possibly to some degree egalitarian. A social economy of personal influence and benefit sharing may prevail, reflected in activities conventionally seen by outsiders as corrupt, such as graft, bribery, and favoritism. Leadership may be diffuse. Organization can look either incredibly complex or anarchistic to an outsider unfamiliar with traditional societies and expecting a neat hierarchy ("Where's the mayor?" "Where's the chief?" "Where's the king?") and codified law. Kinship may play a significant role. Routine decision-making is by consensus, with debate carried out informally, behind the scenes, or in traditional forums, which the intervenor may discover only by chance. Examples of such forums include religious gatherings, funerals, social events, and clubhouses. Even when the forums take a familiar and standard form, such as the zoning board of a small municipality, the real practice of decision-making may follow the face-to-face model.

Human activity systems organized along family lines, such as family farms, present special problems for the analyst/facilitator. A family farm is a farm and a family, an enterprise and a household. The actual balance of familial and business attributes varies from society to society. In many Third World societies, the social economy of family, kin, and community is more emphasized in relation to an egalitarian ideal of sharing work and benefits. North American farmers, however, have been convinced (after a prolonged national debate) that they are entrepreneurs, at least most of the time (see subsequent text). Nevertheless, family dynamics strongly influence decision-making and overall management objectives.

The Mucho Sacata Ranch case might fit in here because some of the concerns expressed by participants seem related to a somewhat chaotic transition from a strictly family operation (before Mr. Bell inherited it) to a business operation.

As was indicated earlier in this chapter, conventional attempts to intervene in small communities, such as in technical assistance projects overseas, frequently run into resistance if not outright refusal. Would-be intervenors may attribute this to mindless conservatism and primitivism. The probable truth of the matter is that *the people involved have their own ideas about their problems and needs, desirable technologies, and what they want to do with them.* Just as in our own country, ordinary and unschooled people in Third World communities can have an extraordinary capacity for innovation, learning, problem solving, and self-development. (The Papua New Guinea co-op case illustrates this point.) It is a fact that unhappily has been overlooked by many development theorists and practitioners. Indeed, *it is the conventional intervenor who is often mindlessly conservative in refusing to find out what people want and need.* The appropriation of rationality by science and technology is another of the myths we have to overcome in order to deal with real-world problems (see Habermas, 1973:265; Mingers, 1980:42).

Reflecting the complexity and dynamism of the situations that may attract a soft systems practitioner, the problematic nature of a situation may be related to the inadequacy of established ways of doing things to deal with changing circumstances. Consider, for example, the situation in a semirural municipality when the balance of power in the town council shifts from farmers to nonfarmers.

Because the analyst/facilitator may lack the necessary linguistic and cultural skills or the social standing to operate successfully in such forums, the key may be to train local facilitators along the lines previously discussed. The expectation is that the people with whom we work most intensively will, in their turn, facilitate in the traditional forums with new understandings.

A common alternative in our own society is that an already established small group, organization, or community is not oriented toward debating change. For example, the floral growers, discussed in previous chapters, belong to a cooperative, as well as to a number of commodity and trade organizations that in the past have had fairly restricted roles. The co-op has been concerned with marketing, the other commodity organizations with exchanging information. One of the analyst/facilitator's early tasks in the finding-out process is to investigate such organizations to determine if they provide an appropriate structure for assuming and carrying out new tasks. Just going through the stages of the soft systems approach will tend to reorient an existing organization or establish a new one by virtue of its collaborative and consensual mode of operation.

Formal Organization

Here, the people involved in debate are associated with business firms and bureaucratic organizations. Stereotypically, large companies and government agencies have a hierarchical organization with clear lines of authority and distinct (even specialized) roles. Some of their characteristics, particularly responsiveness to novel problems, are discussed later in this chapter and in Chapter 8. Here, certain features of formal organizations that may affect responsiveness to change and the conduct of a debate on change are identified.

According to Charles Reich (1970), bureaucracies have a life of their own in the form of imperatives and rules. *The first imperative is self-preservation.* Members try to avoid personal responsibility. Bureaucracies also tend to maintain *any* policy once it is implemented. New organizations are more flexible and open to innovations than old ones are. Too many bureaucratic organizations were created to deal with problems or exploit opportunities that no longer exist (Gardner, 1963).

While the organization as a whole may be large, the number of participants in the soft systems process must be manageable and carefully selected: officials and key employees who have a hierarchical relationship to each other; seniors and juniors, stratified, with owner/decision makers distinguishable from other actors. Features of the egalitarian face-to-face community persist, however, in that key employees of junior rank, due to years on the job or technical competence, enjoy authority and respect that is not reflected in the table of organization. This is the kind of constituency to which Checkland's methodology has been most thoroughly applied.

The first four techniques described in the previous section are well adapted to debating change within formal organizations, and the others may apply as well. For example, using the question-generation technique also reveals information on the structure of a situation. Checkland (1981:178) describes a situation in which current activities in a firm were radically different from the content of a conceptual model the analyst had developed for a particular operation. He points out that there is nothing sacred about a conceptual model and that it would be foolish for the analyst to tell a group of very experienced people what ought to be done. The manager and other participants were able to develop new options as a consequence of using the technique. The specific actions finally implemented were the decisions of the firm's top manager, not the proposals of the analyst.

Note that many of the concerns of our potential clients in food, agriculture, and natural resources involve their interactions with formal organizations, specifically governmental departments and regulatory agencies. These include both federal and state offices that operate in such areas as health, environmental protection and conservation, water, energy, recreation, and agriculture. These kinds of interactions are highlighted later in this chapter in connection with a debate arena labeled "large polities."

Diverging Interest Groups

Here debate participants are members of *groups that have adopted conflicting positions on issues* involving change long before an analyst/facilitator has been engaged. Disputants may include governmental agencies, business firms, public utilities, advocacy groups, and community organizations. Typically, the conflict centers on what is called a distributional dispute, that is, who gets the benefits and who experiences the costs associated with some kind of change. Increasingly, an initiative or decision by a governmental body is at the center of the conflict. Included are situations in which government authorities are considering or have already granted necessary permits, licenses, or variances to a private party, such as a developer, industry, or utility. Even when such a public dispute becomes an issue in electoral politics, substantial electoral victories do not necessarily translate into the power needed to resolve the issue and implement a change (Susskind and Cruikshank, 1987:3). This is because, as noted elsewhere in this book, *voting does not lead to consensus, but rather produces losers as well as winners.*

Mediation is felt by proponents to produce more agreeable results. Thus, two or more parties to a dispute invite another party to intervene when previous attempts to reconcile differences have been unsuccessful. This involves mediation, yet another form of facilitation. In the language of soft systems, mediation may be the preferred form of debate in situations involving multiple ownership, that is, two or more parties with widely divergent interests and Ws each separately have the power to make the system cease to exist.

To mediate literally means "to come between." The role of the mediator is to facilitate discussion and debate among the parties, sometimes face to face, at other times by serving as a go-between, and ultimately to persuade them to agree to a settlement. That settlement may be the mediator's conceptual model, a proposal arising from one of the parties, from a subgroup, or from the discussion of all of the parties, or a combination of these sources. Note that mediation differs from *arbitration,* in which the third party has legal authority to impose binding decisions.

Mediation has a long history in international affairs in connection with boundary and resource-use disputes and peacemaking. Mediation and arbitration are best known for their employment in the labor–management field to settle disputes that might otherwise escalate to prolonged strikes, litigation, and other costly disruptions of the interests of all parties to the dispute. There is an extensive body of experience in the field (e.g., Seide, 1970; Simkin, 1971).

Policy makers and judicial authorities have become increasingly aware of the potential of nonlitigated solutions to other kinds of social and economic conflicts. The cost, frequent failure, and time involved in litigation stimulated research and development on mediation as an alternative method of settling disputes. It has emerged in cases involving torts, breaches of contract, and matrimony, as well as labor relations. Law schools have begun to offer courses on mediation, and university-based programs have emerged. Possibly more relevant to the present concern, since the early 1970s, mediation has been used to help resolve environmental disputes. This development also might provide precedents and guidelines for using mediation in the agricultural, natural resources, and environmental fields.

In a comparative study carried out for the Ford Foundation and the Conservation Foundation, Talbot presented five case studies of disputes ranging from the conflict between environmentalists and electrical utilities over the uses of the Hudson River to a dispute between two municipalities and several state agencies over a proposed garbage dump in Wisconsin, all settled through the use of a mediator.

Talbot feels that two central questions need to be resolved before mediation can be attempted in any field. The first has to do with *feasibility*: What kinds of disputes are amenable to mediation, and what are the relevant constraints? The second has to do with *methodology*: What are the best, most appropriate techniques of dispute settlement in different kinds of situations? The soft systems approach suggests the addition of a third critical question to the agenda, that of *desirability*. Is it realistic to expect (or require) that mediation produce results that are in any sense optimal to all parties? Or must those involved be satisfied with more limited results?

There is growing interest in using mediation in the agricultural sector. For example, Section 5.D. of New Jersey's Right to Farm Act (S-854) (legislation intended to deal with local obstacles to farming in the context of suburbanization) calls for "a period of negotiation" when conflicts between farmers and nonfarmers (including local governments) arise. To date this requirement that disputes be mediated has not been implemented. No machinery has been set up at the local level, no guidelines promulgated, no panel of trained mediators established, nor have research and development been undertaken to provide the prerequisite and sound bases for these necessary features of an agricultural dispute settlement program in the state of New Jersey. This may be because few agricultural disputes have escalated to high levels of conflict. A recent survey (Lisansky, 1986a) indicates that, while disputes are common, the majority are settled informally, that is, without involving formal mediation machinery. Either the parties settle their differences unassisted or prominent local figures, such as county extension agents, town councilors, or members of county agricultural development boards, serve as unofficial facilitators.

The example of agriculture is used here to discuss the three questions regarding mediation that were outlined earlier.

1. *Feasibility:* There is some literature on *types* of agricultural disputes, including Brooke's (1987) and Lisansky's (1986a and b) work in New Jersey and Moyers and associates' (1968) work in Wisconsin. Typical issues include the use of certain agricultural techniques that nonfarmer residents consider to be nuisances, such as noxious and threatening chemical applications, noisy irrigation pumps and bird-frightening devices, and smelly swine and poultry operations. Lisansky (1968b) has shown that most of these disputes are settled informally. More serious, with potential for escalation to higher levels of conflict, are disputes that arise when sudden major changes in land and resource use are waiting to happen: the decision of farmers to develop a portion of their holdings or to sell to a developer, proposals to divert water from agricultural to other uses (as in the Chatham River case study), or the selection by a governmental agency of a farm tract for development of a sanitary landfill.

Expanding some of the suggestions of Talbot (1983:31), mediation can be successfully applied when

a. The parties to the dispute and the interests they represent are limited and the issues circumscribed.

b. All parties to the dispute have a sense of urgency and a greater interest in getting on with their normal business than in damaging each other or achieving some other objective, such as compensation.

c. There is a balance of power among protagonists.

d. The parties are willing and able to implement whatever they agree to. (This is a statement of desirability and feasibility. In this context, feasibility particularly means that the parties' scope of action is not constrained vis-à-vis possible outcomes. For example, do the parties actually have viable options in the face of possible settlements or will their very survival be threatened?)

e. The status, prestige, and neutrality of the mediator are unquestioned by all parties.

2. *Techniques:* There is an extensive literature, including textbooks, on mediation in the labor–management field, which focuses on the institutionalization of mediators and panels of mediators, the use of single mediators or teams, their mode of selection, their training and procedures, the structure of settlements, and modes of establishing the legitimacy of settlements. This information has not been rigorously reviewed, nor have models been screened for their appropriateness in the contexts associated with agricultural disputes except in the field of

agricultural labor. There they have been found to be inadequate due to problems ranging from the operational constraints faced by farmers to issues of intercultural communication to the profound poverty of members of the workforce. Some of the specific techniques employed by mediators are described subsequently in connection with a substantive example, the Hudson River settlement.

The recent best seller, *Getting to Yes* (Fisher and Ury, 1981), is a readily accessible guide to specific techniques and broad approaches for achieving mutually acceptable agreements in all kinds of disputes. A product of the Harvard Negotiation Project, it is notable and should be particularly useful to you because it applies learning concepts similar to those we have adopted in this book. For example, in their chapter entitled "Invent Options for Mutual Gain," they present an approach that, in outline, closely resembles the learning cycle and the soft systems approach (Fisher and Ury, 1981:70). They advocate many of the techniques used in this book in defining conceptual models, including identifying interests (or **W**s), defining purposes in terms of desired future states, separating the invention of options from their debate, and brainstorming. They emphasize that mediation techniques are not mysterious or only for the gifted, but can be learned.

Another product of the Harvard project is Susskind and Cruikshank's *Breaking the Impasse* (1987). It focuses on public disputes, including those pitting government agencies against private parties. It also is oriented toward techniques and uses several cases of environmental disputes. Elsewhere, Susskind and Weinstein (1982) have outlined a nine-step procedure that places mediation well within the reach of the soft systems framework.

a. All parties with a stake in the outcome are identified.
b. Involved groups are appropriately represented.
c. Differences among values and assumptions are discussed and narrowed to fit a workable agenda.
d. A sufficient number of alternatives is considered.
e. The scope of the dispute is agreed upon.
f. Relative values of costs and benefits are agreed upon.
g. The worth of compensating actions is established.
h. Bargains are implemented.
i. The consulting parties are held to their commitments.

3. *Desirability:* Proponents tend to idealize and obscure mediation's functions and objectives and present its track record uncritically. Talbot (1983) is an example. He says, "Conflict, in and of itself, is not the problem." An alternative view is that *the most significant task of mediation is to remove or reduce conflict* so that the parties to a dispute can (1) participate in a collaborative process to settle differences and (2) ultimately get on with their normal routine, only partially constrained by the terms of the settlement. Thus, a test of effectiveness is the abatement of conflict itself, for example, *avoidance of litigation* and the duration of the state of peace. Note that an achieved consensus is the reciprocal of conflict. Mediation does not necessarily result in outcomes that are desirable in all respects to any party. Indeed, settlements can easily entail serious new problems and incremental effects not anticipated by anyone. These probably must be dealt

with using approaches *other* than mediation. Nevertheless, threatening litigation may gain entrance to a decision-making process for a party previously excluded in connection with a mediated settlement. And the early implementation of a broad-based and open planning process that goes out of its way to include all interested parties will do much to avoid the kind of polarization that leads to mediation or litigation. The soft systems approach may also be used to realize the terms of a mediated settlement or to help deal with its consequences.

The growing popularity of mediation has led to the establishment of a nation-wide network of public dispute centers that make negotiation skills and training available at no or low cost (NIDR, 1984). Some are free-standing, others university-based or sponsored by state or local government. They go by many names in various localities. Many states have offices of mediation or public advocates. Private, for-profit centers also exist in some localities to provide alternative and fast ajudication and mediation services for a substantial fee. Analyst/facilitators who feel that they are over their heads in the depth and complexity of a dispute can probably turn to one of these facilities for backup assistance, or they can even build in their participation in advance.

To illustrate these points and also to show how one mediator performed, this chapter turns to a war of conflicting interest groups that broke out in the greater New York area in the early 1960s and was not actually settled for 17 years. Although the specific details are different, in outline it resembles conflicts over water use that have occurred all over the U.S. between such interest groups as farmers, recreation advocates, municipal officials, resource managers, residents, Native Americans, industries, environmental organizations, and governmental regulators. In other words, change the names of the actors and the play is much the same!

In addition, the specific intervention techniques used by mediators are the same ones used by facilitators in other arenas and can be learned by most people. Thus, the image of mediators as lion tamers, stepping into already aggravated debates in which adversaries are sometimes literally ready to kill each other, should not deter you from looking into the field. The mediator in the Hudson River case, Mr. Russell Train, found himself in this kind of extreme situation and ultimately was successful in getting the parties to make a deal.

The Hudson River Settlement

In 1963, the New York State Department of Environmental Conservation granted an operating license to Consolidated Edison Corporation ("Con Ed") for the construction of a pumped storage hydroelectric facility at Storm King Mountain, approximately 40 miles north of New York City. This set in motion a 17-year battle among four regulatory agencies, three public advocacy groups, and five utility corporations over the use of the waters of the Hudson River estuary. This overview, which focuses on the role and activities of the mediator who ultimately helped the adversaries reach a settlement, is based on a case study in Allan Talbot's (1983) survey of environmental mediation. It is presented here because it provides warnings about, and a possible template for, the conduct of debate in other messy natural resource disputes involving competing uses of important regional water sources, including the Chatham River case, presented elsewhere in this book.

In addition to the Storm King hydro proposal, the utilities had three other major new installations that depended on Hudson River water. There were two separate power complexes, each containing two 600-megawatt oil-fired plants, and the Indian Point nuclear installation of three reactors; all required vast amounts of river water for cooling. There were also two older thermal stations with similar needs and impacts.

Both New York State and federal regulatory agencies were embroiled as well. In addition to the state environmental authorities mentioned earlier, the Federal Power Commission was involved in the licensing of the Storm King hydro project, the Nuclear Regulatory Commission was involved with the Indian Point nuclear installation, and the New York regional office of the federal Environmental Protection Agency was involved in the issuance of certain water discharge permits. The federal involvement provided the wedge for kicking the whole matter into the federal court system because, at least at that time, federal regulations mandating environmental reviews were more stringent than those of the state of New York. This illustrates the important point that mediation is often a supplement to, rather than a substitute for, court action. Nevertheless, at various times the State Department of Environmental Conservation was at war with the federal EPA. At one point there was even internal dissension between the wildlife officials and the parks and recreation officials of the same state department.

The advocacy groups included the Scenic Hudson Preservation Conference, the Natural Resources Defense Council, and the Hudson River Fishermen's Association. They had several concerns. One was essentially aesthetic: the scenic impacts of the hydro installation, cooling towers, and other facilities along the Hudson shore. The other major issue involved the cumulative impacts of all of these facilities, especially thermal effects, on aquatic and marine life in the Hudson estuary. It was already seriously threatened by other environmental insults, including long-term dumping of a myriad of chemicals and other wastes. At certain points in this long dispute, advocacy groups were even pitted against each other; e.g., fishermen concerned about impacts of thermal pollution on fish regarded a regulatory decision requiring cooling towers as a victory, whereas other interests concerned with scenic impacts regarded cooling towers to be a blight! It is important to note that these groups were not equal partners. Scenic Hudson, in particular, represented Hudson shore property owners of great wealth and social power. And the Natural Resources Defense Council is a public-interest law firm, not a citizens' group, "whose primary function has been to litigate against the government in environmental cases" (Lieberman, 1983:94).

The federal court system was to settle jurisdictional disputes between agencies and between state and federal levels, interpret legal and regulatory requirements, and ultimately sanction any mediated settlement. Along the way, it made "new law" regarding the standing of citizens' groups in environmental disputes and paved the way for environmental reviews in federal regulatory actions.

At various times, the case traversed an escalatory continuum from community concern to corporate boardroom to state and federal regulators to the courts and then down again. Deescalation involved first the mediation effort and then, as part of the settlement, the federal EPA allowing the state regulatory agencies to issue permits based on the settlement.

Public hearings held by a regulatory agency on cooling for thermal electric installations presented what the lawyer for the Natural Resources Defense Council called "the drama of advocacy science," a battle of experts for both sides. For advocacy groups, the objective was to wrestle a proposal to a standstill. For advocates bent on stopping a proposed change, delay is the strategy of choice. Here the ally of delay was the objective fact that marine biologists knew pitifully little about the life histories of significant fish species such as the striped bass, and so one expert's opinion was as good as another's! Needless to say, the utilities experts argued for minimal impact and attempted to discredit the other side's experts. The utilities' overall objective was to question the group's standing to take part in the case in order to negate their very expensive demand that cooling towers be built as a substitute for using river water directly.

The possibility of mediating this dispute first came up in 1979, approximately 16 years after the controversy had erupted and both sides had incurred legal expenses in the tens of millions of dollars. Then lawyers representing two of the advocacy groups met with lawyers of Con Ed (the lead utility) to discuss the possibility of mediation in recognition of the effective stalemate that had been reached. The idea was to include all pending Hudson River energy disputes, the issue of coolant for the three electricity-generating installations, and Storm King in the negotiations. The management of Con Ed agreed that it was worth trying, and a consensus was reached that a third-party mediator would be required because of the large number of parties and the complexity of the dispute. Russell Train, then president of the World Wildlife Fund and a former administrator of the federal Environmental Protection Agency, was accepted as mediator by all the parties.

What follows is a chronological description of Mr. Train's actions, providing sufficient context to suggest why he did what he did. His actions illustrate some general points about the performance of mediators, many of which highlight propositions outlined earlier about intervention and debate facilitation. *A mediator is a facilitator of debate specializing in situations involving high levels of conflict.* Sometimes extraordinary means — techniques that might not be endorsed in less aggravated situations — may be justifiable. This is because the mediator's or facilitator's job is to help the people involved to define the features of the dispute in terms that are manageable, to move the impossible to the possible. As in a war, the first task is to stop the parties to a dispute from shooting at each other. Once that task is fulfilled, the soft systems approach or one resembling it (e.g., Fisher and Ury, 1981) can be used to help the people involved to debate the basic facts as well as proposals for an improved future — possible settlement terms — to which all can agree. Readers who desire a more detailed account of the substance of the dispute are referred to Talbot's case study (1983) and Tucker's (1977) critique of the whole affair.

The more complicated the dispute, the more prestigious and well-connected the mediator must be. Utter neutrality may even be of lesser importance as long as all parties agree to the particular mediator. Russell Train's appointment in this case seems to exemplify the principle; he had substantial social, political, and perhaps even economic connections with many of the principal parties. According to Talbot (1983:17), Train gained credibility by being so busy that scheduling sessions was difficult. Thus, his willingness to continue when the going got tough encouraged the others to stay with the process.

Mediators begin at the beginning by going through a phase of finding out. Train describes how, in April 1979, he "oozed into the role of mediator" and did "some quiet checking with the parties involved" (Talbot, 1983:14). More than anything, he needed to know right away if everyone was serious about a settlement and comfortable with him as mediator. If the answer to either of these questions was no, then he would be bound to withdraw. Natural Resources Defense Council and Con Ed lined up quickly. The federal EPA, which Train had headed years before when permits for the three installations had been granted, proved more difficult. It would appear that Train accepted their conditional agreement only because he felt that he could in some sense "handle them" later on (this is an interpretation). In the long haul that followed, these three groups were to be the principal players.

Formal negotiations cannot be initiated without ground rules regarding the behavior of the parties and the exchange and discussion of proposals. Formal meetings began in New York City in August 1979, attended by 28 people representing 11 groups. The very first session was devoted to making rules. They included a gag rule, although, interestingly, the press turned out to be indifferent to the whole affair. It was agreed that any offers presented in mediation would be off the record with respect to the ongoing regulatory and judicial proceedings. It was also decided that all major proposals would be submitted to Train in writing and that he would distribute them to all the other participants. This points up another important duty of a mediator in an aggravated situation: *to control all significant communications among the parties and between them and the outside.*

In the early to middle phases of the process, mediators must be alert to breakthroughs of principle and be prepared to ignore substance. In a later meeting, "nonhardware" proposals emerged, whereas earlier proposals, and also regulatory jousting, had focused on hardware such as cooling towers. A relevant example of a nonhardware option is to shut down some stations during fish spawning season. With the *principle* established, some progress was made with proposals involving a mix of hardware and software solutions. This also illustrates another important feature of how mediators do their jobs. *The mediator is always leaping ahead to anticipate the* next *turning point.*

Interminable and indeterminate debate over details, especially technical issues, may be diffused by appointing subcommittees and temporarily removing the issue from the main body of participants. Early optimism about the mixed proposal just referred to was dashed when serious doubts arose over the utilities' estimates of saved fish. Train averted disaster by appointing a separate small technical committee of biologists and other experts, drawn from the staffs of the parties to the dispute. Its brief was to develop an operational test of fish and egg mortality in relation to various proposed water-use technologies and scheduling scenarios. Unhappily, the experts could not agree on anything either, and negotiations were once again stalled. This stalemate actually involved the divergent positions of the utilities and the EPA. In subsequent meetings, the representatives of these parties actually "began trading barbs and insults" (Talbot, 1983:30).

Separate the parties and/or even separate the experts from their clients. To break this stalemate, in a June 1980 meeting, Train requested that the lawyers for Scenic Hudson and NRDC get together with the members of the previously formed technical committee to "work on some new ideas." According to Train, he was not "at all sure what those ideas might be." His purpose "was just to keep the process moving" (Talbot, 1983:20). Through the summer, this group, which included marine biologists working for the different parties, was the core mediation group.

One of the experts on this committee had a hard systems computer model of power-plant operation, fish abundance, river flow, and water temperature. The committee used it to test options and determined that, in order to reduce fish kills to required levels, not only would the thermal plants, which had previously been "sacrificed" in mediation sessions, have to be shut down during certain periods, but the nuclear plants would have to shut down as well. Apparently, when faced with a consensus of the technical experts serving on the technical committee, some of whom were their own, the utilities made an offer involving the nuclear plants.

The mobility of mediators, their ability to cross boundaries and jump lines of authority, is not only appropriate, but critical to facilitation as long as they confine themselves to things the disputing parties cannot do for themselves. The EPA representatives, however, were not satisfied with the utilities' response and, because everyone else seemed to be moving toward agreement, they became increasingly isolated and inflexible. This time, Train met with a high-level Washington-based EPA administrator and persuaded him to send the New York regional staff back to the bargaining table.

Once the mediator smells a settlement, it's time to lock the parties in a room and rush them to a conclusion. The period from September to December 1980 saw the most intensive bargaining period ever. The general outline of an agreement had finally emerged, and now specific details were being haggled over and bargained for. So Train leaned on the parties to accept a wrap-up date, when all arguing would cease and a document of some sort would be signed.

The final agreement included the termination of Storm King, operating modifications of cooling water intakes and thermal outflows to reduce fish kills, development of a fish hatchery, shift of regulatory responsibility from federal to state environmental authorities, establishment of a research fund, and reimbursement of the legal fees of the citizens' groups. As the outline of the mediation process suggested earlier, no party to the settlement achieved an optimal result from its perspective, but there were winners and losers.

The biggest winners were Scenic Hudson and NRDC, who killed the Storm King project, got the area turned over to a parks commission, and also beat the

EPA out of requiring cooling towers. According to at least one critic (Tucker, 1977), this was utterly pernicious, since these groups represented the leisure class of wealthy weekend homeowners instead of full-time residents of affected areas. Con Ed and the other utilities also received immediate gains, although they had been seriously damaged by the prolonged dispute. They were finally able to move on major projects, got the EPA off their backs, and avoided the super-expensive cooling towers. All of this was at the expense of vaguely worded concessions on technology and operations and the Storm King hydro project, which had been rendered obsolete in the interim. The losers were the EPA, the Hudson River Fishermen's Association, and possibly the fish! More recent disputes involving the development of the Hudson's resources continue to hinge on experts' conflicting opinions regarding impacts on fish, particularly the striped bass.

The final agreement was signed on schedule amid the glare of TV lights. Even then, the mood was grim, and Train ruled the signing with an iron hand. He insisted that no one but he would speak until all 11 participants had signed. He feels to this day that a single angry word from anyone beforehand might have destroyed the settlement. Moreover, implementation of the terms of the settlement proceeded more slowly than anticipated because Train dropped out right after the signing. This reinforces the concern *that facilitators not leave their clients at least until an implementation plan, including standards of performance and evaluation, has been created* (see Chapter 8).

In summary, when the people involved in a problematic situation have widely diverging **W**s, interests, and visions of an improved future, and over time their differences have escalated to actual or potential conflict, special techniques to help them improve the situation may be called for. Analyst/facilitators or mediators must help the people involved to back away from the conflict mode in order to establish a semblance of normal discourse and cooperation. The soft systems approach is then a useful tool for gathering basic facts on the situation, defining an improved future, and so on, as long as it is understood that every stage of the approach may be characterized by debate in the generic sense of the term.

Courts

Adjudication, in the form of court trials and lawsuits, is debate and intervention "invited" by only one party to a dispute. That party may be the state, bringing an action on behalf of the people; or it may be a private individual, group, or corporation. In the language of soft systems, ideally two competing definitions of conceptual models are subject to comparison and debate. Alternatively, one conceptual model is presented by one of the competing parties, and it is the task of the other to undermine it to prevent it from meeting the required test of proof. Of course, in practice the definitions are often incomplete and only implicit, with the **W**s most prominently represented in the discourse.

Contrasting with consulting and mediation procedures, litigation is a harsh and unforgiving way of attempting to improve situations (Sax, 1970:107). Determination is characteristic of adversaries, and plaintiffs are frequently bent on vengeance and redress, rather than merely on seeking an improvement in their situations. Moreover, major new areas of litigation, such as health and medicine, consumer safety, and environmental protection, have mushroomed in recent years (Lieberman, 1983). Awards may be large and rulings damaging to normal operations. Accordingly, defendants must at least equal the vigor of the party bringing the action.

Adjudication is also ritual combat, a substitute for real violence, veiled in forms, rules, and authority. One function of judges is to apply the law to novel circumstances in order to maintain an appearance of continuity with the past, represented in the body of common law, relevant statutes, and court precedents. Judges welcome deescalation in the form of out-of-court or court-approved settlements, sanctioned mediation, and the appointment of special masters, who are essentially arbitrators chartered to supervise the implementation of a decision or settlement.

Outside of criminal law, courts have the following concerns that are of particular relevance to the agricultural and natural resources area: (1) damage suits, (2) suits directed at existing problems, and (3) preemptive suits to head off problems before they actually manifest themselves. Preemptive suits, especially, are directed against administrative agencies that are unresponsive to concerns of the public while responding effectively only to their own political necessities (Sax, 1970). Unresponsiveness is examined in more detail in the next section of this chapter. As the Hudson River settlement showed, courts also have a key role in regulating debates in all the other arenas when circumstances are aggravated enough to warrant their intervention. The regulatory arena is especially subject to judicial review, although courts have been reluctant to tamper with the legislative mandates underlying administrative actions (Lieberman, 1983:97).

Although no real examples can be cited, it seems reasonable for a lawyer and client to go through stages 1 to 4 of the soft systems approach to plan their case, whether they are plaintiffs or defendants. A systemic approach might involve explicit definition and modeling of the desired future state of affairs, including attempting to account for the Ws of other parties. It also is feasible for groups designing other kinds of improvements to incorporate an adjudication subsystem to reduce a constraint. For example, in the case of the floral growers' (Chapter 5) desire to provide additional acreage for the industry, litigation against county or state planning boards might be necessary.

Large Polities

In the best of all possible worlds, government intervention occurs only (1) when responses at lower levels prove inadequate to the task due to the magnitude and scope of the problematic situation and (2) if the people experiencing a problem request it. Of course, we do not live in an ideal or even a parsimonious world — some governments and processes are more democratic than others, politicians and bureaucrats in all systems seem to have totalitarian impulses, some parties to a situation have more influence or power than others, some mechanisms of intervention were designed to deal with problems that no longer exist, and so on. One authority has described public planning as involving, not systematic problem solving and rational weighing of fundamental issues, but a process of "muddling through" or the "method of successive limited comparisons" (Lindblom, 1959). Historically, the U.S. political system reflects a flux between centralization and decentralization, as well as great variability among the states with regard to home rule versus centralized state government authority. This has been a topic of debate throughout American history.

It is hard to imagine a true national debate with the whole country as the arena. Certainly the political arena of presidential elections and the rivalry of major political parties are characterized by conflict and competition. The candidates' carefully selected issues are either intended to reflect or capture the interest of voters. Nevertheless, issues do gain currency in the public at large or a significant portion of it. Today, with the electronic media, this can happen quite rapidly, probably because people from one end of the country to the other can share experience while actually carrying on dialogue and debate in much smaller and more localized groups. The period of the past thirty years has seen many social and technological innovations adopted, accepted, or tolerated in this way, including civil rights, artificial fibers (plastic clothing), birth control, nonmarital relationships, anti-Vietnam War sentiment, and energy conservation.

Before radio and television, this process was probably slower and more dependent on itinerant facilitators and intervenors, such as labor organizers, government agents, traveling salesmen, and preachers of diverse religious, moral, and economic persuasions.

A good example is the development of national agricultural policy in the U.S. Current policy (and problems too) is rooted in a protracted debate — followed by an uneasy consensus — regarding the fundamental nature of farming. In essence, the issue has been whether farming is a business or a family production system or life-style. In the long run, society came down on the side of business. As a result, both the **W** of surviving farmers and policy have been framed in these terms ever since.

From the farmers' standpoint, the two competing models of farming entail radically different strategies, management styles, and overall objectives. The strategy of business in the modern world is *growth*; the alternative of no growth, or stagnation, is not stability, but failure (Jacobs, 1969:118–20). Growth is accomplished by debt capitalization and specialization, along with enlarging the scale of operations. This contrasts strikingly with the strategy of domestic production units, *persistence*; in other words, the name of the game is to stay in the game. Persistence involves conservative assessments of change and diversified, as well as down-scale, operations.

According to Durrenberger (1985/86), the farmers' plight is that, starting in the 1920s, they defined themselves as businessmen but continued to behave like household production units. And, starting as far back as the post-Civil War period, national institutions and policies, such as the land-grant college system, favored the business orientation and large-scale growers, while pressuring small- to medium-sized operations into failure. There has been real debate, as exemplified by the competition of the Farmers Union and the Farm Bureau in the 1930s. The issue has been resolved, however, not through debate, but by the progressive elimination of one of the competing sides by means of low commodity prices and growth-oriented advice on technology and management. And because in the real world no issue is ever so clear-cut, many have been uneasy with this ever since. They believe the policy has hurt small, poorly capitalized family farmers the most, in part by redefining the operational meaning of *small, commercial, part-time,* and so on.

At the level of national governments and their major subdivisions, such as states and provinces, the concern is with at least two kinds of arenas. The *policy-making arena* is associated with parliamentary or law-making bodies such as state legislatures and the U.S. Congress. The *administrative, regulatory, service, and enforcement* arena includes executive agencies and commissions that at least nominally report to national chief executives, as well as to state governors and members of their cabinets. Some notice must also be taken of the political arena of elections, politics, and influence that, even though characterized by debates about change, is not otherwise known for consciousness, coherence, or defensibility. This complexity should not put large polities out of reach of the soft systems approach as long as certain limits are observed. The client group has to be of workable size. The transformation statement might be limited to implementing within an executive bureaucracy a policy already determined or to helping legislative staff contribute to the enactment of a policy. All the debating techniques described earlier in this chapter are potentially appropriate here. Nevertheless, much more goes into the debate at this high level, and it needs to be examined in a holistic way.

Note that the proposed use of the soft systems approach in government should not be confused with the already established uses of *other* systems approaches, such as operations research, systems engineering, management science, and cost-benefit analysis. These are of the hard systems type discussed elsewhere in this book. In the present context, the problem with the use of these techniques is that *in practice* they involve *designed systems* rather than *mutual learning.*

There are always people somewhere in any modern nation or state who are experiencing difficulties that *might* be addressed by their governments but are not. Some students of public policy have asserted that it is just as interesting and productive to study "unpolitics"—inaction, the absence of debate, or nondecision-making on issues that objectively exist—as to study action in the public policy arena (Crenson 1971). Indeed, inaction is as characteristic an act of government as action. This is because officials have a vested interest in keeping their own bureaucratic situation on an even keel, of changing how they do business as little as possible. Another source of inaction is that people have an interest in husbanding their autonomy by keeping government off their backs *most of the time.* This is because government-sponsored interventions tend to be uninvited, authoritative, inappropriate, and hard to reverse. They may justifiably be viewed as a last resort.

Nevertheless, situations arise in which people not only need to mobilize resources beyond their own but, having recognized the need, demand intervention by government. What often ensues, however, is not mutual learning and constructive debate between the people involved and officials, but rather a pattern of nondebate, inaction, and efforts to convince people that their concerns are imaginary or that the state has a monopoly on concern in the area in question. This seems especially generic to problems and situations that appear to be novel and for which there is no explicit policy or official pattern of response.

Thus with issues that are still below the surface, in what is sometimes described as a policy vacuum, government officials and executives of companies that bear some responsibility for a situation adhere to a scenario that repeats itself over and over again:

1. Authorities ignore the situation as long as possible.
2. If pressed, they deny responsibility and attempt to shift blame to the victims.
3. In response to mounting pressure, they may rush to be seen doing something (anything) on the basis of preliminary information on the situation.
4. Their preferred style, however, is to defer action, based on the need to do more studies according to their own priorities (and to ignore the priorities of the people concerned).
5. They then minimize the extent of the problem relative to the resources they have at their disposal, while also controlling all information about it, including the existence of other victims, a reflexive coverup that slows the spread of awareness and political consciousness.
6. Outside intervenors, such as experts employed by concerned citizens or righteous professionals, who attempt to refute the emerging official description of the problem do so at the risk of physical threats or being labeled dissidents and professionally incompetent when, in fact, they are whistle blowers.

The behavior of victims of a serious, spreading problem situation also seems to follow a typical scenario, which interacts with industrial and official behavior in unfortunate ways:

1. Victims respond minimally at the outset and try to deal with novel or unprecedented problems as if they were familiar ones.
2. They view their situations as isolated and accept official acounts (if any), even guilt and blame.
3. Their capacity to act effectively may be inhibited by physical and emotional trauma as well as by limited resources.
4. In any event, action usually depends on awareness that they, as victims, are not alone.
5. With regard to the foregoing, expertise is no guarantee against victimization, although sophisticated victims may have an enhanced ability to learn their way through a situation.
6. Victims achieve their greatest success when they band together to form an advocacy group or make use of existing local institutions to take action.
7. This may involve escalation, including entering the judicial arena.

Advocacy groups have a mixed character. Important features may include (1) how long they have persisted after a catalyzing situation or the emergence of the issue that forms their reason for being, (2) their evolving organizational structure, and (3) whether or not they have moved beyond advocacy to take on other roles. Advocacy groups have an important role to play in forcing debates about change, broadening the participatory base of debates, and mobilizing people and resources. New local groups form around a specific situation, such as a hazardous waste dump, a nuclear power plant, or a resource development proposal. They raise funds, organize demonstrations, lobby local political leaders, hire experts such as lawyers, and engage in litigation. Beginning as totally independent cells, they may also form alliances with similar groups for specific actions and join together in regional or even national clearinghouse organizations that serve as information exchanges. By the same token, they may also fade away once closure of some kind has been achieved regarding the situation that led to their original formation.

More mature organizations, such as the Sierra Club, other national environmental groups, and the National Farm Bureau Federation, have the characteristics of formal organizations, although local chapters may have a great deal of autonomy. They are more likely to have permanent, salaried staff for managerial functions and to provide expertise in science and technology, lobbying, law, public relations, and so on. They are able to sustain efforts across a broad front on issues affecting their constituencies rather than being a single-issue "adhocracy." They also tend to establish close working relationships with legislative and executive agencies and relevant science organizations and to operate backstage as well as in the sunshine. It is not unusual for ranking members of advocacy groups to make a career of moving among these institutions, e.g., from club to agency to university, and so on.

National organizations or their local cells become derailed from time to time when they take on ancillary functions that develop into serious structural constraints on their advocacy roles. The latter seems to be what happened to the Michigan Farm Bureau in the case described subsequently. From being the farmers' advocate, it branched out into farm supplies. When those operations went bad, it dealt with its constituency like any company trying to minimize its liability. This left some Michigan farmers with no effective advocate.

In a study of the circumstances surrounding the passage of laws aimed at controlling air pollution, a problem of great scope and magnitude, Crenson claims that "The only way to prevent air pollution legislation from being enacted is to prevent *the issue* from coming up in the first place" (1971:87; emphasis ours). A technique commonly employed to keep the cat in the bag for as long as possible (in hopes that the issue will go away by itself) is to avoid sharing knowledge, especially knowledge of the existence of other concerned people or victims, and to maintain that the victims' situations are unique and that somehow they brought them on themselves. The cat can be let out of the bag, and the action or policy vacuum broken regarding such an issue, by

1. Continued expansion of the problem.
2. Increased awareness of the (possible) expansion of the problem.
3. Increased awareness of localized action *somewhere* to deal with the problem.

4. Increased knowledge of how to deal with the particular kind of problem based on experience and models developed in a pioneering locality.
5. The ultimate decision by policy makers to implement a general solution affecting all localities where local solutions have not been implemented or have not been effective, as well as localities that haven't even experienced the problem in the first place.

The political activity surrounding this emergence also seems to follow a pattern with these features:

1. The people who first promote the debate are outsiders to relevant bureaucracies or political bodies, for example, leaders of local ad hoc advocacy groups.
2. Conventional political party organizations are detached from the issue.
3. Support for action on the issue does not become consolidated in formal organizations but rather is carried forward by individuals and a myriad of small groups and grass-roots organizations, which form alliances and coalitions.
4. The most organized support comes from individual professionals and career bureaucrats in relevant agencies and from the news media.
5. Relevant business and industry, after initial and even prolonged opposition, sooner or later become active participants in policy making to shape it to their interests.

Imagine serving as analyst/facilitator to a party caught up in a situation that fits the trajectory described by this model. For example, you might be a consultant or a practitioner of some sort—a physician, vet, extension agent, agency employee, lawyer, or citizen—serving or feeling a responsibility to one or more people who are attempting to cope with a problematic situation: residents with a toxic waste dump in their neighborhood, farmers confronting a natural or technological hazard. If you understand the roots of governmental behavior, you may be able to help clients break through more readily. It is particularly important for the analyst/facilitator to understand the overall state of development of the issue. For example, when confronted with a case of hazardous waste, environmental and health agencies around the country behaved differently (although not as well as one might hope) after the Love Canal episode than they had before.

Because they seem to play an important role in promoting debate by disseminating information about common problems, journalists have a critical role to play. Their nose for news can be sharpened through understanding of the local situation and the performance of government agencies. Employees of government agencies can themselves benefit from exposure to the big picture if it makes them more critical of their own and colleagues' biases. Such insiders do, however, run the risk of being seen as disloyal and labeled as whistle blowers.

The following case, involving widespread contamination of the human food chain with an exotic chemical, illustrates many of these points. It shows an advocacy organization deflected from its role by conflict of interests. The fact that a key victim, possessing excellent scientific credentials, was hardly better off than less sophisticated victims tends to give a lie to the claim frequently made by government scientists that people are to blame for their own ignorance. It is a revealing and ironic situation that highlights *the frequent dominance of* **W**s *over science*. Professionals-in-training take heed!

The Contamination of Michigan's Food Chain

In the spring of 1973, a truck belonging to St. Louis Freight Lines, a common carrier, loaded a ton of polybrominated biphenyl (PBB) from the Michigan Chemical Corporation (MCC) and delivered it to a feed mill of Farm Bureau Services (FBS), a subsidiary of Michigan Farm Bureau (MFB), the farmers' own advocacy and support organization. FBS, with 87 mills and more than 200 dealers, was the largest agricultural feed supplier in the state.

MCC is a subsidiary of Northwest Industries, in turn a subsidiary of Velsicol Chemical Corporation. According to Epstein et al. (1982:49), Velsicol is one of the most heavily penalized firms in the chemical industry. This kind of corporate structure has functioned in the past to compartmentalize legal responsibility and buffer parent companies from potentially serious damage claims and other losses.

The delivery was a catastrophic mistake. The feed plant expected a ton of magnesium oxide (also known as milk of magnesia), an innocuous substance used to enhace the digestion of livestock and marketed by MCC under the trade name Nutrimaster. MCC was also a supplier of PBB, a fire retardant used in industry, under the trade name Firemaster. At this time, due to a paper strike, many of its chemical products were packed in 50-pound plain brown sacks, distinguished only by crudely stenciled labels.

Workers at the feed plant routinely mixed the contents of the bags into succeeding batches of feed, which were sold all over the state. Other feed mixtures prepared in the same machines were also tainted. FBS shipped bags of supposed magnesium oxide to their various mills and dealers as well. Soon entire herds of farm animals were dead or dying. Dairy animals, the usual consumers of the augmented feed, were especially hard hit, but sheep, beef cattle, and poultry also were affected. Ultimately, as many as 100,000 cattle died or were destroyed, as well as millions of chickens and thousands of swine and sheep. While animals were dying, livestock continued to be slaughtered and animal products, including milk, meat, and eggs, marketed throughout the region, to be eaten by urban dwellers as well as farm families who consumed their own produce. In the end, medical authorities estimated that all 9 million people of the state of Michigan had measurable levels of PBB in their tissues. The worst symptoms occurred in farm families.

PBB is a member of a novel family of chemicals called *halogenated hydrocarbons*. They are made from basic hydrocarbon, or petroleum fraction molecules, by adding atoms of elements known as halogens: chlorine, bromine, or iodine. Bromine is the element that contains the desired fire-retardant character that is added to the hydrocarbon benzene to produce PBB. In order to reduce the fire death rate in the U.S., use of fire retardants in consumer goods is mandatory in industry. Unfortunately, many of the substances used for this purpose are suspected carcinogens and mutagens, as well as being extremely toxic.

Measuring toxicity is hardly straightforward. One of many complicating factors is that there is no such thing as a "pure" chemical. Manufacturing gives rise to chemical variations known as *isomers,* small differences in molecules, so that there are more than 100 different PBBs, and these are thought to have varying toxicities. Moreover, products include accidental contaminants as well as purposeful additives, which may increase their toxicity while also making diagnosis more difficult. When authorities finally got around to testing for PBB, they only looked for one isomer, and some contaminants were entirely overlooked.

In the Michigan case, the health of hundreds of thousands of people who consumed large quantities of PBB-contaminated animal-food products was seriously affected. PBB and other halogenated hydrocarbons are novel toxicological agents because they do not act in organisms the way familiar poisons do. PBB accumulates in the tissues of exposed people and animals alike, and it is thought that repeated small doses are potentially more injurious than a few massive exposures. Symptoms included extreme fatigue, dizziness, headache, neurological deterioration, skin problems such as chloracne, stomach ailments, unusual nail growth, hair loss, and increased susceptibility to disease (Epstein et al., 1982:49n). Many scientists and technical people, including veterinarians, physicians, and university and government experts, confronted these problems with fatal ignorance, compounded by arrogance. Farm families, the early victims, faced the situation initially with bewilderment, fear, and guilt, and later with hurt, anger, and a sense of betrayal. Their physical ailments were difficult to separate from the emotional effects of the economic impacts.

This brief description of the course of the disaster is principally based on *Bitter Harvest* by Fredric and Sandra Halbert (1978), themselves victims, and *The Poisoning of Michigan* by Joyce Egginton (1980), a journalist. Other books and articles also have been consulted (Chen, 1978; Coyer and Schwerin, 1981; Peterson, 1978).

The mysterious symptoms did not begin to appear in the herds of FBS customers until the summer of 1973, as farmers exhausted their supplies of home-grown silage and substituted purchased feed. An early victim with a singular role to play was Rick Halbert, a third-generation Michigan dairy farmer, who held an M.S. in chemical engineering from Michigan State. Between August and October 1973, he and his father, who operated a neighboring dairy farm, bought 65 tons of magnesium oxide-enriched feed from FBS. Ironically, the feed mixture itself had been produced to Mr. Halbert's specifications. For some time after problems emerged, he was convinced that it was custom-blended for him alone and that therefore his problem was unique. The fact was that FBS adopted his idea and distributed the mixture widely in the state.

For the Halberts, "The story . . . began to unfold with a vague uneasiness about dropping milk production" (Halbert and Halbert, 1978:7). Their first sick cows were observed on September 20, with the first death in October. Symptoms included loss of appetite; sudden drop in milk production; grotesque curling hoof growth; dull, matted coat; thick, wrinkled skin; loss of balance; stillborn and grossly deformed calves; reduced disease resistance; mastitis and metritis; uterine problems; and teeth grinding due to abdominal pain.

Rick Halbert suspected feed problems right away. He even zeroed in on the additive and *queried FBS directly about it*. "Are you sure you used magnesium oxide?" FBS was explicitly reassuring. He noted the resemblance of symptoms to incidents involving other chemical contaminants, including pesticides, and one incident involving chlorinated napthalenes (in the end, brominated napthalines were found as a contaminant of PBB!). At the request of the Halberts' vet, two diagnostic veterinarians from the Michigan Department of Agriculture (MDA) visited the Halbert farm and gathered samples for testing routine bacterial, viral, and fungal problems. A dairy scientist from MSU, who also visited the farm, concurred that it seemed to be a nonspecific feed problem and recommended that they change *every* item in the cows' diet. The general health of the herd improved somewhat after that, although milk production continued to plummet, from 33,000 to 8000 pounds, and later a general deterioration of the herd set in.

In the remaining four months of 1973, the Halberts lost $80,000. More important, due to sharply reduced milk production, they lost their milk base, the annual quota assigned by the Federal Milk Market Administration. This meant that even if the family's herd returned to former levels of production, the difference between the old and the new quotas would have to be sold at a punishingly low price (Halbert and Halbert, 1978:54–59).

Note that at this point everyone, except possibly Rick Halbert himself, treated the situation as a routinely familiar and isolated problem. Apparently, none of the Halberts' neighbors used the same feed, and when he asked FBS directly if there had been other complaints, as late as February 1974 they claimed his was the only one. This was a falsehood that was to be maintained in various guises throughout the crisis, as FBS and government agencies consistently minimized the magnitude and scope of the problem as a rationale for inaction. Later it was revealed that *all* the hundreds of farmer victims had been told that their problems were unique and that *their own* poor husbandry practices were to blame.

> They thought it was like Watergate and they called it Cattlegate. But there was no big coverup. It was not what was done that was wrong, but what was not done by a number of people in authority who did not realize the magnitude of the problem. [Harry Iwasko, state official, cited in Egginton (1980:84)]

Halbert quickly learned that "little but frustration was to be gained in making a formal approach to a government institution, even though this was usually the place with the expertise" (Egginton, 1980:59). The Michigan Department of Agriculture and Michigan State University were unresponsive, as were the federal Department of Agriculture and the Detroit office of the Food and Drug Administration. He had contacted all directly and was brushed off on the basis that this was one farmer's problem and thus highly localized. Some private labs out of state could do some of the work, but at a price (FBS commissioned such work but held the results privately). Under these circumstances, Halbert developed his own technique for penetrating state and federal agencies, taking circuitous routes around them until he found an *individual* who was interested in the case and willing to do some research. In this way, he convinced USDA research scientists at Ames (reproduction specialist), Baton Rouge (mycologist), and Beltsville (toxicologist) to pick up a piece of the problem. It was at Ames where samples were subjected to gas chromatography because of Rick's suspicions about pesticides. *By accident,* the lab left the machine on over lunchtime, and the peaks that were later found to represent the heavy PBB molecules were first discovered.

Finally, a Beltsville livestock toxicologist, to whom Rick Halbert had speculatively sent a sample of feed, identified PBB. This was in April 1974, eight months after the sickness had first appeared in the Halbert cattle. By this time, more than 90 percent of the PBB was already in the environment; in livestock; in food products on grocery shelves, refrigerators, and freezers; and in human tissue. This was the cost of bureaucratic delay.

Although Rick Halbert thought that now his and other farmers' problems were over, they were really only beginning. The Food and Drug Administration responded to the Beltsville report by ordering a check of feed stocks in the state. MDA introduced a quarantine program, barring shipment of milk from herds in which contamination levels exceeded one part per million (ppm) of PBB. In the years to come, however, no ban was ever placed on marketing meat from quarantined or otherwise contaminated herds, not even after a program to dispose of contaminated cattle by burial was introduced. Little was done to prevent recontamination, or even re-recontamination, of feed plants, farms, and barns. Indeed, a "Silent Spring" situation developed on many farms, with a wide variety of fauna disappearing, even earthworms.

Undoubtedly, part of the problem was the absence of press notice and coverage. Studies of news reporting on a wide range of environmental disasters show that government spokespeople are used disproportionately as news sources (Media Institute, 1985). Since the government was denying the existence of a problem, there was *no news.* It is also the case that the big, aggressive newspapers are located in cities, and hence reporters are removed from rural issues.

The first news story on the disaster appeared on an inside page of *The Wall Street Journal* on May 8, 1974. Michigan papers were largely silent until as late as 1977. The story of what had actually occurred inside the feed plant back in 1973 was broken by a trade paper, the *Michigan Farmer,* in November 1975. The Michigan Farm Bureau immediately canceled its $45,000-per-year advertising account. It was not until early 1977 that the two biggest papers in the state, the *Detroit Free Press* and the *Detroit News,* produced any in-depth reporting at all. This was probably because the authorities' success in preventing the consuming public from making the connection between farmers' difficulties and their food supply was beginning to unravel.

State health authorities were virtually inactive. The state Department of Public Health carried out one poorly conceived study comparing families living on officially quarantined farms with other farm families but, since the criteria for quarantine were so arbitrary, it is likely that many unquarantined farms were highly contaminated too. Nevertheless, for several years this study was used as a rationale for not carrying out more thorough work.

The first disturbing health information came from surveys carried out by a citizens' group. It was not until 1977 that the state health department carried out a serious study, this time of human maternal milk. They were shocked to find PBB in 96 percent of the women sampled and levels up to four times that allowed, 1.22 ppm, in cows' milk sent to market in certain rural areas. Finally, in the summer of 1976, a legislative aide contacted Dr. Irving Selikoff, head of the Environmental Sciences Laboratory of New York's Mount Sinai Hospital. Almost two years *earlier* Selikoff had agreed to send a team to Michigan, but an invitation was never issued by authorities. Now such authority was forthcoming from the legislative leadership, and Selikoff's large team of specialists arrived to begin their work in an unused wing of a Grand Rapids hospital in 1976. Preliminary results, announced in January 1977, found that there was a variety of PBB isomers in the subjects; that 40 percent had some neurological disorders, with memory lapse the most dramatic symptom; and that many had skin, breathing, and joint problems. Later analysis also confirmed that many suffered from damage to their immuno-suppressive system. Selikoff urged that the so-called tolerance level be reduced to as close to zero as possible. Later, with state support, the study was extended to include urban residents, and Selikoff's team was the author of the conclusion that all 9 million Michigan residents carried PBB in their systems.

An exchange between Selikoff and a Michigan State University scientist at a conference in October 1977 says it all regarding the four-year stance and **W** of official agricultural science during the catastrophe. The MSU scientist: "I've got to stick up for the Department of Agriculture. . . ."

Replied Selikoff: "You have to stick up for the truth. You have no other options."

The reader can compare this case with other incidents involving chemical contamination to see if they follow similar patterns along the lines described earlier in this chapter. A critical element in this unfolding situation was the absence of awareness, common expressions of concern, and the development of public discussion and debate. Hence, timely and effective action by agencies with appropriate capacities was also lacking. Without intending to blame them directly, the silence of the press was critical. This was fostered by the success of the Michigan Farm Bureau and Farm Bureau Services in controlling both the flow of information and state governmental processes, including regulatory performance. The farmers certainly had strong grievances against all of the institutions that they thought had been set up to serve them, including their advocacy organization, as well as the various state and federal agencies that existed to deal with this kind of issue (Coyer and Schwerin, 1981). It is also important to note that the *novelty* of this situation contributed to the way the various parties performed. Notwithstanding his technical credentials and real expertise, Rick Halbert's claims lacked credibility precisely because they appeared to the authorities he contacted to be without precedent. For their part, the authorities responded to what is now understood to be a novel problem as if it were an old, familiar one — "a farmer with some sick cows."

Halbert's role and the Michigan PBB situation resemble in some respects that of Lois Gibbs and the notorious Love Canal toxic waste dump case in Lewiston, New York (Brown, 1981; Gibbs, 1982; Levine, 1982), but there are also some important differences. Halbert pursued a technological fix; because of a combination of his own **W** and lack of information on the scope of the problem, he was unable to see it "whole." Somewhat later, a citizens' advocacy group in Michigan seems to have looked more broadly at the situation, organizing, engaging the media, and launching its own health survey.

Viewed in isolation, the official response in Michigan appears to have been painfully slow and inadequate. Yet general and official awareness developed in the span of a year, and most of the other events just described occurred within three years after the onset. In the Love Canal case, residents began noticing seepage of suspicious materials into their homes and other frightening occurrences in the late 1950s and had to wait almost 20 years for any official response. On that scale, Gibbs took the stage late in the game to play the key role of forming a community group that designed strategies initially aimed at pressing authorities for more effective action, including evacuation, but finally aimed at initiating an increasingly public debate. Once engaged, the media played a critical role, particularly in expanding the arena in which debate was occurring, until it was a truly national debate, engaging the Congress, most state legislatures, and a myriad of localities.

In Michigan, the PBB contamination and the state's handling of it became little more than an issue in a gubernatorial election. Yet, the Michigan incident exposed some 9 million people to a toxic chemical, while Love Canal directly exposed a few hundred residents near the site and as many as a hundred thousand people dependent on local groundwater supplies. The Halberts' book was made into a Hollywood film, *Bitter Harvest,* but the incident, as well as the film, seems largely forgotten. Love Canal has come to symbolize the threat posed by chemicals in the environment. Ultimately the activities of Gibbs and her neighbors made it possible for anyone in the United States to press a credible claim regarding toxic waste, while incidents involving milk contamination have continued to occur around the country and tend to be treated as unique. Gibbs went on to found the Citizens' Clearinghouse for Hazardous Waste, a national communications networking organization devoted to facilitating citizen action. It produces two newsletters and publishes an extensive list of informational, technical, and how-to materials.

Legislative Action

Laws enacted at the federal, state, and local levels to deal with the whole range of issues faced by our society are the result of a process of factual investigation, the development of alternatives, and their public and private debate. For some, effective policy making seems to consist of getting a policy enacted rather than of getting a policy enacted that works! As is discussed here, the business of politics is the continuous debate on issues related to desirability. Feasibility, on the other hand, is the concern of executive agencies. This way of dividing up the tasks of government signals significant departures from the premises of the soft systems approach that, as was the case in the discussion of mediation, may be needed in the context of high-level conflict.

Throughout the book, discussion has been peppered with references to science and experts. In the present context of government action, questions such as the following need to be posed: "How much does science contribute to the debates leading to policy making?" "Isn't there an objective, scientific basis for enacting legislation to solve or mitigate pressing problems that have physical or biological components?" The tentative answers to these questions are respectively "unclear or marginal, at best" and "no."

Governmental action and policy making are a messy and not entirely rational or objective process. New policies are necessarily built on a foundation of old ones, and overall consistency has to be maintained if the laws enacted are to withstand judicial tests. As individuals, legislators attempt to balance conflicting interests and values, including their own, as well as those of others with standing to participate in a debate. Most of the debate in legislative bodies on proposed legislation focuses on what has been labeled (and defined as) desirability. Another way of saying this is that policy makers make decisions on the basis of values. Economics and political influence loom large, and well-endowed interest groups spend large amounts of money on legislative lobbying at the state and federal levels. Science enters here, as it did in the Hudson River case, in that interest groups muster their own experts to testify at legislative hearings. In many ways, their conflicting input on messy real-world problems tends to cancel science out of the process except when evidence on one side of an issue is over-whelming and beyond dispute, as in the Michigan case *after* the Selikoff studies. Or else, science information consistent with value judgments seems to prevail, which amounts to the same thing. As often as not, there are no relevant scientific findings sitting on the shelf waiting to be dusted off and put in evidence. The life history of the striped bass, which figures in a number of public disputes and policy debates in the Northeast, is an example. The assessment of health risks associated with chemicals and other substances is another (e.g., Wilkinson, 1987).

Science and other systematic knowledge has more to do with what has been labeled feasibility. Questions of feasibility come into policy making, but they are much more significant when executive agencies set about implementing laws passed in the legislature. It is as if two or three or even more separate groups of participants were associated with the different phases of the soft systems approach. A typical scenario at the state and federal levels would have legislative staff members — that is, unelected employees of individual legislators, of legislative committees, or of the whole legislative body — carry out the finding-out tasks. Then the legislators themselves, or at least selected members of a relevant committee, are brought in to complete the finding-out phase in legislative hearings where witnesses appear. Then the legislators and staffs work to conceptualize proposed transformations, not in systems terms, but in legal ones. As the process moves on, proceedings become less and less consensual and perhaps also less coherent. Ultimately, members of legislative committees vote on whether or not to deliver their conceptualization to the whole legislative body. If they do so, then the whole body will vote. If the vote is positive, then the piece of legislation will be passed on to the other house (in a bicameral legislature) or to the executive branch if both houses have already passed it. The executive, the president or the state governor, as the case may be, then can approve it by signing it into law or disapprove it. If signed into law, the new conceptualization falls in the lap of an executive agency. Agencies, such as a state agriculture department or a federal regulatory agency, have the task of translating the laws passed by the legislature or Congress and accepted by the president or state governor into *practical proposals for change* and then implementing them.

This is where both scientific expertise and the soft systems approach might have a useful role to play. Since the legislation itself presents the agency with an abstract conceptualization, the clear task is to subject the proposals embodied in it to a reality test corresponding to stages 5 and 6 of the soft systems approach. To carry out the equivalent of stage 5 requires a picture of reality and other basic documentation. Some of this may be part of the legislative record, but independent investigations are frequently carried out. It is common for state and federal agencies to enter into research contracts with universities for this kind of work if it is beyond their own capacity. Executive agencies also have a constituency of interests they serve and otherwise interact with. For example, state environmental or conservation departments deal with industry, recreational groups, environmental advocacy organizations, and scientists as well as the public at large. To carry out the equivalent of stage 6, it is normal practice for executive agencies to work informally with representatives of such interests and also to conduct public hearings on how to translate legislative mandates into regulations or programs, as well as to evaluate and revise past initiatives. Beyond this, it appears to be feasible to carry out stages 3 and 4 as well by using legislative mandates as bases for developing transformation statements and treating other aspects of law and established practices in government as environmental constraints.

Obviously, this discussion has an artificial or imaginative quality about it because the governmental bodies involved *do not* use the soft systems approach. The intention here has been to use the soft systems approach as a standard of consciousness, coherence, and defensibility against which actual procedures may be compared — the direct overlay method. When this is done, not surprisingly, there are certain correspondences as well as serious deviations from the standard. The correspondences offer the promise that a professional trained in the approach can in fact use his or her competencies in situations involving governmental operations. The deviations ensure that there is plenty of work to do.

Some of these points can be illustrated in a discussion of a complex legislative proposal that was debated in the New Jersey state legislature several years ago.

Pesticide Legislation in New Jersey

"A historic struggle" is how a New Jersey writer hyperbolically described the debate of a 1984 state senate bill to restrict the use of agricultural chemicals (Ripton, 1985). The bill was introduced in February 1984 by Senator Ray Lesniak, who represents an urban district, albeit a heavily polluted one, with a wary and activist constituency. It would have imposed the most stringent pesticide controls in the nation, since it borrowed language from laws on the books in other states, which had followed a variety of approaches. Specific provisions:

1. Ban nonagricultural aerial applications of broad-spectrum pesticides, e.g., the use of Sevin for gypsy moth control.
2. Require a minimum 24-hour notification of neighbors within 500 feet of an agricultural site targeted for aerial, air blast, or hydraulic spray application of pesticides.
3. Protect farm workers with the strictest safety regulations in the U.S.
4. Severely restrict the ability of the State Department of Environmental Protection to register pesticides.
5. Establish a test program to determine the extent of pesticide contamination of the environment.
6. Broaden representation on the state Pesticide Control Council.
7. Increase funding for the Rutgers University-based Integrated Pest Management (IPM) program, which had already helped participating farmers to reduce chemical use sharply.

New Jersey is the most urbanized state in the U.S. and experiences the whole range of pollution problems, mostly associated with industrial waste and auto emissions. Yet, agriculture is still a $3-billion-a-year industry. Suburban and leapfrog development has taken farmland at an accelerating rate and placed non-farmer exurbanites cheek by jowl with working farms. Some 4 million pounds of pesticides are applied annually to crops in the state; another 85 thousand are used for mosquito control, and 40 pounds per year of the deadly Chlordane were used (in highly diluted mixtures) for termite control until the EPA banned its use. In New Jersey, pesticides are turning up in landfills, private wells, public water supplies, and bioassays of human tissue!

The bill was not introduced in response to a catalyzing event or rising grass-roots clamor. Rather, it emerged from long-term staff research and input from public-interest groups. One influential document, entitled *The Poisoned Garden: Pesticide Use in the Garden State*, was compiled by an advocacy group, New Jersey Citizen Action. Expert witnesses, such as toxicologists and other health experts, also appeared.

It must be understood, however, that when trying to quantify health risks associated with pesticides and other substances, the experts are dealing in *trans-science*; they are asked questions that exceed the capabilities of the science of toxicology to answer (Wilkinson, 1987:39). Chronic adverse health effects of suspect chemicals cannot be verified by direct experimentation on humans. Extrapolations from animal studies are suspect and constantly debated. Members of the legislative branch of government, both state and federal, as well as informed lay people, are upset by the absence of a scientific basis for policy making. Lacking solid epidemiological or toxicological data on health problems related to pesticide exposure, advocates of the bill nevertheless mustered much anecdotal testimony in legislative hearings.

A powerful coalition was ranged against the bill. It included the farm lobby, led by the New Jersey Farm Bureau and supported by the State Department of Agriculture. Farmers were particularly upset by the bill's notification provisions and the farm worker protection requirements. Possibly more powerful was the opposition of the chemical industry. This interest was represented by the Alliance for Environmental Concerns, which employed as its principal lobbyist Philip Alampi, who had served 26 years as New Jersey's Secretary of Agriculture. The chemical industry placed defeat of the New Jersey proposal at the top of its national legislative agenda because the bill would set a pattern that, it was feared, many other states would emulate. Accordingly, it poured very substantial resources into the campaign, including hiring a leading New York City public relations firm. The industry position was, and continues to be, that the state Department of Environmental Protection and federal regulators already have the power to address all the concerns contained in the Lesniak bill.

Although the bill and legislative debate received a moderate amount of media coverage, it never took off as a focus of broad public concern the way other issues have in the state, such as toxic waste and water quality. The bill's opponents managed to keep it bottled up in committee for most of 1985, where it finally died. It may be that only the occurrence of a large-scale disaster could have overcome the influence and resources ranged against it. Nevertheless, many of the bill's provisions turned up in new pesticide regulations, which were collaboratively drafted by Department of Environmental Protection staff during the period when the legislative debate took place. Participants in this backstage debate were another state agency and an advocacy group, both with significant ties to the agricultural community.

Summary

This chapter has focused on three aspects of the topic of debating desirable and feasible change. It defined and discussed the notion of debate itself and emphasized the full-and-open-discussion aspect rather than the combative aspect. It is to be hoped that by this stage of the approach the people involved will have established amicable and constructive working relationships with each other as well as with the analyst/facilitator. The debate phase of the soft systems approach is a special kind of reality test. Whereas the comparison stage involved going back to the basic facts developed in stages 1 and 2, stage 6 looks forward to see if the formal model of an improved future developed in stage 4 is really what people want and if it is within their capacity to implement. These two issues respectively bear the labels *desirability* and *feasibility*.

Since this is the point in the soft systems approach at which the analyst/facilitator and possibly a select group of participants come back to the larger body of clients with some results, this chapter briefly reviewed the subject of intervention, which had been detailed in Chapter 2. Once again, emphasis was placed on the rationale underlying facilitative interventions in contrast to authoritative ones. Your purpose as a facilitator is to work yourself out of a job by helping people to be self-reliant in pursuit of improving their situations. This is the opposite of conventional intervention, aimed at making people dependent on your expertise and your designed solutions to their problems. Experts and expertise have a role to play in situation improvement, but they are secondary to the mutual learning of analyst/facilitator and the people involved. That is what the soft systems approach is all about. As an analyst/facilitator, you must truly know the people involved, establish effective two-way communication, and then work alongside them for the purposes they select.

It has not been possible to provide a cookbook, a set of tried-and-true recipes that will always turn out the same. You cannot eat franks and beans every night! Instead, this chapter has presented a selection of principles and guidelines, illustrated with concrete examples that might point you in appropriate directions for dealing with the real-world situations in which you will be professionally involved in the future. In addition, the bibliography for this and other chapters points you toward sources of guidance on debate techniques that have only been mentioned in passing.

Finally, this chapter also provided some facts and insights concerning the contexts in which you might find yourself intervening, the *debate arenas*. For this, it was important to talk about the kinds of groups you might work with as an analyst/facilitator. Discussed were features of small face-to-face communities, businesses, multiparty conflicts, the courts, bureaucracies, and government. This was not a civics lesson. An analyst/facilitator needs to know how our social institutions really work, not how they are supposed to work. Again, this account is not the final word. To be effective, you must constantly expand your base of information and understanding and improve your assimilative tools. You do this by reading books and newspapers, taking courses in fields seemingly remote from your own—particularly courses that initially make you feel uncomfortable—reflecting on what you have found out, discussing things with peers, and going out in the world to grapple with real situations.

Many of the inquiry activities discussed in this chapter continue in Chapter 8. The tasks involved in implementing the proposals for change that have survived the tests of comparison and debate will be presented there.

REFERENCES

Amy, Douglas J. "Environmental Mediation: A New Approach with Some Old Problems." *Citizen Participation* 4(2):10–11, 24, 1982.

Bingham, Gail. "Does Negotiation Hold Promise for Regulatory Reform?" *Resolve* (Fall):1, 3–6, 1981.

Boulding, K. E. "Review of Checkland: Systems Thinking, Systems Practice." *Journal of Applied Systems Analysis* 9:137–38, 1982.

Brighton, D., J. Northrup, and C. Motyka. *Shaping the Forest's Future*. Rutland, VT: Green Mountain National Forest, 1987.

Brooks, F. A. "Conflicts Between Commercial Farmers and Exurbanites: Trespass at the Urban Fringe." In *Farm Work and Fieldwork: American Agricul-*

ture in Anthropological Perspective. M. Chibnik, ed. Ithaca, NY: Cornell University Press, 1987

Brown, Michael. *Laying Waste: The Poisoning of America by Toxic Chemicals.* New York: Pocket Books, 1981.

Carson, Rachel. *Silent Spring.* New York: Houghton Mifflin, 1962.

Checkland, P. *Systems Thinking, Systems Practice.* Chichester: Wiley, 1987.

Chen, Edwin. *PBB: An American Tragedy.* Englewood Cliffs, NJ: Prentice-Hall, 1978.

Coyer, Brian W., and Don S. Schwerin. "Bureaucratic Regulation and Farmer Protest in the Michigan PBB Contamination Case." *Rural Sociology* 46(4):703–723, 1981.

Crenson, M. A. *The Un-Politics of Air Pollution: A Study of Non-Decisionmaking in the Cities.* Baltimore: The Johns Hopkins University Press, 1971.

Crowfoot, James E. "Negotiations: An Effective Tool for Citizen Organizations?" *The NRAG Papers* (Fall)(3):24–44, 1980.

Durrenberger, E. P. "Notes on the Cultural-Historical Background to the Middlewestern Farm Crisis." *Culture and Agriculture* 28(Winter):15–17, 1985–86.

Egginton, J. *The Poisoning of Michigan.* New York: Norton, 1980.

Epstein, S. S., L. O. Brown, and C. Pope. *Hazardous Waste in America.* San Francisco: Sierra Club Books, 1982.

Fisher, R., and W. Ury. *Getting to Yes: Negotiating Agreements Without Giving In.* New York: Penguin, 1981.

Gagnon, J., and C. Greenblatt. "Health Care Planning and Education via Gaming Simulation." In U.S. Department of HEW, *The Hemophilia Games: An Experiment in Health Education Planning.* Washington, DC: U.S. Government Printing Office, 1975.

Gardner, J. *Self-Renewal.* Evanston: Harper & Row, 1963.

Gibbs, Lois. *Love Canal: My Story.* Albany: State University of New York Press, 1982.

Greenblatt, C., and R. D. Duke. *Game-Generating Games: A Trilogy of Issue-Oriented Games for Community and Classroom.* Beverly Hills, CA: Sage, 1979.

Habermas, J. *Theory and Practice.* Boston: Beacon, 1973.

Halbert, Fredric, and Sharon Halbert. *Bitter Harvest.* Grand Rapids: William B. Erdman, 1978.

Harter, Philip J. *Negotiating Regulations: A Cure for the Malaise?* Technical report. Washington, DC: Administrative Conference of the United States, 1982.

Institute for Alternative Futures. *Alternative Futures for Vermont: A Strategic Planning Workbook for Forests, Parks, and Recreation.* Alexandria, VA: The Institute for Alternative Futures, 1982.

Jacobs, Jane. *The Economy of Cities.* New York: Random House, 1969.

Lake, Laura M. *Environmental Mediation: The Search for Consensus.* Boulder: Westview, 1980.

Levine, Adeline G. *Love Canal: Science, Politics and People.* Lexington, MA: Lexington Books, 1982.

Lieberman, J. K. *The Litigious Society*. New York: Basic Books, 1983.

Lindblom, C. E. "The Science of 'Muddling Through.'" *Public Administration Review* (Spring):79–88, 1959.

Lisansky, J. M. "Farming in an Urbanizing Environment: Agricultural Land Use Conflicts and Right to Farm." *Human Organization* 45(4):363–71, 1986a.

Lisansky, J. M. *A Survey of the Regulatory Environment for New Jersey Agriculture*. New Brunswick, NJ: Agricultural Experiment Station Publication No. R-26102-1-86, 1986b.

Media Institute. *Chemical Risks: Fears, Facts, and the Media*. Media Research Series. Washington, DC: Media Institute, 1985.

Miller, Allen. "Psychosocial Origins of Conflict Over Pest Control Strategies." *Agricultural Ecosystems and Environment* 12:235–251, 1985.

Mingers, J. C. "Towards an Appropriate Social Theory for Applied Systems Thinking: Critical Theory and Soft Systems Methodology." *Journal of Applied Systems Analysis* 7:41–49, 1980.

Moyers, H., E. Nelson, and J. Duncan. *A Preliminary Report: A Study of Attitudes Towards Agricultural and Natural Resources Concerns in Wisconsin*. Madison: College of Agriculture and Life Sciences, University of Wisconsin, 1968.

National Institute for Dispute Resolution (NIDR). *Dispute Resolution Resource Directory*. Washington, DC: National Institute for Dispute Resolution, 1984.

New Jersey Citizen Action. *The Poisoned Garden: Pesticide Use in the Garden State*. Hackensack, NJ: New Jersey Citizens Action Coalition, 1984.

Peterson, Iver. "Michigan PBB: Not a Comedy but Plenty of Errors." *The New York Times* 7/2/78:E16, 1978.

Reich, C. A. *The Greening of America*. New York: Bantam, 1970.

Ripton, John. "The Politics of Pesticides." *New Jersey Reporter* (April):18–26, 1985.

Rundle, W. L. "Teaching Negotiating Skills: A Simulation Game for Low Level Radwaste Facility Siting." *Environmental Impact Assessment Review* 6(3):255–263, 1986.

Sax, J. L. *Defending the Environment: A Handbook for Citizens' Action*. New York: Random House, 1970.

Schuck, Peter H. "Litigation, Bargaining and Regulation." *Regulation* 3 (July–August):26–34, 1979.

Seide, K. *Dictionary of Arbitration and Its Terms: Labor, Commercial, International, Concise Encyclopedia of Peaceful Dispute Settlement*. Dobbs Ferry, NY: American Arbitration Association, 1970.

Shor, I., and P. Freire. *A Pedagogy for Liberation: Dialogues on the Transformation of Education*. South Hadley, MA: Bergin & Garvey, 1987.

Simkin, W. E. *Mediation and the Dynamics of Collective Bargaining*. Washington, DC: Bureau of National Affairs, 1971.

Susskind, Lawrence, and Jeffrey Cruikshank. *Breaking the Impasse: Consensual Approaches to Resolving Public Disputes*. New York: Basic, 1987.

Susskind, L., and A. Weinstein. "How to Resolve Environmental Disputes Out of Court." *Technology Review* 85(1), 1982.

Talbot, Allen T. *Settling Things: Six Case Studies in Environmental Mediation*. Washington, DC: The Conservation Foundation, 1983.

Tucker, W. "Environmentalism and the Leisure Class." *Harpers* 255 (December):49–56, 73–80, 1977.

Wilkinson, C. F. "The Science and Politics of Pesticides." In *Silent Spring Revisited*. Gino J. Marco, Robert M. Hollingwoth, and William Durham (eds.). Washington, DC: American Chemical Society, 1987.

CHAPTER 8

Implementation: Stage 7 of the Soft Systems Approach

This chapter is about planning for and then taking action to improve situations. To implement means "to put in practice." It is the endpoint of a complete cycle of inquiry in the soft systems approach. It is also the beginning of change in a human activity system because it uses many of the proposals arrived at in the previous stages of comparison and debate for action to improve problematic situations. According to Checkland (1981:19), however, particular kinds of results are not guaranteed; the process is highly idiosyncratic because it is open to the personal styles and objectives of analyst/facilitators and the people involved and *how these interact with each other*, as well as with elements of structure and the established ways of doing things. The distinctions between structural, processual, and climatic or attitudinal change are discussed later in this chapter. Realistically speaking, stage 7 continues the tasks of stage 6 (and possibly other stages too), in that inadequacies may still turn up.

According to the soft systems approach, implementation does *not* mean installing some kind of expert-designed system. It is true that before the debate phase, a conceptual model may resemble an expert-designed system. What remains after the debate, however, is a body of *agreed-to changes* arrived at by means of a consensual process that involved the people who will carry out the changes and live with the results. In principle, changes developed in this way should be easier to put in place than those of a system designed in the expert mode discussed in Chapter 7.

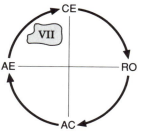

The main tasks of the implementation stage are (1) to design an implementation plan, (2) to carry out the specific and highly varied actions of that plan, (3) to communicate the specifics to all affected parties, including, but not limited to, actors who have not previously been directly involved in the process, (4) to monitor performance and the environment and evaluate results, and (5) to modify aspects of the plan if information accrues requiring it.

The role of the analyst/facilitator in the implementation stage is not clear-cut. Remember that the bottom line is that the analyst/facilitator is always an actor in the problematic situation (Naughton, 1981:65) until the job is completed. The question of when it is appropriate for an analyst/facilitator to leave and other aspects of the role in stage 7 are discussed later in this chapter.

Preliminarily, two extreme positions (and one that falls in between) can be described. On the one extreme, when the debate phase is complete and prospective changes are clearly defined, the job of the analyst/facilitator is done. He or she phases out of the process or, as in the case of a consulting job, terminates the relationship, after delivering the final report. The other extreme, perhaps corresponding to the position of the analyst/facilitator who is a regular employee of the organization involved in a problematical situation, is characterized by full involvement in implementation. Falling somewhere in between these extremes is the analyst/facilitator who stays to use the soft systems approach to monitor and, if necessary, modify the changes that are otherwise carried out by the people involved, perhaps to function as an expert at critical points, and to complete his or her own learning cycle.

Accommodative learning competencies are particularly needed in the implementation stage. You will recall that accommodators excel at action, carrying out plans, working with people, and getting involved in new situations. More than likely, during the comparison and debate stages, accommodators were the people who urged their partners on to take risks and seize opportunities. Convergers can now take the lead in formulating the implementation plan that the accommodators (among the people involved) put into action.

Implementation Plans

As the beginning of this chapter indicated, people move from stage 6 with a very detailed picture of an improved future state, consisting of one or more conceptual models that have been thoroughly tested by means of comparison and debate. In other words, they start Stage 7 with a very complete mission statement. In addition, the people involved have already expressed ideas about *how* to operationalize the human activity system(s). As a matter of course, these ideas too will have been discussed and recorded. This does not amount to a concerted plan of action, however. It doesn't identify specific acts, the order in which they are supposed to occur, who is responsible for what actions, where the required resources will come from, how details will be communicated to the larger constituency of interested or affected people, and how everyone will know when the future is here.

Convergence on a detailed plan of action covering this ground is the initial objective of stage 7. Undoubtedly it will involve cycling back to comparison and debate in order to sort through all the suggestions for *how* that have already been put on the table. There is a very extensive literature in the fields of business management and human resource development, called strategic planning, that can provide a wealth of tips and examples (see especially Benton, 1978: Pfeiffer, Goodstein, and Nolan, 1987). Drawing on some of these materials, the discussion here will focus on a *minimum* set of items that must be included in an implementation plan if the people are to coordinate their activities successfully, fulfill their responsibilities adequately, and communicate effectively. These are the activities and elements that *must* be included in any implementation plan:

1. *Benchmark activities* (sometimes referred to as *strategic actions*). What are the key activities that must be carried out to realize each of the subsystems or "whats" developed in stages 4, 5, and 6? (HOW?)
2. *Performance measures.* What operational measurements will be used to evaluate the effectiveness of benchmark activities? (HOW WELL?)
3. *Responsible actors.* What individuals or groups are responsible for carrying out each benchmark activity? (WHO?)
4. *Timetables.* When is each benchmark activity to be completed, including the order in which each activity is to be carried out? (WHEN?)
5. *Needed resources.* What financial and other resources are needed to accomplish each benchmark activity, and where will they come from? (WANTED?)
6. *Budget.* How are monetary resources allocated among the benchmark activities and implementation phases, and how are these monitored and controlled?
7. *Overall leadership.* Who has the individual or collective authority to make important decisions regarding coordination, monitoring, and revision of the implementation plan and to lead people through change?
8. *Communications.* How will those responsible for particular actions or phases of the plan communicate with the larger constituency of actors and other concerned parties in order to assure cooperation and coordination and to avoid conflict?

As noted in other parts of this book, the food, agricultural, and natural resources sector, especially at the primary producer level, suffers from a lack of organization. This is most apparent at the local level. As a result, political influence and power, the ability to gain access to resources and markets, and cooperation between individuals and among groups are all low. Where they exist and function, co-ops and growers' associations provide mechanisms for local cooperation and action on common concerns. These kinds of organizations, however, are often disorganized and underfunded. If the changes envisaged are to be carried out in part by modifying existing organizations, then creation of an implementation plan — perhaps as the capstone to earlier activities of the process — will strengthen the organization and people's capacity for cooperative action.

A sample segment of an implementation plan developed by the floral growers (whose modeling efforts were described in Chapter 5) to operationalize one of the functions specified in their human activity system model is presented later in this chapter in Table 8.1. This illustration provides a clear indication of how specific an implementation plan needs to be in describing all the elements defined in the foregoing discussion.

The Facilitator's Role

The beginning of this chapter briefly described three possible roles for the analyst/facilitator in the implementation phase of the soft systems approach: (1) withdrawal after the conclusion of debate, (2) full involvement in and beyond implementation, and (3) monitoring and indirect assistance. This section will flesh these out and discuss their implications.

If the facilitator is to withdraw at the conclusion of the debate phase, Checkland's option, then the overall quality of the intervention from the beginning through to the submission of a final report is crucial. The latter must be sufficiently detailed, unambiguous, and of a quality that leads to productive planning. This option may be most appropriate for the formal organizations described in Chapter 7, the business firms and governmental agencies involving a well-organized, unified clientele group and ample human and technological resources. It may be decidedly inappropriate where the people involved are only weakly organized, as in small agricultural cooperatives or community groups, and particularly when there is potential or established conflict.

It will be recalled from the previous chapter that implementation of the Hudson River settlement took much longer than anyone expected. According to the lawyer for Scenic Hudson, this was because "Russ Train left the case right after the signing and wasn't around to push and prod everyone" (Talbot, 1983:24). This statement reflects Train's intervention style, heavily weighted toward authoritativeness. One can only speculate that this was a necessary reponse to the exigencies of the situation, the longstanding and intractable nature of the dispute, and so on. It is also symptomatic of the absence of mutual learning and the consistent denial of the client group's autonomy. And that is the central issue — that the overall objective of intervention is successful learning to foster and/or reinforce the autonomy and self-reliance of the client group. In fairness to Train, it was never an objective of mediation to bring about a merger of Con Edison, the advocacy groups, and the regulatory bodies.

The full-involvement option may reflect a situation with different *structural* characteristics in which the analyst/facilitator is a regular full-time salaried employee of the owner and a coworker of the other actors involved or a member of some other relevant reference group, such as a community. It should be emphasized that, although perils exist, there is nothing the matter with an organization possessing and using its own in-house capability of learning itself out of a problematic situation. Indeed, the possession of this capacity is an asset.

You should, however, be alert to the pitfalls. One is that the stakes for an insider thrust into the role of analyst/facilitator are higher because the personal cost of failure is high. This will become more burdensome as the soft systems approach unfolds phase by phase in a cumulative way. Both the analytical and the action phases of the process can become distorted by deep involvement in the situation and the outcome, if the analyst/facilitator is not on guard. The benefits of being enabled to remain on the scene to monitor, evaluate, and, when necessary, to cycle through the soft systems approach (see the following) again should outweigh the risk of discord and personal loss.

In another scenario, the analyst/facilitator, originally an outsider, works his or her way into a job. This presents difficulties because it contravenes the learning and empowerment functions, which are at the heart of the approach. In this respect, the purpose of visiting experts is to work themselves *out* of jobs by training their replacements. This is particularly apt in developing-country contexts.

The third or middle option, in which the analyst/facilitator monitors the implementation phase being carried out by the clientele group, has much to recommend it. This is especially true for an insider who is concerned about strengthening the organization's capacity for self-renewal. It applies particularly to those situations discussed previously in which initially the group of participants was poorly organized and continues to lack important resources. This option depends a lot on the facilitator's style. *Hands off* is the watchword. The more the people involved do for themselves, the better their learning. The analyst/facilitator's value is his or her ability (1) to offer advice, encouragement, and approval; (2) to cycle through the soft systems approach again if necessary, even reconvening a debate, if some phases of implementation are seen to be seriously unraveling; and (3) to function as a resource person. A few final cautions are appropriate here.

Not everyone is able to envision an improved future or think through an implementation process. As in previous stages of the soft systems approach, facilitators must ensure that enough people who possess the required abilities are involved in implementation activities. Otherwise, implementation will not happen. Alternatively, it will take much longer to happen because more learning will have to accompany the process of making changes. No matter how concrete and detailed, an implementation plan will not be meaningful to everyone. It will be too abstract for some and will lack credibility for others. Written and oral communications must be clear, detailed, and consistent throughout the implementation phase — the time for debate on the improved state being sought is largely over, and the objective is to take action and help people stay in tune with the effort. People can easily become confused, lose their sense of direction, forget their own roles and those of others, ignore deadlines, and otherwise deviate from the means originally agreed upon for reaching the desired improved state. Effective leadership, coordination, and communications competencies are needed by the people selected to manage the implementation phase.

Not everyone will immediately accept change. Some people will wait and watch while others try it out before they themselves move to change their ways. Rogers (1981) surveyed thousands of change initiatives that had been recorded in journals and the documents of development agencies. Most of these initiatives were in the health, agricultural, and natural resource development areas. His conclusions consisted of a series of communications propositions about how people respond to change and how facilitators of change effectively communicate ideas and practices. One of the things he found was that farmers and agribusiness managers around the world do not all accept change at the same rate. While his choice of descriptive labels for the five types of responders — laggards, late majority, early majority, early adopters, and innovators — has received heavy criticism, his work has helped us understand that, although the rate at which people adopt change is variable, it does have a pattern. A key lesson for facilitators from Rogers' work is that it is unrealistic to expect all actors, owners, or customers to greet any proposal for change with equal enthusiasm or initially to participate in implementation at the same level.

In the beginning, only a small number may be willing to assume the risks associated with trying something new. Others will wait and see. Early in the implementation phase, there will be few tangible things for people to grasp. Only those few individuals who are able to envision the change sufficiently and positively will see its potential. The modeling phase of the soft systems approach helps more people see what the change is about. Yet even with modeling and debating, not everyone is able to translate and understand abstractions. As other people's experience becomes available, they are better able to see and evaluate the changes. Rogers' work contains many other insights on how to communicate ideas and take people through a change process. You should make his work part of your professional toolkit as you prepare for your careers.

The timing of implementation activities is important, particularly the time to begin. Inexperienced facilitators tend to defer action until everyone appears to be willing to begin. Experienced facilitators encourage rapid initiation, working with the people who are most willing to take risks. Those people will be the examples whom others will watch. The caution here is to avoid a pioneering group that is too unbalanced, involving an unrepresentative slice of the total population that the change is actually intended to help. On the one hand, in the case of the floral growers, to start with heavy representation of the large-scale growers might permanently exclude smaller operators. On the other hand, starting with younger family farmers in a Third World community situation might be all right because they will "grow old" with the project and be succeeded by younger participants in their turn. To begin the implementation without broad or at least representative participation can actually jeopardize success. A critical judgment is required.

Kinds of Changes

According to Checkland (1981:180), *three kinds of changes may emerge from the modeling, comparison, and debate stages: structural, procedural, and attitudinal (or climatic).* Structural changes have to do with the frameworks of institutions and organizational groupings. These features of social reality are fairly durable. They do not change easily, routinely, and in the short run. This does not imply that structure is unchangeable. Rather, structural changes take longer, are riven with conflict, and require directed will and effort, even power, to accomplish. Procedural changes affect the dynamic elements of the situation, such as how communications and reporting will be carried out, how resources will be used, and how products and services will be measured and controlled. According to Checkland, structural and procedural changes are relatively easy to specify and implement. Implementation difficulties arise because structural and procedural changes have unanticipated consequences. Here the watchword is *all solutions to problems give rise to further problems.* Attitudinal changes refer to elements of the **W** characteristic of the people involved, especially owners and other important actors, and their immediate feelings about the unmodified and modified situation, the element of *climate*. Attitudinal change is also difficult to attack directly. Before proceeding, the three kinds of changes need to be discussed in greater detail.

Structural Change

As a result of socialization and education in our own society, we come to situations equipped with a priori notions of what structure is; it's a part of our **W**. As Americans, we see structure (if we think about it at all) in fairly rigid, even legalistic, terms as a neat hierarchy of larger and larger units and divisions, tables of organization, clear lines of communications, authority and power, proper division of obligations and responsibilities, and so on. (There are related procedures too, such as reports, deadlines, etc., which will be discussed subsequently.) Indeed, a lot of these matters are encoded in constitutions, formal policies and procedures, laws, by-laws, contracts, and other binding agreements. For example, the United States of America can be depicted as consisting of a hierarchy of municipalities, counties (or parishes), states, and the nation as a whole, all with political offices, governmental bodies and agencies, laws, and responsibilities both to citizens and higher-level authorities. Corporations are thought to be organized along similar lines. This, we are taught to think, is what structure and organization are all about. But how do individuals, with all their variability, and especially vaguely defined neighborhoods, communities, movements, catchments, and regions, fit into this kind of structure? Does the absence of clear hierarchies imply anarchy? Or chaos? Or, more to the point, absence of structure?

Some social scientists and other observers would say no to all of the preceding questions. There are kinds of structures in which readiness for change is, in some sense, built in. Anthropologists have described them in tribal societies as segmentary and acephalous, village-sized groups without rulers or an overarching hierarchy (e.g., Bohannon, 1954). This kind of organization has also been found to characterize social movements in modern industrial nations such as our own. For example, Gerlach and Hine (1970, 1973) focus on such social, political, educational, and religious movements as black power, environmentalism, and Pentecostalism.

The structure of these organizations consists of semiautonomous cells. New cells form continuously as groups break off from parent groups because of disagreements of various kinds. New cells can also spring up independently. Some cells may be temporary organizations formed in response to a local problem, dissolving when that problem goes away, and reviving again when a new issue emerges. Or a cell can loosely ally itself with others on a regional or national (or even international) basis without formal organizational bonds. What ties cells together are means of exchanging information, such as meetings, demonstrations, and newsletters; the fact that individual people may belong to more than one cell; that leaders of individual cells have personal ties; and that there may be itinerant speakers, preachers, or other leaders who make the rounds from group to group. This kind of structure is consistent with the face-to-face pattern described in the previous chapter.

Several authorities have found some of these features side by side or even within established bureaucracies in formal organizations. For example, John Gardner has described the self-renewing organization as one that constantly changes its structure in response to changing problems and needs (1963:26). In discussing what he refers to as the "new ad-hocracy," Alvin Toffler (1970:124–151) points to the emergence in large corporations of increasing modularism, "sideways movement," and "project" or "task force" management. The latter is a structural technique whereby, drawing on personnel from diverse departments of a firm or different governmental agencies, teams are assembled to solve specific short-term problems.

In general, segmentary organizations can respond quickly and in ways that reflect real needs and concerns of members. In particular, they can deal with new situations; can be expanded rapidly; are capable of innovation; and, while individual cells are relatively easy to demobilize, the general movement or overall organization is more difficult to attack. Hierarchically organized, bureaucratic organizations have more staying power and command greater resources but are slow to respond, do poorly with novel situations, and can be shut down by lopping off the head.

All this points to the usefulness of viewing structure as consisting of both *segmentary* and *hierarchical* elements. This helps to shed some light on the history and function of the U.S. political system, real variations that exist in the political organizations of the various states, and the way the system actually changes. The issue emerges again and again in food, agriculture, and natural resources when jurisdictions at different hierarchical levels engage in a debate regarding the proper locus of governmental policy and management.

For example, some readers are probably familiar with the concept of states' rights. This is the assertion of the sovereignty of the individual state in opposition to assertions of the power of the federal government. The issue has come up in connection with a wide range of issues, including civil rights; the ownership of off-shore energy and mineral resources; nuclear power plant safety; agricultural chemical regulation; and, most recently, control of biotechnology. Such disputes reflect the tension between segmentary tendencies and hierarchical ones. The same kind of tension is played out within individual states, shedding light on the difference between states with very strong central governments, such as New York, and those with historically weak ones, such as New Jersey. The latter situation produces what goes by the name of home rule, again an example of the segmentary tendency, with the state's more than 500 municipalities exercising real power and constantly resisting assertions of state government authority.

Pressure to centralize increases as home-rule states become more densely populated, developed, and industrialized. For example, one of the first things to happen in municipalities that have rapidly suburbanized with a growing residential population is that farmers and agriculture-related businesses lose control of local government. Local government increases real property taxes in an effort to provide expanded municipal services. Farmers are particularly hard hit because they have so much land to tax. Such municipalities may also try to regulate the remaining farmers in ways tending to make them think of quitting agriculture. Authorities may limit or forbid certain livestock operations because of the odor they produce, restrict the movement of slow-moving farm vehicles along public roads, take water resources for urban purposes, forbid the use of noise makers, object to agricultural chemical application, and subject farm utility structures to urban building codes. And farmers who decide to develop their land are faced with local zoning and planning boards that are increasingly hostile to development, out of a desire to conserve "green space," that is, the farms!

In New Jersey, state government has intervened in a number of ways aimed at reducing local sovereignty in order to help farmers. One of the first was farmland assessment legislation, which said that local authorities could only tax the *agricultural value* of farmland rather than its *development value*. More recently the state has enacted a right-to-farm law, aimed at preventing municipalities from outlawing standard agricultural practices. These are all examples of *procedural* changes (see subsequent text), but their cumulative effect is in the direction of structural change, that is, greater centralization.

Thus, an example of a *structural change* in a home-rule state might be the decision to make the state legislature full-time rather than part-time. This is because, with a part-time legislature, many state legislators are simultaneously *local* government officials at the municipal or county level. As such, they tend to mirror local concerns more closely and are subject to review by their electorates with much greater frequency and in detail. A change to a full-time legislature would by degrees tend to remove legislators from some of these concerns and make it easier for state agencies to get legislation allowing them to preempt local laws. Another structural change would be to invest county government with new authority. This is most likely when new kinds of concerns and activities arise, such as county-wide environmental controls, growth management, waste water treatment, solid waste management, and the like. Ultimately, such changes also reduce the sovereignty of municipalities and may also make government less directly responsive to citizens' concerns. That is a cost of centralization that some may consider undesirable.

Easily overlooked are the less obvious aspects of structure, the beliefs, values, and sentiments underlying people's behavior in groups of various kinds. Proposals to change "obvious" features of structure can run into difficulties precisely because some people perceive them as contrary to deeply held and unquestioned values. In other words, structural features and **W**s in a situation are related. Racism as an institution in the United States consists not only in the existence of a relatively advantaged white majority and relatively disenfranchised black, Indian, Hispanic, and Asian minorities, but also in an elaborate structure of beliefs—a mental map—that shapes, sometimes quite unconsciously, the behavior of individuals to resist structural change.

During stages 3 and 4, a group going through the soft systems approach may create a new organization; new role relationships; new lines of communication, responsibility, and authority — all implying structural changes. For example, in the case of the floral growers (Chapter 5), although one of their models maintains the separate existence of the three established growers' associations, by establishing a super organization in the Advisory Board and by vesting it with considerable authority for an initiative intended to revolutionize the industry, the group has planted the seeds of structural change. It also provides a context in which conflicts may arise based, in part, on the old divisions. These kinds of problems are likely to arise first during implementation.

Procedural Change

A procedure is a technique, a particular way of doing something, a series of steps followed in a regular order. Here the concern is with procedural change in many spheres. At its core, technological change *is* procedural change, although it may imply or necessitate structural and attitudinal change. What is being altered in technological change is the sequence of steps or the steps themselves.

For Checkland (1981:180), *procedures* are "all the activities which go on within the (relatively) static structures." Procedures involve communications and control of the flow of information, materials, and energy. It has to do with the hows rather than the structural whos or whats of organizational roles, physical environment, and mechanical components. During stages 3 and 4, procedural changes were at issue when you discussed who the owners and actors in your system would be and identified the future subsystems, inputs, outputs, and boundary conditions.

The procedures within an organization may be rigid and automatic, at one extreme, or ad hoc or nonexistent at the other extreme. Correlated with elements of structure, at least to some degree, procedures determine how fast an organization can respond to change and how appropriate responses are to a situation. Bureaucracies move slowly but respond to lots of situations. Ad hoc task groups move rapidly but only in relation to narrow objectives. The same distinction applies to manufacturing: Assembly lines are relatively inflexible; they can only stop and start, but they deal in large quantities. Contrastingly, individual craftspeople can vary their product infinitely but in small volume.

The national debate in the 1930s regarding whether farming was a way of life or a business implied procedural changes at the outset. When more and more farmers decided to operate as businesses, they changed how they did things: they increased their use of debt capital, decided to expand rather than stand pat (growth versus equilibration), pursued mechanization, purchased more inputs, and so on. Ultimately, this had structural implications as a concomitant of growth, including the rise of corporate agriculture, an increase in average farm size, and the sharp reduction in the numbers of farms and farmers. The disappearance of black farmers in the South is a particularly significant structural change. Today, many family farms are chartered as corporations even as they remain structurally, procedurally, and attitudinally separated from corporate farming organized along industrial lines. But there is more convergence on the way.

The relationship between structure and procedure is not precise. For example, up to a point it is possible to reorganize work procedures around an assembly line by changing rules, such as who is allowed to shut down the line. That's a procedural change. Once you permit workers to exchange roles or tinker with the arrangement of machines on the line, however, you are dealing with structural change. Similarly, the federal Voting Rights Act, a procedural change, has dramatically improved the role of minorities in elective office, a structural change. This spotlights the useful tip that easier-to-introduce procedural changes ultimately *cause* structural changes of a kind that could not be implemented directly, except possibly at gunpoint.

For those involved in producing an implementation plan, this means, at a minimum, that they place procedural changes high up on the priority list for early action. Their implementation and successful operation can provide the context for implementing structural changes later on, and this can be provided for in the plan as well.

Attitudinal or Climatic Change

Attitudes are positive or negative judgments associated with an issue. At least four functions can be attributed to them. First, as a component of **W**s, attitudes aid in understanding the world. Second, although they strongly bias one's view of the world, attitudes also protect self-esteem. Third, attitudes motivate attempts to increase rewards and minimize punishments in situations. Fourth, attitudes may express fundamental values. It is generally believed that this ordering reflects the ease or difficulty of inducing attitudinal change. Thus, it is thought to be easier to change understandings of the world than to alter fundamental values.

For some authorities, attitude has three components: cognition, affect, and intention (e.g., Thurstone, 1931; Triandis, 1964, 1971). *Cognition* is the connection people make between one object or category and another, their sense of the relatedness of things. *Affect* is how people feel about something. *Intention* is the sense of future action, including the possibility of adopting a particular change. At a given time, in members of a group, these components may be consistent or in conflict with each other. When in conflict, psychologists call it *cognitive dissonance* (Festinger, 1957). Dissonance gives rise to pressures in the individual to either reduce it or to avoid increasing it. This pressure is what has been referred to in earlier chapters as unease about a situation. Pressure arising from cognitive dissonance or unease can manifest itself in changes of behavior, changes in cognition, and exploratory responses to new information (Katz, 1960). The latter refers to diverging behavior, tentative and circumspect attempts to make sense out of something.

Some literature in the field of economic development makes it sound as if it were necessary to induce unease in order to prompt change. This is only logical if the change agent assumes from the outset that changing how local people do things is indeed necessary (to say nothing of desirable and feasible). Colonial administrators took for granted the need for change. A typical mechanism for inducing dissonance was to impose taxes. This created the necessity for indigenous people to enter the modern sector to earn money with which to pay their taxes. There is also a hypothetical model of agricultural change that says, in effect, that people will adopt new techniques that require more work when they experience the effects of population pressure on resources (Boserup, 1965). Individuals experience pressure on resources when their quality of life declines, giving rise to cognitive dissonance. As the soft systems approach unfolds, phase by phase, there should be a flux of attitude, sometimes involving increased dissonance and other times in the direction of permanently reduced dissonance.

Proposals for change, whatever their source, produce dissonance. Proposals that have testable results built in produce less dissonance (Triandis, 1971:335). This was one of the functions of the use of a computer simulation in the Hudson River settlement case, discussed in the previous chapter. Similarly, agreement with other people, as in the debate phase, reduces dissonance. Accordingly, Checkland feels that attitudinal changes are more likely to come about when the people involved share the experience of living through structural and procedural changes than when a facilitator attempts to change attitudes per se. More important is for the analyst/facilitator to monitor attitudinal changes to see how things are going, for example, to sense when the actors agree that improvement has been achieved, reflecting a reduction in cognitive dissonance. As Checkland notes, the client knows the problem is solved when, in retrospect, the answer appears to have been obvious!

The kinds of attitudinal changes in the people involved that one might find correlated with structural and procedural changes are changes in influence and authority, expectations, sense of appropriateness and propriety, standards of conduct and performance, and the like.

The description of **W**s in the CATWOE exercise of stage 3 is, in a sense, an exercise in attitude clarification or modification for the people involved. Most people do not closely examine their attitudes, at least not routinely. The CATWOE, modeling, comparison, and debate activities all facilitate the formation of new attitudes among participants in the process, not only by clarifying existing **W**s, but by means of new experience. For example, the cooperation required on the part of the floral growers, otherwise members of competing organizations based on historical differences of perspective and interest, can by itself reduce the attitudinal barriers to change.

It is not possible to list, let alone discuss in detail, the specific skills, technologies, and other competencies that people might need to tap or learn in order to operationalize a given human activity system according to an implementation plan. A list might include legislative lobbying, fundraising, communications, community organizing, workshop education, bottom-line and social-cost accounting, and computer networking. Communications is focused on here because it is a generic issue in the probable situation of smaller, local, and less well-organized groups.

Communicating Change

Establishing effective communications with all parties involved in a problematic situation is vital to the execution of an implementation plan because the emphasis shifts from face-to-face meetings and the small group dynamics that have been characteristic of earlier tasks to bringing on board a much larger group of diverse and previously unengaged people. Here the task may be conveying the changes that a smaller representative group feels are desirable and feasible to potential participants (actors, owners, customers) and, in some cases, to the community at large. The groundwork for some of this will have to be laid during the other phases of the process by ensuring that participants have represented all of the **W**s. The analyst/facilitator and the emerging leadership of the overall effort now must be alert to the informal structure of the larger group and the modes selected for conveying information and coordinating activities. Often, because co-ops and other small local groups lack a well-developed formal structure, the informal networks of friendship and social relations that tie together the people involved will have to be utilized.

Research on communications networks (e.g., Lawrence and Lorsch, 1969) indicates that there are two extreme types, the *closed* and the *open*. You are undoubtedly familiar with the *closed* network from attending large lecture courses in school. The key person in the node, such as a lecturer or boss, conveys the message to the other, passive, and silent members of the group. No questions are asked and communications are one-way. The other extreme, the *open* network, may still be characterized by a key person in a central role, but information flows are two-way (or multiple-way), not only between the leader and individual members of the groups, but among the members too. The closed network is faster than the open one but considerably less efficient, accurate, and adaptable to ever-changing tasks. It should also be obvious that the closed network is associated with the authoritative and the open one with the facilitative modes of intervention.

It's also important to keep an eye on subgroups that may emerge around the activities related to particular subsystems. They can function like cliques, with plenty of two-way communications within but only one-way communications with each other. It has long been recognized by rural development authorities that formal, closed communications modes are effective for initially introducing knowledge of an innovation to prospective participants, but they are not a likely means to convince people to adopt the innovation (Arensberg and Niehoff, 1971:89–90). Modes of two-way communication, feedback, and demonstrations are critical.

Key actors in the implementation phase need also to be alert to the existence of needed communication capabilities within the larger pool of involved people. For example, studies have discovered the existence of information gatekeepers (Allen, 1985). These are individuals who, because of their wide-ranging social or professional activities and contacts, as well as their reading and special interests, can help link the organization and the external world. Like news editors, gatekeepers decide what is important. Small, face-to-face communities also contain opinion leaders, who have a larger than average role in shaping a consensus on issues of concern and hence are key participants in informal communications networks.

In order to reach potential participants and members of the larger community who may be affected, implementation also requires careful selection of communications *media*. This could include use of mass media, as well as locally produced newsletters and other written material, workshops, and various kinds of demonstrations. The general rule of thumb is the earlier in the process and/or the smaller the number of people to be reached, the better it is to use open, face-to-face communication techniques. Again, note in passing that some of these show-and-tell techniques will already have been mastered by participants in the previous stages of the approach. Workshops and demonstrations are particularly useful for promoting the adoption of processual changes among participants.

The mass media are noninteractive and can therefore suffer from the same limitations as the closed communications mode just discussed. Newsletters and other written materials produced and disseminated by the organization can be open media, or integral parts of open networks, if they are carefully responsive to their constituents, for example, by printing everything they receive from their readership. A new communications resource is the community-access channels that most cable television operators are legally required to provide. If used in a newsletter mode, as just described, community programming can be made to function as an open communications medium.

The selection of communications media, as well as other means for producing change, depends on resources. Co-ops and other local organizations concerned with change typically operate with limited resources — limited personnel, time, money, experience, expertise, and political influence. The kinds of collaborative, consensual, or educational approaches that have been emphasized require fewer resources than conflict approaches such as mediation or ajudication. The latter may be unavoidable, however, if structural change involving features of a situation that people do not presently control, appears to be the only way to go. Proposals for structural changes frequently meet with opposition, often well organized, and resources must be directed to addressing that opposition. Examples of such situations are those that can be improved only by taking on a governmental agency, a major processor or distributor, a prospective land or natural resource developer, and the like. Conflict approaches tend to be costly precisely because the other side usually has what appears to be unlimited resources that the group bringing the action has to try in some way to match.

Despite their costliness, proposals for structural change and the use of conflict approaches have become much more common since the 1950s and 1960s, when they were widely reported by the media and legitimized in the context of the civil rights movement and the rise of community activism. There is a substantial literature on the conceptual issues involved in conflict, as well as mixed conflict and collaborative approaches (e.g., Alinsky, 1971; Brager and Holloway, 1978). A variety of handbooks providing practical tactics and techniques have also resulted from experience in community organizing, consumer affairs, and welfare reform (Huenefeld, 1970; Kahn, 1970, 1982; Nader and Ross, 1971; Oppenheimer and Lakey, 1964).

Case Studies of Implementation

To flesh out the points raised in this chapter, the book concludes with two concrete cases of implementation. The first case involves the floral growers' associations whose earlier modeling efforts were presented in some detail in Chapter 5. It is an ideal case because it has been possible to follow it through one complete cycle of the soft systems approach (although one would expect also to monitor it in the future). The second case returns to the experience of one of the authors several years ago in attempting to organize a fruit and vegetable cooperative for the small-scale producers of a village in Papua New Guinea. This case was introduced in Chapter 5, and background materials are available in the appendix of this book. The experience shows how bugaboos, such as a priori assumptions, poorly defined Ws, failure to conduct a debate, and other violations of the soft systems approach, are finally expressed in implementation and how the errors are recovered by the people themselves.

The Floral Growers' Associations

The following is a sample portion of an implementation plan developed by the floral growers. The growers' modeling efforts and a portion of the narrative that accompanied the first version of their model were presented in Chapter 5. That narrative refers to the information-gathering and -disseminating subsystem. The same subsystem will be used to illustrate what an implementation plan might look like. To review, the people identified five major human activities in the information-gathering and -disseminating subsystem. These are as follows:

1. Collecting data on imports, exports, production, marketing, and distribution.
2. Assessing the needs of consumers and growers.
3. Every month, communicating these needs to researchers of Bliss State University's department of horticulture and the State Department of Agriculture as well as to growers who wish to participate in the research themselves.
4. Receiving information based on this research.
5. Putting this information in forms usable by growers and consumers.

Table 8.1 presents the implementation plan for the first function, collecting import, export, production, marketing, and distribution data. For the plan to be put in practice, the people involved had to develop an implementation plan for each major function identified under all subsystems. The growers worked in small groups to develop the initial implementation plan. Each group took a dif-

TABLE 8.1 Sample Section of the Floral Growers' Implementation Plan for the Information Collection Function

Activity to Be Implemented: An information-gathering and disseminating subsystem that involves the following five major functions:

Function A: Collects import, export, production, marketing, and distribution data.

Function B: Assesses consumer and grower needs.

Function C: Communicates monthly these needs to researchers of Bliss State University's department of horticulture, the State Department of Agriculture, and growers who wish to participate in on-farm research themselves.

Function D: Receives from researchers information based on their work.

Function E: Prepares information in forms usable to growers and consumers.

Implementation strategy:

Function A: An information-gathering and dissemination system that collects import, export, production, marketing, and distribution data. In order to operationalize Function A, the following strategic actions are planned:

1. *Strategic action:* Develop detailed listing of the kinds of data that the State Department of Agriculture (DOA) and the university already collect on a routine basis.

Performance measures:

Advisory board reviews qualifications of prospective volunteers related to data gathering and processing experience and other relevant aspects of their backgrounds.

A preliminary list of all routinely gathered data available from relevant sources is assembled and distributed to key actors and advisors, including university agricultural economists, for their comments on possible gaps and oversights.

Date of completion:

Volunteers selected, January 30,1988.

Listing completed, March 30, 1988.

Needed resources:

Data bases from DOA, the university, commercial and government banks, and other agencies.

Computer equipment and related supplies.

Who is responsible:

One member of the advisory board will be responsible for soliciting the membership of the three growers' associations and the community at large in search of prospective volunteer committee members. The final committee will consist of three growers and two members of the community who have interest, as well as data-gathering and management abilities. The advisory board makes the final selection among applicants.

Reporting procedures:

The advisory board member who volunteers to be responsible for this strategic action shall report on progress at each month's advisory board meeting and at all three growers' association meetings.

Budget:

Computer equipment is available, so initial cost of operationalizing this function is minimal.

Estimated cost is $300 for supplies, transfer of initial data bases, and circulation of listing of data bases available.

2. *Strategic Action:* Conduct survey with growers to determine the kinds of information needed, how often, and in what form.

Performance measure:

Phone survey instruments will be directed at all growers affiliated with three growers' associations. A minimum 75 percent response rate is required. Otherwise, followup calls will be conducted until minimum response is achieved.

Date of completion:

April 30, 1988.

Needed resources:

List of members' names and phone numbers from the three associations.

Four volunteers to conduct phone survey.

University faculty member or extension specialist to assist staff in designing survey and in training volunteers.

Computer equipment and spreadsheet software such as Lotus 1-2-3 or dBASE.

$600 for printing and distribution.

Who is responsible:

One grower and his wife, who volunteer to operationalize this function.

An extension agent, who would provide technical assistance.

University students interested in doing phone survey or volunteers from growers' associations and community.

One advisory board member as a liaison with cooperating institutions and organizations.

Reporting procedure:

Grower and wife report progress on monthly basis to liaison advisory board member.

Grower and wife provide summary of findings to all three associations at quarterly meeting.

Liaison board member reports to associations regarding progress at quarterly meetings.

Budget:

All labor is volunteered.

Computer equipment and programs are available through cooperative extension.

$600 required for printing and dissemination of results.

3. *Strategic Action:* Form partnership with State Department of Agriculture and university so that data-gathering services don't duplicate.

Performance measure:

Steering committee reviews partnership agreements, data sharing procedures, and data collecting responsibilities.

Compliance with agreements and adequacy and sustainability of procedures reviewed on semimonthly basis, with initial emphasis on volunteer staff burdens and morale.

Date of completion:

Initial agreements, procedures, and assignments established by May 30,1988.

Needed resources:

One advisory board member.

Volunteers responsible for operationalizing strategic actions 1 and 2.

Who is responsible:

Advisory board member.

Reporting relationships:

Advisory board member reports to associations and works with key board members and volunteers affiliated with strategic actions 1 and 2.

Results of partnership agreements reported to members and officers of the three growers' associations.

Budget:

No costs anticipated.

4. *Strategic action:* Develop procedures and processes for data sharing between DOA, university, private businesses (banks in particular), and data storage.

TABLE 8.1 (*cont'd*)

Performance measures:

During shakedown phase, unit operators will meet monthly to discuss problems and deal with bugs until they agree that procedures for data sharing (transfer from one unit to another) have been routinized and are working satisfactorily.

A mail survey to a sample of growers will evaluate usefulness of format for communicating data, and revisions will be made.

Grower sample will be resurveyed after use of revised format, if necessary.

Information specialist from cooperative extension will evaluate data storage system.

Dates of completion:

Data sharing procedures, March 30, 1988.

Data storage system developed, April 30, 1988.

Storage system tested by May 30, 1988.

Data storage system revised, August 30, 1988.

Needed resources:

Computer with 60-megabyte hard disk drive.

dBASE or other database-management software.

One volunteer familiar with construction of useful databases.

One advisory board member to volunteer to perform liaison for implementing strategic action 6.

Six interns from university student body.

Who is responsible:

Advisory board member acts as liaison so that board is kept informed of progress and concerns.

One volunteer to supervise/manage the creation and merging of data bases.

Six interns to help collect and enter data.

Interns are responsible to volunteer manager of data base.

Volunteer manager of data base reports to advisory board member responsible to board and associations.

Reporting procedures:

Interns and data base manager hold biweekly meetings.

Data base manager informs advisory board member of progress on a quarterly basis.

Advisory board member reports progress at board meetings.

When strategic action 4 is operationalized, report is made to floral associations.

Budget:

We are assuming that the data bases maintained by the State Department of Agriculture, the university, the State Department of Business and Economic Planning, and selected banks will be available free of charge.

Computer equipment is available.

Licensed copy of software package is available from extension.

$500 for storage backup tapes.

$600 to test usefulness of way in which data are displayed.

5. *Strategic Action*: Develop a simple manual of procedures that explains the routine operations of this unit so that new staff will have some documentation to help them get started.

Note: This manual is to be a guidebook for how to collect, store, and share data. It is to be practical and outline key procedures established during the first year of operations. Each advisory board member responsible for a strategic action area is to write up procedures related to that strategic action.

Performance measure:

A rough draft of manual will be circulated to staff working on this unit for comments and recommendations to be included in first draft. First draft will be circulated to board members, cooperative extension, and partners in data-sharing network for comments and recommendations to be incorporated in the final version.

Date of completion:

Staff comment period, June 30, 1988.

First draft done by August 1988.

Final draft done by December 1988.

Needed resources:

Computer with word-processing software package.

One volunteer to write manual.

Who is responsible:

A member of advisory board and one volunteer writer.

Advisory board member responsible for circulating first draft of manual to selected group and communicating comments and suggestions for revision to writer.

Reporting procedures:

Volunteers coordinating each strategic action are to write procedures related to their areas and give them to coordinator of strategic action 5.

Coordinator of strategic action 5 reports to advisory board when manual completed.

Budget:

None required.

6. *Strategic Action:* Formulate internship/co-op agreements with university so that interns are available on a routine basis, including the development of a job description.

Performance measures:

Student learning contract reviewed and signed by strategic action coordinator and faculty sponsor.

Faculty sponsor reviews student journals and other evidence of performance specified in learning contract and communicates with strategic action coordinator.

Date of completion:

Initial university units having appropriately qualified students identified by February 1988.

Agreements with university departments/colleges completed by April 1988.

Internship job description completed by March 1988.

Eleven interns with draft learning contracts available on a routine basis beginning January 1989.

Who is responsible:

Chair of advisory board.

Coordinators of strategic actions 2 and 4.

Reporting procedures:

Chair of advisory board reports on progress at advisory board meetings.

Report made to associations when agreements are reached.

Budget:

None.

7. *Strategic Action:* Develop staffing plan, including job descriptions and reporting relationships; how staff will be trained and appointed; how long their term of service is; and recruiting procedures.

Note: By *staffing plan* we mean the operational plan that identifies how people are going to be selected and trained so that the strategic function of collecting data is sustained beyond the initial phase of implementation.

Performance measures:

Within six months of implementation, advisory board members serving in operational

TABLE 8.1 (*cont'd*)
roles will meet individually and in groups with function staff to evaluate the adequacy of
job descriptions, possible overlap of responsibilities, reporting relationships, and
arrangements for recruiting and training new volunteer staff.

Date of completion:

December 1988.

Needed resources:

Advisory board members who participated in the implementation of strategic actions
1–6.

One volunteer to write up plan.

Who is responsible:

One advisory board member who will volunteer to coordinate this strategic action.

Reporting procedures:

When plan is completed, coordinator will make presentation to total advisory board
and to floral associations and will thereafter report routinely on staffing issues.

Budget:

None required

---

ferent function and developed some preliminary ideas of how it might be implemented. All agreed that the first year was to be an implementation year; it would
be devoted to getting the activity up and running. During that first year as well,
plans would be made regarding how the function would be sustained.

An implementation plan can be developed in a number of ways: (1) one person can volunteer to do the first draft in order to facilitate the initial discussion
by a larger group of participants; (2) a small group can be assigned this task; or
(3) if it is important that most of the participants remain involved in order to
ensure their commitment later on, then the larger group can be divided into small
working groups, each with its own assignment.

The floral growers selected a process similar to number 3. First, however, the
group as a whole met to establish guidelines and limits that would apply across
the board to all the functions they were to work on. The principal guideline they
developed was that since initially there would be little money available, any
strategic actions included in a plan would have to be at no or very low cost. For
example, wherever possible, volunteer labor was to be specified in the plan if
the group knew that volunteers or other sources of free labor were available or
could be found.

Then they worked through all the subsystems in a series of meetings (one
subsystem a meeting, each meeting lasting two hours with plenty of food available!). At each meeting they divided into working groups. For a given meeting,
the number of working groups corresponded to the number of key functions
listed under the subsystem statements (as shown in the model). The first task of
a given working group was to list the strategic actions they felt were needed in
order to operationalize the function they were assigned. Next they preliminarily
developed a sequence for these strategic actions (this changed several times as
developmental discussions continued). Performance measures, completion dates,
resource needs, assignment of individual responsibilities, reporting procedures,
and budget estimates were developed for each strategic action.

A Papua New Guinea Co-op

Chapter 5 included a retrospective soft systems analysis of a self-development initiative that one of the authors had observed in a community in Papua New Guinea. Based on the case materials available in the appendix, a conceptual model for the people's original plan was defined and formulated. The discussion went on to show how the people involved had revised it over a period of approximately ten years. Also presented was the naive researcher's attempted intervention to improve the business subsystem of the overall model. The reader should review the case materials and the discussion of those modeling activities (in Chapter 5) before turning to the subsequent presentation of the implementation phase.

What follows is a historical or retrospective soft systems analysis (Checkland, 1981:194ff) of a qualified success with special attention devoted to the intervenor's failures and excesses along the way. It also provides a down-to-earth look at what some critics of conventional international development assistance activities call bottom–up development.

Implementation of the designed system began with a variety of actions by the intervenor that involved things that community people could not do themselves. He wrote letters to the camp managers of the two mining operations in the larger region, Ok Tedi and Frieda River, as well as to merchants in a coastal town that served as the region's commercial center. He discussed air freight rates with the Missionary Aviation Fellowship (MAF)—the missionary airline serving bush airstrips in the area—and determined that very favorable backload rates might make some outlets feasible. He also negotiated with the Australian Baptist Mission Society to permit nonmissionary aircraft to use the airstrip to pick up produce, under the assumption that buyers such as the mining operations might find it cheaper to use fixed-wing aircraft rather than helicopters. He queried the extension services, both the locally based agents at the district headquarters at Telefomin and in the provincial capital, regarding the availability of model co-op by-laws and other educational materials as well as flats and produce bags (but received no reply).

The intervenor also worked on an accounting procedure that would permit managers to weigh and record individual growers' deliveries and then, after receipt of payment from buyers, to carry out simple arithmetic to generate individual growers' payments. Weighing was also necessary to calculate aircraft loads. He even loaned spring scales to the enterprise, intending that such instruments would go on the shopping list for capitalization in the future.

One trial shipment and sale to a market at the district headquarters at Telefomin was organized. The idea was to use some community members who were working there to unload the produce, move it to the site, and conduct the sales. For reasons not understood, the trial failed, and much produce was wasted. An unanticipated spinoff was that the MAF pilot, on his own, took several stalks of bananas to Oksapmin, the site of the only functioning co-op in the region, and they were well received. This formed the basis for more enduring business relationships later on, when Oksapmin served as a conduit for some Miyanmin produce, especially fruit and bananas, to the Ok Tedi mine. Later, too, arrangements were made through an Overseas Volunteer Service extension worker to send two Miyanmin managers to Oksapmin for an apprenticeship.

The foregoing reports the up side of the intervention. The implementation of what was to become the most remunerative kind of operation seemed to be a disaster at the time. In response to one solicitation letter, the manager of the Frieda mining camp dropped in by helicopter one afternoon and cut a deal involving semiweekly deliveries of fruit and vegetables on a set schedule (since there was no radio contact with the outside). Some details were vague, such as whether a chopper or fixed-wing aircraft would be involved, but the buyer provided a breakdown of what was needed. People began bringing produce in two days before the scheduled delivery. It was thus possible to begin to use the accounting procedure, described previously, in a leisurely way. The intervenor supervised weighing the two price classes of produce and observed the entry of the information in the books, as well as the placement of the produce on the weighed pile, with a small paper label indicating weight and grower.

Everything looked fine until the morning of the first delivery, when the management was overwhelmed by optimistic growers and a mountain of commodities. The intervenor felt that he had to be everywhere at once, and at one point the leading manager, who occupied the position by virtue of his experience running the small local co-op store, went and hid in the bush! When the total order specified by the buyer had been reached, it was necessary to announce that no more produce would be accepted. The chopper finally arrived and disgorged a large, fine-meshed cargo net, which was promptly filled with the carefully weighed and recorded delivery. The intervenor was too busy to notice the restiveness of the crowd, but the sight of the chopper hovering to pick up the loaded net triggered something. There was a mad rush of disappointed would-be sellers to heave their produce into the net too.

This raises the obvious question for a change agent in this kind of situation: "How do you organize so many *very* small-scale growers?" The answer was also provided: "You don't!" In effect, you let them do it themselves!

Payment for the first delivery was received several weeks later. In the interim, the managers and the intervenor had worked through the books, adding the names of last-minute vendors and assigning them a nominal share of expected proceeds. At the people's request, the intervenor disbursed the money one Sunday morning after church. By local standards, it was an immense amount, although individual returns were modest. Nevertheless, people came forward to collect their money glowing with pride. The feeling in the group was palpable.

Feedback on the mix of commodities this customer desired in the next order was also received. The buyer wanted more sweet potatoes and lemons! This information was disseminated to growers. Otherwise, the intervenor vowed to keep his hands off the next delivery, except for making sure that the helicopter was not overloaded, a safety concern.

On the next occasion, people severely scaled back the intervenor's conceptual model. Names of individual growers contributing to the order were recorded, but weighing of produce was sporadic and ultimately abandoned under the press of business. Otherwise, the day came off well, the order was picked up in a timely way, and the buyer paid off on the spot. The division of this payment was particularly revealing; the total was *divided by the number of growers*! And when the proceeds were disbursed on this basis, there was universal satisfaction.

Superficially, this situation resembles the fate of many marginally successful development interventions. The recipients of the designed package partially disassembled it and then used parts of it in their own way. Even more understanding can be wrested from the case by cycling through the soft systems approach. This can start with a revised (sub)system definition and discussion of some facts that support the revision. Clearly, the (sub)system definition and conceptual model presented earlier in Chapter 5 cannot explain what happened here. A closer look at the overall modernization-providing system presented in Chapter 5 gives some guidance. For the time being, the business subsystem might be more realistically expressed as a system whose output is cash production. This cash is, in turn, a vital input to the education production susbsystem. In the larger scheme, the output of education is, in turn, the improved future for the next generation. So a revised (sub)system definition might look like this:

> A community-owned, cash-producing system engaged in growing and selling vegetables and fruit in order to improve people's ability to pay school fees and related extraordinary expenses of modern society.

The revised conceptual model, reflecting this system definition, is presented in Figure 8.1. It was subsequently followed with no complaints or difficulties. This forcefully brought home the radical difference in **W** between the intervenor,

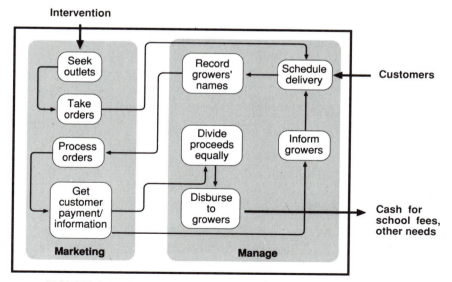

FIGURE 8.1 Conceptual model of changes actually implemented.

who was socialized in a stratified, capitalist society, and the Miyanmin people, who lived in an egalitarian, communal society.

The people's basic objectives and priorities became clear in time. Late in February 1981, a local financial crisis was precipitated by the start of the new school year, when tuition fees were due. Primary school fees are minimal, the equivalent of three U.S. dollars or so. Parents of children who had been selected for academic high school, however, were faced with what was for them an astronomical sum, in excess of 150 U.S. dollars, plus air fares and incidental expenses. A few families had the money because fathers had spent several years away from the community as agricultural contract laborers. Others had only a portion, and the intervenor offered to make up the difference. Two families reluctantly accepted help. Friends told him that the reluctant families were "ashamed," and the desire for self-sufficiency was expressed repeatedly in debates, which occurred in church in subsequent weeks. Thus the "real" purpose of business development emerged, at least for the time being. It was to be able to buy educational services for their kids. In other words, there is a very strong future orientation among these people and an authentic vision of the role and purpose of education. Interestingly, people were not interested in just *any* education.

Children who graduate from primary school in good standing may attend a secondary vocational school, for free. In the late 1970s, the West Sepik vocational school, in the provincial capital of Vanimo, had been restructured around a model curriculum developed by the World Bank. The problem addressed by the curriculum was that, due to the poor job-market characteristic of the modern sector of many Third World economies, high school graduates faced chronic unemployment and ended up hanging out on the streets of urban centers. Hence, this vocational curriculum sought to teach skills that would be "useful in villages" to encourage graduates to return to their home communities. Of course, the contradiction was that the vocational education provided by traditional societies before they were disrupted by outside intervention was originally more than adequate to this task.

The first Miyanmin youngsters had attended this newly renovated vocational school in 1980. They wrote letters home describing what they were being taught. Parents were scornful and angry: "Cabbages? I don't want my kid learning how to grow cabbages! I want him to learn business or mechanics or construction." In accordance with their vision, they pulled their children out of the school.

When the intervenor left to return to the United States in June 1981, community income from the business had grown from a nominal zero to 450 dollars per month. It has continued to prosper. Now the only Miyanmin kids needing assistance with school fees are residents of remote villages who lack access to this community enterprise.

The foregoing case of an attempt to improve a cooperative enterprise in a small Third World community illustrates many points raised earlier in this chapter and elsewhere in the book. The first is that procedural changes, which are what the case finally boiled down to, can produce satisfactory results. Note, however, that these changes were implemented against a background of profound structural and attitudinal changes effected in previous decades.

A second lesson has to do with the character of the intervention. Initially, it was pursued in a more or less authoritative mode, although it clearly was invited by the people. The intervenor already knew intellectually that authoritative behavior was not appropriate in Miyanmin society, but he had to have this knowledge reinforced by people's assertion of their customary autonomy. The intervenor quickly backed off.

The third issue is the benefit of adhering to the participant mutual-learning mode from beginning to end, including using the debate mode to develop specific changes. The standard anthropology research technique of participant observation does involve mutual learning, but it is not part of an anthropologist's **W** to view the world in intervention terms. In practice, the intervenor and the people involved ended up learning side by side only because there was no other way to get the job done. Had the intervenor known of the soft systems approach then, things would have gone more smoothly, although the short-term outcome would have been similar.

This case also reinforces the importance of getting the system definition and CATWOE more on target with those who are undertaking change before rushing to start changing things. Note the changes in system definition, in specific transformations, and especially in the **W** from one cycle to the next. A lot of the intervenor's evident misapprehension had to do with the desirability of the system and, even more dramatically, of the subsystem. Desirability, of course, refers to knowing what people really wanted, this in turn based on adequately capturing their **W** as part of the task of system and subsystem definition.

Summary

This chapter has described the five main tasks of the implementation stage of the soft systems approach and discussed the alternative roles of the analyst/facilitator in implementation. The first task is to develop an implementation plan. In a highly detailed fashion, an implementation plan spells out the "how," "who," and "when," as well as the budget, resources, communications, and monitoring required for each distinct strategic action. It was not possible to be specific about these elements of the plan because the possible actions, even within a given plan, are so varied. In addition, the distinction between structural, procedural (or processual), and climatic (or attitudinal) change was presented. Structure refers to relatively durable social and environmental features of a situation. Changing them directly is difficult, riven with conflict, but not impossible. Procedure refers to how things are done within a given structure. Procedural changes are more readily implemented and cumulatively may lead to structural change. Climate or attitude is related to the **W**s characteristic of a situation and how these affect immediate feelings as a situation changes. Climate is also difficult to attack directly but tends to change by virtue of participation in the soft systems approach and as changes are implemented and seen to improve a situation.

Communications was discussed in some detail, both the theory and specific applications appropriate to the industrial, agricultural, and environmental contexts in which you are likely to find yourselves. Communications competencies are essential in all phases of the soft systems approach, but they come to the fore in the implementation phase, when the number of people involved greatly expands.

The fact that the application of the soft systems approach does not really end with implementation is signaled by the requirement that performance measures be built into the implementation plan for each strategic action. In other words, once an agreed-upon change is implemented, someone must be responsible to monitor its consequences, and there must be explicit criteria or procedures for doing so, to ensure that the action is working as intended and that it is not having unintended and undesirable side effects. The discovery of problems is an indication that another cycle of the soft systems approach may be required in order to design modifications of a given subsystem's function.

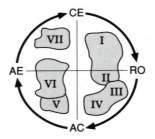

The key question regarding the role of an analyst/facilitator in the implementation phase is: When is the job complete? Or when is it practical and ethically appropriate to leave? Three alternatives were discussed, each of them potentially sound in different contexts. Analyst/facilitators who have spent most of their professional lives working in formal organizations, such as business firms, advocate withdrawing after the debate phase and the submission of a final report to the client. This may be entirely appropriate in that context, but working with small, unorganized, and resource-poor groups may require full involvement in implementation and subsequent monitoring activities. There are numerous variations between the two extremes.

The chapter concluded with two substantial cases of implementation. The first, involving a group of floral growers, tended to show the virtues of taking the soft systems approach full cycle, from describing a situation through implementation. Although some potential sticking points were discussed, implementation seemed likely to proceed on a firm conceptual and empirical basis. The second case, involving an attempt to intervene in the development of a co-op in a small community in Papua New Guinea, illustrated the usefulness of the soft systems approach by default. Because the intervenor rushed to implementation without having cycled through the approach, many problems arose, which were corrected mainly by the people involved taking charge. In this case, the soft systems approach was useful for looking back on that situation to draw valuable lessons. These include the need to enter into a situation with an open mind, to establish an authentic mutual learning approach with the people involved, to explore the **W**s involved in a situation fully, to conceptualize a vision of the future in a formal way, and to subject that picture to a reality test through debate and other means — before introducing changes.

REFERENCES

Alinsky, Saul. *Rules for Radicals*. New York: Vintage, 1971.

Allen, T. "Communications in the Research and Development Laboratory." In *Organizational Psychology: Readings in Human Behavior in Organizations*, 4th Ed. D. A. Kolb, I. M. Rubin, and J. D. McIntyre (Eds.). Englewood-Cliffs, NJ: Prentice-Hall, 1985.

Arensberg, Conrad M., and Arthur H. Niehoff. *Introducing Social Change: A Manual for Community Development*. Chicago: Aldine-Atherton, 1971.

Benton, Lewis, ed. *Management for the Future*. New York: McGraw-Hill, 1978.

Bohannon, P. "The Migration and Expansion of the Tiv." *Africa* 24(1):2–16, 1954.

Boserup, Esther. *The Conditions of Agricultural Growth: The Economics of Agrarian Change Under Population Pressure*. Chicago: Aldine, 1965.

Brager, G., and S. Holloway. *Changing Human Service Organizations: Politics and Practices*. New York: The Free Press, 1978.

Checkland, P. *Systems Thinking, Systems Practice*. Chichester: Wiley, 1981.

Festinger, L. *A Theory of Cognitive Dissonance*. Stanford: Stanford University Press, 1957.

Gambling, Trevor. *Societal Accounting*. London: George Allen & Unwin, 1974.

Gardner, J. *Self-Renewal*. Evanston: Harper & Row, 1963.

Gerlach, L. P., and V. H. Hine. *People, Power, Change: Movements of Social Transformation*. Englewood Cliffs, NJ: Prentice-Hall, 1970.

Gerlach, L. P., and V. H. Hine. *Lifeway Leap: The Dynamics of Change in America*. Minneapolis: University of Minnesota Press, 1973.

Huenefeld, J. *The Community Activist's Handbook*. Boston: Beacon, 1970.

Kahn, S. *How People Get Power*. New York: McGraw-Hill, 1970.

Kahn, S. *Organizing*. New York: McGraw-Hill, 1982.

Katz, Alfred H., and Eugene I. Bender. *The Strength in Us: Self-Help Groups in the Modern World*. New York: Franklin Watts/New Viewpoints, 1976.

Katz, D. "The Functional Approach to the Study of Attitudes." *Public Opinion Quarterly* 24:163–204, 1960.

Lawrence, P., and J. Lorsch. *Developing Organizations: Diagnosis and Action*. Reading, MA: Addison-Wesley, 1969.

Marris, P., and M. Rein. *Dilemmas of Social Reform*. Chicago: Aldine-Atherton, 1973.

Nader, R., and D. Ross. *Action for Change*. New York: Grossman, 1971.

Naughton, J. "Theory and Practice in Systems Research." *Journal of Applied Systems Analysis* 8:61–70, 1981.

Oppenheimer, M., and G. Lakey. *A Manual for Direct Action*. Chicago: Quadrangle, 1964.

Pfeiffer, J. William, Leonard D. Goodstein, and Timothy Nolan. *Applied Strategic Planning: A How to Do It Guide*. San Diego, CA: University Associates, 1987.

Rogers, Everett M. *Communication of Innovation: A Cross-Cultural Approach,* 3rd ed. New York: The Free Press, 1981.

Rothman, J. "Three Models of Community Organization Practice: Their Mixing and Phasing." In *Strategies of Community Organization,* 3rd. ed. F. M. Cox, J. L. Erlich, J. Rothman, and J. Tropman (Eds.). Itasca, IL: Peacock, 1979.

Rothman, J., J. L. Erlich, and J. G. Teresa. *Promoting Innovation and Change in Organizations and Communities*. New York: Wiley, 1976.

Susskind, L., and J. Cruikshank. *Breaking the Impasse: Consensual Approaches to Resolving Public Disputes*. New York: Basic, 1987.

Talbot, A. T. *Settling Things: Three Case Studies in Environmental Mediation*. Washington, DC: The Conservation Foundation, 1983.

Thurstone, L. L. "The Measurement of Social Attitudes." *Journal of Abnormal and Social Psychology* 26:249–269, 1931.

Toffler, A. *Future Shock*. New York: Bantam, 1970.

Triandis, Harry C. "Exploratory Factor Analyses of the Behavioral Component of Social Attitudes." *Journal of Abnormal and Social Psychology* 68:420–430, 1964.

Triandis, Harry C. "Social Science and Development: Social Psychology." In *Behavioral Change in Agriculture*. J. P. Leagans and C. P. Loomis (Eds.). Ithaca, NY: Cornell University Press, 1971.

Appendix

Student Case Materials

Three sets of case materials have been assembled to provide depth of detail to the cases that run through the book beginning in Chapter 4. The possibilities for analysis have not been exhausted, however. Instructors and students are encouraged to review one or more of these cases in order to explore alternatives not considered here and to gain experience in working through the various stages of the soft systems approach. These materials also indicate the kinds of information sources that might be useful in stage 1 of the soft systems inquiry process, as well as illustrating some of the products of stage 2.

Case 1. Chatham River

Prepared by Dr. Robert Sowell and Mr. David Miller, North Carolina State University, with contributions by Dr. Lovel Jarvis, University of California, Davis, and Dr. George E. B. Morren, Jr., Rutgers University.

Note: Names of places and principal parties and some details have been changed to protect confidentiality.

Outline of Materials

    I. Overview of case
    II. Transcript of public hearings conducted by state Office of Water Resources
    III. Sample model output; river simulation study (Figure A.1)
    IV. Composite mind map
    V. Summary of interview with riparian/farmer
    VI. On water law
    VII. Relevant court decisions
    VIII. Synthesis statement

I. Overview Of Chatham River Case

In 1981 the town council of Springville, in the upper coastal plain of North Carolina, recommended that the town augment its water supply by constructing a pumping plant for withdrawing water from the Chatham River. Two factors prompted this recommendation. First, growing domestic demand for water had already strained the existing system to its limits. In addition, Springville, like many other southeastern towns, was attempting to attract light industry, and the town council felt that the inadequacy of the local water supply hurt Springville's chances for success. Reasons for seeking manufacturing plants in or near the town were several, but the principal one was the feeling, shared by the vast majority of the area's inhabitants, that manufacturing would bring employment opportunities to complement the uncertain agricultural economy upon which most people relied. Town leaders hoped that an assured water supply would help bring jobs to the area and that the expanded tax base created by industry would absorb much of the cost of the new pumping and treatment plants.

At the time the town council decided to expand the water supply, it was pumping water from three wells located within the city limits, and it initially considered simply increasing the number of its wells. However, engineering studies indicated that it would be less costly for the town to treat water pumped from the Chatham River, which formed the western boundary of the town, than to install and operate additional wells.

Announcement of these findings and of the town's intention to withdraw water from the Chatham generated opposition from others who felt that they had a right to use the river's water. This opposition may have been exacerbated by the relative rarity of water rights disputes in the Southeast. The groups opposing the town council's proposal to withdraw water from the Chatham felt strongly about their rights, but there were few legal precedents defining what these rights were or how they might best be defended. What has evolved up to this point is more a hardening of attitudes than any serious attempt at resolution.

II. State of North Carolina, Office of Water Resources

Excerpts from the proceedings of a public hearing relating to an application filed by the board of commissioners of the town of Springville requesting approval of their intent to pump water from the Chatham River. J. J. Whitfield, director, presiding

Mayor Hall, mayor of the town of Springville, North Carolina: On the behalf of the town of Springville I would like to thank you for calling this meeting here to consider our problem. As you probably know, Sunderland County is one of the fastest-growing counties in the state. Springville, located approximately six miles from the city of Jefferson, is also beginning to experience a high rate of growth.

At the present we have three wells in Springville supplying our water. In 1983 Mr. Paul M. Van Camp, a civil sanitary engineer for the town of Townsend, made a survey of the town's water supply. In his report he said that well no. 1 was pumping approximately 57 gallons per minute; well no. 2, 14.5 gallons per minute; and well no. 3, 64 gallons per minute. This comes out to approximately 135 to 140 gallons per minute. In this particular area well water is hard to obtain. We recently had a well dug in town and all we could get there was 75 gallons per minute. The town currently uses approximately 200,000 gallons of water per day. We anticipate the population of Springville will triple in the next 20 years. Therefore, these are the reasons that we hope to withdraw water from the Chatham River.

We have potential of more housing developments and industry in Springville. In fact, right now we have one textile firm that is looking at a site in town. We anticipate this company will locate here. Of course they would consume a great amount of water each day. This is another reason that we request approval to withdraw water from the river. I would be pleased to try to answer any questions that you have at this time.

Chairman Whitfield: Any questions?

Mr. Hubbard: I have one, Mr. Chairman. Mayor Hall, in the resolution requesting this hearing, it was stated that the town would develop this as a source of public water supply. Have you any information as to how soon this development might occur?

Mayor Hall: No, Mr. Hubbard, we do not have any data, but we have discussed this with engineers around the state. As you know, we are small and something like this would probably require outside help. We store approximately 1,000,000 gallons of water in Springville. When I left this morning, we had a half a tank of water. So you can see the problem we are likely to have when it comes to July. Last year our tank got down below the quarter mark in July. Talking about doing some sweating, the people of Springville really prayed for rain.

Mr. Hubbard: Mr. Hall, may I ask one or two other questions? Have you employed an engineer, consulting engineer, to prepare any preliminary estimates or plans of the development of this stream as a source of public water supply?

Mayor Hall: As of yet, Mr. Hubbard, we have not because we felt like our first step should be to see if our plans met with the support of the county and the city.

Mr. Hubbard: What I am trying to bring out for the benefit of the Committee, Mr. Hall, is whether or not you have any immediate plans, in the next four or five years, to go ahead with this project or whether this is something projected in your long-range planning.

Mayor Hall: Mr. Hubbard, as you know, you can never tell what future town councils will do. But with the present water situation and the projected growth rate in Springville, I don't see where the town would have any choice other than to consider this a source of water. I, and the town council members with whom I have talked, would like to act in the very near future.

Mr. Hubbard: Anyone else have a further question of Mayor Hall? Thank you, Mayor Hall.

Mayor Hall: Thank you, sir.

Chairman Whitfield: I would like now to recognize Mr. L. A. Priest, the town clerk at Springville, for such comments. Mr. Priest.

Mr. L. A. Priest, town clerk, Springville, North Carolina: I don't know a great deal I can add to what Mayor Hall has said. I have been affiliated with the town of Springville now for 21 years and know that our water system was installed in 1963. We have a rock formation that restricts drilling to depths of about two hundred feet. We do have a stream that we feel could be an asset to the community. We are growing, as Mayor Hall says. We have over 500 customers now. With an average of 3.8 persons per home, we can now provide water for almost 2,000 people. We showed a population of 1,100 in the 1980 census. Building permits selling as they are now, I feel sure that by 1990 we will hit 3,000 and we cannot supply the needed water with wells.

Mr. Hubbard raised a question concerning timing of the project. This is not something for the distant future. We have an emergency situation now.

Now we know that we are going to have a lot of obstacles to overcome, because revenue for small towns comes slow and we are going to have to get a lot of help. We might get some federal help. We are going to have to have a filtering plant in the town of Springville in the very near future. It is something we must do now, and the reason we have not hired an engineer was because we didn't know what type of treatment plant to ask for or what you would recommend. We felt that our first step was to find out where you stand on our withdrawing and treating water from the river. With your interests in mind, we would then do what we could about getting an engineer and finding out what we can do to get a water system for the town of Springville.

Mr. Hubbard: Thank you, Mr. Priest. I am glad you emphasized the urgency of the situation. I was trying to get the information into the records that this is something you anticipate doing within the relatively near future and not a long-range plan.

Mr. Priest: That is right, sir, that is right.

Chairman Whitfield: Any question that the members of the committee would like to ask Mr. Priest?

Mr. Clary: Mr. Chairman, sir, may I ask Mr. Priest a question, please?

Chairman Whitfield: Mr. Clary.

Mr. Clary: Mr. Priest, as you look to the future, have you considered the possibility that the city of Jefferson will extend its system so it is in reaching distance of Springville?

Mr. Priest: That is a slight possibility, but I don't believe that it's something we can depend on now. Jefferson has its problems, too, and whenever we put Springville's problems and Jefferson's problems together, you have got a lot of problems, so I don't believe that is something we can depend on getting us out of an emergency right now.

Mr. Hubbard: Mr. Priest, may I ask you one other question? Does the town keep records which would indicate the amount of water that you are using on an average day?

Mr. Priest: Yes, sir, they are available at any time.

Mr. Hubbard: Do you recall what your average daily use is?

Mr. Priest: Our average daily use at this time is 175,000 to 200,000 gallons per day.

Mr. Hubbard: Per day?

Mr. Priest: Per day.

Mr. Hubbard: That requires the pumping of your wells . . .

Mr. Priest: Constantly.

Mr. Hubbard: Normally, constantly, I would think.

Chairman Whitfield: Any further questions of Mr. Priest? Thank you very much, Mr. Priest.

Mr. Priest: Thank you.

Mayor Hall: Mr. Chairman, we have with us, also, Mr. A. N. Bradford, member of the Springville Town Council.

Chairman Whitfield: Mr. Bradford, we would be glad to have your comments.

Mr. A. N. Bradford, member of the town council, town of Springville, North Carolina: Thank you, Chairman Whitfield. I would like to fill in a few background comments.

Chairman Whitfield: Go right ahead, sir.

Mr. Bradford: Concerning the emergency nature of this request, last year we had occasion to ask for assistance and advice because we thought we were going to have to use the stream at that time. Mayor Hall contacted the Corps of Engineers in Wilmington and asked them to provide for an emergency pump and other equipment and advice as needed. We got that desperate at one time. We haven't been exactly idle and trying to put everything on someone else's shoulders. We renovated two wells last year, and we were happy with the results. We also drilled a well last year. We haven't put in a pump unit yet, but we expect to do so in the forthcoming budget year, as soon as we have the money available. But with the present growth rate, which seems to be accelerating very sharply lately, I would say within five years or less we would very likely need to use water from the river. In the past we have considered it as a source of recreation and as our town's outstanding beauty spot, but we are going to have to use it for more serious purposes in the near future. Thank you.

Chairman Whitfield: Thank you, Mr. Bradford.

III. Sample Model Output

The accompanying tables are sample outputs of a computer model commissioned by the State Department of Water Resources to assess the impacts of Springville's proposed abstractions from the Chatham River. The actual model and runs were conducted by the department of biological and agricultural engineering, North Carolina State University, Raleigh, under contract OWR/SU/NCAES-84–22. Professor Robert Sowell was the overall project director. The results are provided for purposes of scientific communication only and do not represent actual predictions of flows, levels, environmental impacts, and public benefits and costs by the Office of Water Resources.

Printouts provided as figures below include the following sections:

   I. Rainfall vs. streamflow.
   II. Crop returns vs. irrigated area, crop mix, and market price.
   III. Irrigation water needed for optimal returns for crops.
   IV. Domestic and industrial needs.
   V. Reduced streamflow vs. requirements for fish habitat.

```
        I      A      II B II    C    II  D  II    E    II    F   II    G   II   H   I
 1                               CASE: CHATHAM RIVER
 2    SECTION I.  Rainfall vs. Streamflow -------------------- Rows 21-40
 3                INPUTS: C35, D35, E35, F35, monthly rainfall for May, June
 4                     July, and August Respectively.
 5    SECTION II. Crop returns vs. irrigated area, crop mix, and market price
 6                --------------------------------------------- Rows 41-60
 7                INPUTS: C51, irrigated land area
 8                        B52, B53, and B54, portion of area in corn, tobacco,
 9                        and vegetables, respectively. (input as decimal)
10    SECTION III. Irrigation water needed for optimal returns from crops.
11                 NO INPUTS: -------------------------------- Rows 61-80
12    SECTION IV. Domestic and industrial water needs. -------- Rows 81-100
13                INPUTS: C87, Projected growth rate. (input as decimal)
14                        H86, Domestic water consumption.
15    SECTION V.  Reduced streamflow vs. flow requirements for fish habitat.
16                --------------------------------------------- Rows 101-120
17                INPUTS: D109, E109, F109, and G109, domestic and industry
18                       water - base on SECTION IV.
19                        D118, E118, F118, and G118, minimum flow required
20                        for fish habitat.
21    ===========================================================================
22    ************    Rainfall Frequency Distribution
23    SECTION I.  *   (likelihood of being less than x, in)
24    Rainfall vs.*        May       June       July      August
25    Streamflow  *      ~~~~~~~    ~~~~~~~    ~~~~~~~    ~~~~~~~
26    ************    10%  1.12      2.13       3.11       2.87
27                    20%  2.13      2.62       4.11       3.62
28                    30%  2.62      3.11       4.63       4.63
29                    40%  2.87      3.62       5.71       5.12
30                    50%  3.37      4.63       6.12       5.61
31                    60%  3.62      5.12       7.36       6.12
32                    70%  4.11      5.61       8.37       7.13
33    ------------------------------------------------------------------
34    Input Monthly
35    Rainfall -------------->   2.00      5.00       6.00       3.00
36    ------------------------------------------------------------------
37    Monthly streamflow
38    (cfs) -------------->      532       462        442        88
39    (mgd) -------------->      344       299        286        57
40    ===========================================================================
```

```
40 =================================================================================
41 =================================================================================
42 ******************
43 SECTION II.         *
44 Crop Returns vs.    *                    +++++Yields+++++    +Gross Returns+
45   Irrigated Area  *        Expected    Water              Water
46   Crop Mix          *        Market      Not   Rainfall    Not    Rainfall
47   Market Price      *        Price    Limiting  Only    Limiting   Only
48 ******************           $/       units/   units/     $/        $/
49                              unit      acre     acre      acre      acre
50 Total Acres of Land --) 10000 ~~~~~~   ~~~~~~   ~~~~~~    ~~~~~~    ~~~~~~
51   % in Corn ------) .25   2500  1.95     110       72       215       139
52   % in Tobacco ---) .25   2500  1.55    2300     1725      3565      2674
53   % in Vegetable -) .5    5000  4.00     250      100      1000       400
54 Cost of Irrigation Water $/ac ft 25
55
56                                          [A]      [B]              [A]-[B]
57 Total Gross Returns for Area ($1000)--14449     9033               5416
58 Total Net Returns for Area ($1000)---- 3633      349               3283
59
60 =================================================================================
61 =================================================================================
62 ****************************************************************
63 SECTION III. Irrigation water needed for maximum yields. *
64 ****************************************************************
65                              Acre-Feet of Water
66                              ~~~~~~~~~~~~~~~~~~~
67                         May   June   July   August
68     Corn -------------- 182    0     284    441
69     Tobacco ----------- 182    0     284      0
70     Vegetables -------- 571    0       0      0
71        TOTALS --------- 935    0     568    441      1944  -
72                          -
73 Total Water Needed
74 per Month (million gal) 305    0     185    144
75
76 Average Daily Water
77 Needed (mgd) --------- 9.83   .00   5.97   4.64
78
79
80 =================================================================================
```

```
80 =================================================================================
81 =================================================================================
82 **************************************************
83 SECTION IV. Domestic and Industrial Water Needs. *
84 **************************************************
85 Current Year -------------------) 1986    Water Consumption
86 Current Population ------------) 5000      (gal/person/day) --------) 150
87 Projected Growth Rate ---------) .02       Water Available from
88                                            wells (1000 gal/day) --) 800
89 +++++++++++++++++++++++++++++++++++++++++++++++++++++++++++++++++++++++++++++++++
90                            Water Required           River Water
91              Popu-        (million gallons/day)        Needed
92        Year  lation    Domestic  Industry  Total       (mgd)
93        ~~~~  ~~~~~~    ~~~~~~~~  ~~~~~~~~  ~~~~~~     ~~~~~~~~~~~
94        1990   5412       .81      .00      .81          .01
95        2000   6597       .99     9.00     9.99         9.19
96        2005   7284      1.09    12.00    13.09        12.22
97        2010   8042      1.21    18.00    19.21        18.45
98        2015   8879      1.33    21.00    22.33        21.56
99        2020   9803      1.47    27.00    28.47        27.60
100 =================================================================================
```

```
101 ================================================================================
102 ****************************************************************
103 SECTION V. Streamflow vs Flow Requirements for Fish Habitat.  *
104 ****************************************************************
105
106                                     May   June   July  August
107 Withdrawals from River (mgd)       ~~~~~ ~~~~~~ ~~~~~ ~~~~~~~
108   Agriculture ---------------)      9.83    .00  5.97   4.64
109   Domestic & Industry -------)     10.00  10.00 10.00  10.00
110 Total Withdrawal (mgd) -----)      19.83  10.00 15.97  14.64
111 Equivalent Streamflow (cfs)-)      30.69  15.47 24.71  22.65
112
113 Natural Streamflow (cfs) ---)        532    462   442     88
114 Streamflow after
115  Withdrawals (cfs) ---------)        501    447   417     65
116
117 Minimum Streamflow Required
118  for Fish Habitat (cfs) ----)        200    100   100    100
119
120 ================================================================================
```

FIGURE A.1 Sample model output I, II, III, IV, and V.

Note: This DBase III spreadsheet simulation model is available for the IBM PC and compatible computers running MS-DOS 2.11 or later versions. The simulation is divided into five sections. Section I computes stream flows based on rainfall distribution. Section II computes crop yields and gross and net returns from crops based on a variety of scenarios involving a mix of three crops and use of irrigation versus rainfall. Section III computes the irrigation water required to maximize yields given the monthly rainfall input. Section IV is on domestic and industrial water use in relation to population growth. Section V computes river flows in relation to rainfall and all withdrawals in order to allow comparison with minimum flow requirements for fish habitat.

The simulation is available for $5.00 from

> Professor Kathleen Wilson
> Department of Urban and Regional Planning
> College of Social Sciences
> Porteus Hall 107
> University of Hawaii at Manoa
> Honolulu, HI 96822

IV. Composite Mind Map
(See facing page.)

V. Riparian-Farmer Interview

SCENARIO

Chet Waller farms 400 acres in the coastal plain of North Carolina, an area drained by numerous small rivers. The area possesses a humid, subtropical climate of hot summers and mild winters: rainfall is normally high, around

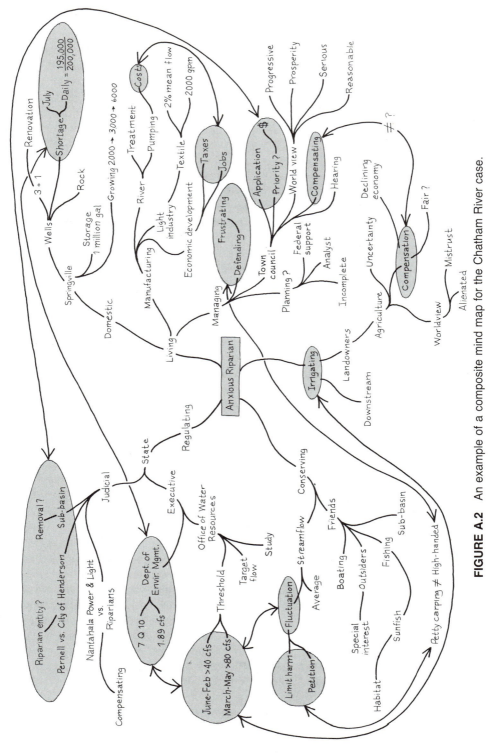

FIGURE A.2 An example of a composite mind map for the Chatham River case.

50 inches annually, and evenly distributed. The growing season is 240 days. Mr. Waller specializes in field crops, switching around between soybeans, corn, and other feed grains, depending on the market. He also grows to- bacco, renting the quotas of a widow and another neighboring family. His home farm borders on the Chatham River, and he is one of numerous family farmers in the area who are worried by Springville's announced plan to begin abstractions to meet municipal water needs. He has agreed to speak to you about these concerns.

There been Wallers here since before Independence. Come over as inden- tured workers in 1680. My grandpappy told me we used to grow peanuts and cotton on this piece of land right here, but cotton moved west round about World War I. Course, I wasn't born yet and don't recall seeing much of that round here.

Except for the tabacca, I'm growing mostly corn on my own land here. Grew some soybeans up to a couple of years ago but lost the whole lot during a drought. Besides, without the export market, prices have been real bad. Corn is good now because I can get it direct to the hog farmers over to Rich- lands. Just got by, the last couple a years, with no rain. Fact is, it was the tabacca that pulled us through. I kept it going by irrigating, took the plunge in June of '85, just went out and bought pumps, pipe, sprayers, and laid it in amongst the plants that first year the best I could.

Course, don't know what's going to happen with tabacca. Can see a time when the quotas aren't going to be worth anything. And I hear how Reynolds is importing lots of low-quality leaf from South America. They're not gonna give us a break nohow and neither is the government. You see, we can't take a couple, three bad years anymore the way we used to. Everyone else round here was near wiped out. Now that irrigation stuff is stacked up beside the barn, but you never know and I been thinking about changing my layout for corn just in case.

Ya see, when I was small, the corn did pretty good, even when it was dry. You know how it was. You could run between the rows the way we used to plant. Sure we used a lot of fertilizer then; had to, with what the cotton done to the land round here. Now we plant so close the bugs can't fly up the rows. No rain; it dries out quick. I reckon I can lay out those fields pretty good so I can get the sprayers in if I have to and not lose any production.

You bet that river's important. Never had to think about it before; just a nice place when we was small, swimming, fishing, you know. My pappy kept some animals — good for them. Now this Springville thing has got me to thinking. True, I just got into irrigating. And don't much need it most of the time. Flow during the drought was pretty good; starts up in the mountains, people say. But what happens next time we don't get no rain? And if Springville starts pumping at the same time, who knows. . . .

Compensation from the town? Heard about that but don't know what it means. What am I supposed to do, put the money in the bank and buy water if it don't rain? And why should I go up and talk to the town council?

Haven't even been in Springville since they closed the movie house!

Don't really trust these ecology people. Used to come around hollerin' about spraying and birds and such. But they got a point about the fish and if we have to fight this thing, every little bit helps. That's what the president of our growers' association says, anyway. He talks to 'em, I don't hardly ever see any of those folks round here. Yeah, the association hired a lawyer from the capital supposed to know about water rights. Went in with another bunch of boys from up north. Goin' to cost a bucket; this thing's fixin to really go through. Don't know nothing about the law myself. I just know we might need the water ourselves and it's always been here.

VI. Water Law Summary

Who Owns the Water? A Brief Overview of Water Law

Conflicts over water are not new. The word *rival* comes from the Latin *rivalis,* "one using the same brook as another" (Worster, 1985:105). Rivalry among current and prospective water users is a response to both environmental and economic change.

Ownership and use of resources are legal matters. The water-law picture in the United States is muddied by regional and even interstate variations and the fact that in some jurisdictions there is not law but, rather, a body of custom that is only partially tested in the courts. The major divide is between the arid west and the semihumid east. In the former, characteristic of regions in which water is scarce, water law is formed by such concepts as absolute property, equal footing for uses, and transferable ownership rights, summed up in the doctrine of *prior appropriation.* In the latter, where water has been plentiful, water law reflects such concepts as usufruct, beneficial use, and reasonable use, which figure in various bodies of law and custom under the *riparian doctrine.* The two doctrines take diametrically opposed positions on the same question of who owns the water.

Riparian Doctrine. "Rights to [make reasonable] use of water in a stream are created by ownership of land which is riparian [i.e., land touching the river bank and within the watershed] to the stream. The water right is an incident of landownership, and cannot be lost by mere disuse" (Sax, 1968).

There are many variants of riparian rights. The original version was based on the English common law that colonists brought with them to settle America. Originally it had more to do with customary usages of common property resources. No one "owns" a river. It should be left free to take its course. According to custom, all members of the reference community have a right to *use* the water flow for "natural" purposes like washing, drinking, and watering livestock, but they do not own it. This is a *usufructory* right rather than a *property* right, and it cannot be sold, bartered, or traded. One person cannot seize a river and develop it—say by damming or diversion—especially for an "unnatural" purpose, to enrich himself at the expense of others.

Some observers see the original version of riparian rights as biased against economic development. This may explain why a variant of the doctrine evolved that still pertains in some jurisdictions. Under this modified doctrine, all riparian *owners* are granted equal use of the flowing stream; that is, a property right is involved, but it is tied to land ownership and location, with water still shared. Under the riparian doctrine, when water is put to new "reasonable" uses, existing users may be required to make way for them by ceasing or reducing current uses. The *may be* depends on ad hoc, case-by-case interpretations by the courts or flexible lawmaking by legislative bodies. For example, some state legislatures have designated agriculture to be a preferential user. State legislatures and the Congress have also created regional and interstate authorities to control water use and development, which have tended also to take the rights of riparian landowners. Related to this, governmental bodies and certain public utilities have the right of eminent domain, according to which they may, following payment of "just compensation," take by condemnation private property for public purposes, such as highways, dams, and water-supply facilities.

Nevertheless, in many jurisdictions, water may not be diverted from the stream to nonriparian lands or other drainages, nor can rights or water be sold, nor can a riparian landowner lose rights by failing to exercise them.

Prior Appropriation Doctrine. This is a deviation from the old common law, which was abrogated in many (but not all) western states. It is based on claims of first arrival, in which potential large-scale users, such as miners or irrigators, band together into camps or districts that collectively hold the right to abstract and use water against secondary claimants, even later arriving riparians. Thus (following Anderson, 1983:30), under this doctrine, the law (1) granted to the first appropriator an exclusive right to the water and granted water rights to later appropriators on the condition that prior rights were met; (2) permitted the diversion of water from the stream so that it could be used on nonriparian lands; (3) forced the appropriator of water to forfeit his right if the water was not used; and (4) allowed for the transfer and exchange of rights in water between individuals.

References

Anderson, Terry L. *Water Crisis: Ending the Policy Drought.* Baltimore: The Johns Hopkins University Press, 1983.

Sax, Joseph L. *Water Law, Planning and Policy.* Indianapolis: Bobbs-Merrill, 1968.

Worster, Donald, *Rivers of Empire: Water, Aridity and Growth of the American West.* New York: Pantheon, 1985.

VII. Court Decisions

Nantahala Power & Light Co. v. *Moss*
Same v. *Norton*
Same v. *Davis et al.* (two cases)
No. 26
Supreme Court of North Carolina
Oct. 29, 1941
17 S.E. 2d 10

Facts

The petitioner owned a dam site on the west fork of the Tuckaseegee River upstream from the lands of the several respondents. It also owned a large body of land above the dam site for reservoir purposes. The petitioner also owned all the land on both sides of the river below the dam to the property of the respondents, a distance of about one-half mile, including what is known as High Falls. Below the lands of the petitioner each respondent owned a tract of land running up to the thread of the stream, including riparian rights in the waters of the river as follows: (1) On the right-hand side going down, the J. W. Davis heirs owned 11 acres, known as the Davis "little" tract, having a frontage of 450 feet on the stream; (2) the respondent Ida Moss owned 372 acres with 11,731 feet frontage on the stream on the right-hand side; (3) across the river on the left-hand side the respondent W. C. Norton owned a tract containing 525 acres fronting approximately 4726 feet; (4) next on the left-hand side was situated the J. W. Davis heirs, "big" tract. Within the Davis "big" tract and fronting on the river was the Reed and Warren tract with a frontage of 2417 feet, belonging to the petitioner. The Davis "big" tract, had a frontage of 2,155 feet within the boundary of the Reed and Warren tract, its frontage, due to the location of the Reed and Warren tract, being broken into three sections.

The petitioner established its power house downstream a considerable distance below the property of the respondents and constructed a tunnel or tube from the dam to the power house to convey the water. This diverted all of the water entering the river above the petitioner's dam, but other waters entering the stream furnished water to the lands of the respondents for all ordinary domestic and stock raising purposes.

These several proceedings were instituted for the purpose of fixing the compensation to be paid to the several respondents for the diversion of such waters through the instrumentality of the tunnel or tube extending from the dam site above the property of the respondents to the power plant considerably below such property.

Consolidated proceedings were brought by the Nantahala Power & Light Co. against the respondents for condemnation of the rights to divert the waters of the west fork of the Tuckaseegee River from the lands of the various respondents and for a right of way 240 feet long for a tunnel over the lands of the respondent Ida Moss. From the judgment the petitioner appealed.

Issue

Are the respondents entitled to compensation for the diversion of the water of the Tuckaseegee River on the basis of advantages thereby accruing to the petitioner?

Holding

The just compensation rule merely requires that the owner of the property taken shall be paid for what is taken from him. "It deals with persons, not with tracts of land and the question is, What has the owner lost? not, What has the taker gained?" *Boston Chamber of Commerce* v. *Boston,* 217 U.S. 189, 30 S.Ct. 459, 460; 54 C.Ed. 725. The value of the property to the condemnor for his particular use is not to be considered. *Western Carolina Power Co.* v. *Hayes,* 193 N.C. 104, 136 S.E. 353.

The very purpose underlying the authority to take by eminent domain is to prevent the owner who is aware of the necessity of the taker from making the most of the necessity and from demanding the highest price such necessity impels. Hence "holdup" or "strategic" values created by the necessity of the taker are not the true criterion. *United States* v. *Chandler-Dunbar W.P. Co.*

The trial court erred in instructing the jury to consider the theory that the respondents were entitled to compensation for the diversion of waters up the Tuckaseegee River on the basis of advantages thereby accruing to the petitioner.

New trial.

Pernell v. *City of Henderson*
No. 161
Supreme Court of North Carolina
Sept 24, 1941
16 S.E. 2d 449

Facts

The plaintiff had for some time owned and operated a gristmill on a stream known as Sandy Creek near Henderson, a city of some 7,600 inhabitants. The city had constructed and maintained dams and reservoirs on the tributaries of this stream above the mill site, from which it pumped a supply of water through mains to the city and distributed it to the inhabitants and users through a water system in the usual way. The plaintiff claimed that this diversion of the water from the natural flow of the stream had so diminished it that the value of his mill site had been destroyed or greatly reduced and his operation of the mill rendered unprofitable. He further alleged that his injury was constantly increased by the rapid growth of the city and its increasing needs. He alleged that the defendant had expressed its intention of continuing the diversion and that it would continue to his injury and damage.

For a second cause of action, the plaintiff complained that for some years prior to January 1, 1940, while he was owner and in occupation of the premises, the defendant created and continuously maintained a nuisance by emptying raw sewage into an upper tributary of the stream on which his mill was located, which sewage flowed down the stream and entered his pond, silting and filling it up so as to greatly reduce its capacity, and causing foul odors about the mill and premises, which could be endured only for a short time, and which caused his customers to complain; and that his premises thereby became unhealthy and were otherwise damaged by the noxious qualities of the sewage, in which respects he alleged that he was endamaged in a substantial amount.

Issue

Does a municipality have a riparian right superior to that of an industry declaring the municipality's use reasonable although it substantially diminishes the quality and quantity of water flowing to downstream riparian industries?

Holding

In its ruling the court quoted Farnham: "The rule giving an individual the right to consume water for his domestic needs is founded upon the needs of the single individual and the possible effect which his use will have on the rights of others, and cannot be expanded so as to render a collection of persons numbering thousands, and perhaps hundreds of thousands, organized into a political unit, a riparian owner, and give this unit the right of the natural unit. The rule, therefore, is firmly established that a municipal corporation cannot, as riparian owner, claim the right to supply the needs of its inhabitants from the stream." Farnham, *Water and Water Rights,* Vol. 1, p. 611.

Conceding that those who own the banks of a stream may, for their own convenience, contrive and use facilities and devices for distribution of water amongst themselves for such purposes, withdrawing from the flow needful quantities, that situation is not presented by the typical construction and use of a water supply system by a municipality, as in the case at bar, which impounds the water in suitable reservoirs, pipes it in large quantities into the city, and distributes and sells it to consumers for any purpose whatever for which it may be used. It could hardly be contended that these users were riparian owners, or that they could invest the city, as representative, or in the role of *parens patriae*, with rights in that respect which they themselves did not have.

The exigencies involved in supplying its inhabitants with water did not confer on the city an exonerating preference over the lower riparian owner who desired to use the water for purposes of manufacturing.

The judgment overruling the demurrers is affirmed.

VIII. Synthesis Statement

Interests involved in this dispute have jelled into three fairly distinct groups: (1) townspeople and other people in the area who support Springville's intention of using water from the Chatham River to meet domestic water needs and to promote economic development of the area, (2) people from the area and from other parts of the state who oppose the town's plan because of possible damage it may do to instream uses of the river, and (3) riparian landowners living downstream of Springville who oppose the town's plan because it may limit their ability to pump water from the river.

To understand this situation better, various documentary materials were reviewed and interviews held with representatives of each of the three groups. Some of these primary materials have been reproduced elsewhere in this appendix. Here is a summary of the major themes of concern they expressed.

The Town Council Viewpoint

Town council members contend that Springville is a riparian entity whose location along the banks of the stream entitles it to make reasonable use of the stream's flow. They hold that there can be no question that use of streamwater for domestic and industrial uses would benefit the economic welfare of the region. Because the town council believes that the economic utility of using the Chatham River for municipal water supply far outweighs any costs to the environment or to downstream riparians, council members regard their opponents' complaints as little more than petty carping and are frustrated by the feeling that they are always being forced to defend their point of view. Representatives of the town point out that Springville's planned diversion of 2,000 gpm (gallons per minute) is only 2 percent of average streamflow in the Chatham, so that on most days the effects of diversions on streamflow would be imperceptible.

Not incidentally, they point out that many of the people who object to the project on environmental grounds are outsiders who talk more about fish habitat than they do about people in Springville being able to make a living. In sum, the council members view themselves as progressives interested in the prosperity of the area and the well-being of the area's citizens. They characterize their opposition as representing special interests who stand to be marginally harmed by Springville's plan.

Finally, the town council members express a willingness to enter into negotiations for compensation of downstream users who care to bring their case before the town council. The town has also agreed to abide by the State Department of Environmental Management's regulation that water not be withdrawn from the river if the withdrawal would reduce streamflow below the 7 Q 10 (the lowest seven-day flow in a ten-year period.)

The Friends of the Chatham Viewpoint

People opposed to Springville's planned diversion of water from the Chatham because of interest in protecting instream uses of the river have organized into a group called Friends of the Chatham. While they do not dispute the town's assessment that municipal withdrawals would be a negligible proportion of mean streamflow, they point out that the Chatham has wide fluctuations in flow, including periods of low flow, when Springville's diversion would substantially diminish flow in the river. The Friends of the Chatham contend that diversions during these periods of low flow would jeopardize the stream's value as an aquatic habitat as well as interfere with boating, fishing, and other recreational activites. This group used a study by the State Office of Water Resources to buttress its claims for the value of the stream as wildlife habitat.

The key point addressed in this study is whether or not flows low enough to disturb aquatic habitat ever occur in the Chatham River. The report begins by describing the Chatham River as a "prime fishery for redbreast sunfish . . . with habitat quality being considered excellent."

Four methods were then applied in the study to set minimum flow thresholds in the vicinity of the U.S. 608 bridge just downstream of Springville. Of the four techniques, the Office of Water Resources found the target-flow method to be the most satisfactory. Judged by this index, June-through-February streamflows lower than 40 cfs at the U.S. 608 bridge were determined to be detrimental to aquatic habitat, while flows lower than 80 cfs during the March-through-May spawning season were felt to be detrimental.

Two miles downstream of the U.S. 608 bridge is a streamgauge, which has recorded flows on the Chatham since 1952. Based upon this streamflow record, while mean flow in the river is 220 cfs, the 7 Q 10 is only 1.89 cfs, much lower than the Office of Water Resources recommendation. The Friends argue that the Department of Environmental Management's insistence on using the 7 Q 10 is based on their interest in maintaining a minimum streamflow necessary to dilute pollutants, an issue that they say is largely irrelevant in considering the Chatham River. For this reason, the Friends have petitioned the state to reconsider its minimum-flow requirements for the Chatham to bring them into conformance with Office of Water Resource's target flow.

While these environmental considerations are what have brought this group together in opposition to the Springville diversion, some fairly lively antagonisms have grown between the Friends and the town to fuel the dispute. The Friends think Springville is being high-handed in not offering to discuss the size of their withdrawal from the stream or alternatives to withdrawing water during periods of low flow, but in offering to negotiate compensation for harm done. The Friends say they are not interested in compensation for harm but in limiting harm. The Friends also have not warmed to the tone of pro-withdrawal people in discussing the town's riparian right to reasonable use of the water. Opponents of the withdrawal point out that some of the town lies outside the subbasin from which water would be withdrawn. There is some question of the legality of a plan to withdraw water from one subbasin to sell to users in a downstream subbasin. Furthermore, implicit in the town's arguments favoring the withdrawal is the premise that because the municipal users would be willing to pay more for water than would downstream riparians or supporters of instream use, the municipal withdrawal should be granted priority. The Friends think that if water is especially valuable to the intended beneficiaries of the Springville diversion, then the town should be able to justify installing wells to eliminate the need to withdraw water from the river during periods of low flow.

The Downstream Riparian Viewpoint

The major offstream users of streamwater are farmers, some of whom are irrigators. They share the Friends of the Chatham's distaste for what they perceive
as the town's arrogance in approaching this issue. While in principle they do not
object to being compensated for their loss of irrigation water during certain critical periods, they have no confidence that the level of compensation they could
exact from the city would measure up to the value that they place on access to
stream water. Like the Friends, the farmers consider the offer of compensation
to be as much a crumb tossed off to silence their barking as a serious offer to
pay them what the loss of their water would be worth. Indeed, some would prefer to have the water and are worried about the adequacy of irrigation supplies
in the event of a drought.

While both the growers and the Friends oppose Springville's use of streamwater, each recognizes that this alliance is to some extent a marriage of convenience. Growers recognize that the minimum streamflow levels recommended by
the Friends could be enforced against irrigators as well as against the town. The
Friends realize that the growers want to halt Springville's withdrawals in part so
that they themselves can withdraw the water farther downstream.

Case 2. Mucho Sacata Ranch

Prepared by Jerry Stuth, Wayne Hamilton, and Donald Vietor, Texas Agricultural Experiment Station

Note: The names of the ranch and the people involved have been changed to
ensure confidentiality.

Outline of Materials

    I. Narrative account of ranch operations and management situation
    II. Pictorial representation summarizing situation;
       Figure A.3; first-generation pictorial description of Mucho Sacata Ranch
       and other holdings of sole proprietor
    III. List of assets: facilities, buildings, and machinery
    IV. Resource assessment; Table A.1
    V. Profit centers:
       (1.) Cow–calf operations
       (2.) Irrigated forage sorghum/feedlot operations
       (3.) Commercial hunt program
       (4.) Combined year's net profit and capitalization
    VI. Sketch Map of Ranch's Land Utilization; Figure A.4
    VII. Regional Rainfall Profile; Figure A.5
    VIII. Cost-Return Projections, Texas Agricultural Experiment Station
       (1.) Forage Sorghum; Figure A.6
       (2.) Cow-Calf; Figure A.7

I. Ranch Operations and Management Situation

Mr. Bell operates several farm and ranch properties, including Mucho Sacata Ranch, which is located approximately 12 miles southeast of Uvalde, Texas, in the Rio Grande Plains resource area. He recently hired a general manager named Mr. Book, who lives on the largest ranch of Mr. Bell's holdings. Mr. Book's job is to provide day-to-day management and make recommendations for all of Mr. Bell's agricultural enterprises. For two years Mucho Sacata Ranch has been run by a young manager, Andy. Mr. Bell had hired him based on observation of Andy's performance on one of his other ranches while he was still an undergraduate at Texas A&M University.

Mr. Bell is getting on in years and wants to detach himself from many management duties. He intends to pursue other investment interests and spend more time on the affairs of the National Agriculture Association, of which he is a prominent member.

Mr. Bell was already middle-aged when he inherited the operation from his family. He left a successful and very lucrative career in agribusiness to move to one of the ranches. He initially employed the old foreman who had worked for his family, but the complexity and growth of combined operations of commercial livestock, a feedlot, and commercial hunting operations, scattered over his holdings, exceeded the foreman's management skills. Mr. Bell was uncertain of his own ability to apply the good management techniques that were needed to maximize profits in the ever-tightening cost–price squeeze and to innovate for the future. He particularly wanted a person who would give the business a new dimension in changing times, not just sound knowledge of fiscal control and accounting, someone who had a good farm and ranch business background.

He finally found Mr. Book, his current general manager. Mr. Book has two agricultural degrees and had been a successful active partner in a consulting business specializing in efficiency analysis and development of livestock programs.

When he came into the organization, Mr. Book faced the challenge of "getting this organization going and trying to get down to Mucho Sacata and get things in order." Andy, manager of Mucho Sacata, is unaware of Mr. Bell's concern. He thinks that the cattle operation is running smoothly and that his commercial hunting program is generating above-average income. In fact, Andy had just mailed Mr. Bell a range inventory plan and livestock evaluation report, which revealed that the ranch is stocked 20 percent over rates recommended in the soil conservation guides for this region. The overstocking effects were confirmed in records from preceding years, which showed a low average calf crop and weaning weights substantially lower than some area producers were getting. Andy informed Mr. Book during the on-site evaluation that the principal contributing factors to low livestock production, besides overstocking, were inferior quality of livestock and inadequate seasonal nutrient supplementation. He had even called Mr. Bell's accountant last week and worked out an economical strategy. Andy's wife helped put together their financial records for Mr. Bell's accountant and typed up the report. Andy had explained the plan to Mr. Book with a great deal of pride, knowing that he had done the job expected of him in running Mucho Sacata, topped off by the innovative and functional plan he had put together.

Andy received no immediate response from Mr. Book. In fact, Andy felt that Mr. Book may have been irritated with him, but he could not put his finger on the problem. Approximately one week after Mr. Book made his on-site inspection, Andy received a terse note from Mr. Bell's office that said that the plan was being returned to him for submission through proper channels.

On top of this, problems were developing with four ranch hands and their families, who were already on the payroll when Andy took over management of Mucho Sacata. At first, Andy was proud of how he had won their confidence. He tried, as much as possible, to help them financially by giving the school-age children of the families odd jobs such as mowing yards and keeping the headquarters looking nice. Although things were going well enough for Andy, he had the feeling that there was growing discontent among his regular employees. He could not put his finger on it, but it seemed that their attitude was deteriorating. He particularly noticed that it was taking more effort on his part to accomplish the same amount of work.

One possible factor that Andy considered was the relations of his people with town folks. He noticed that right after the last cattle working, it seemed as if his people began to complain about a few things that previously seemed all right to them. Andy also knew that town women frequently visited the ranch wives. Evidently, there was some talk going on that had upset his workers and their families.

Andy knew that some town folks earned more money working at the new cannery and grain elevator than ranch employees. Some had nicer homes and cars and dressed better. On the one hand, Andy thought, "We can't compete with businesses' pay scale in town. There's just a limit to what we can pay and stay in the black on a ranch operation." On the other hand, Andy didn't want to lose his help. He knew he couldn't replace his qualified and experienced people very easily, if at all. He began to formulate a plan he hoped would relieve the situation.

Andy decided he would increase the meat allowance from ranch beef by 25 percent per family. This should very nearly take care of all their meat needs. The families were already getting their housing and utilities. Andy felt that if they would consider all these things, the salary comparisons wouldn't be too bad. Another way he felt the ranch could help would be to offer a group medical plan in which the ranch would pay for insurance for the household head and allow him to add dependents at a group rate. Andy went one step further and set up a proposed accident and term life insurance program of $10,000 for the men. This cost was to be borne entirely by the ranch, with the employee owning the policy and naming his beneficiary. The insurance programs could benefit the ranch too, thought Andy, as it would avoid the necessity of paying doctor bills. It would also relieve the ranch of financial responsibility to the families of men accidentally incapacitated or killed.

Andy added up the total value of the benefits he proposed, and it amounted to a sizable pay increase. He felt that his plan would go a long way toward fostering better feelings among his people. Andy presented the whole package, this time to Mr. Book, who was enthusiastic about the idea and supported the plan completely. As soon as the logistics of the insurance were finalized, Andy put the plan into effect. He was a little surprised that there was only a small positive reaction from the employees.

Several months passed, and labor relations on the ranch continued going downhill. One of the men had quit and moved his family to town. Another had asked Andy for a $100-a-month raise. Andy had refused the raise, and he had heard that the man was talking to another ranch about a job. Andy felt frustrated that his efforts to satisfy his labor force had failed. He really couldn't understand what was wrong. Mr. Book's expectations for improved worker productivity also were not being met.

Andy was inwardly pleased that he was going to do the culling on the crossbred cow herd this go-round. He had felt that Mr. Bell had hung on to too many aged and nonproducing cows in the herd to give them "one more chance." Andy knew that the calf-crop percent could be raised considerably by a stiff culling program. Since Mr. Book allowed him to be responsible for the business of this ranch, Andy wanted to make this decision to repay the ranch through his good work.

Andy began to call Mr. Book about once a week to report on ranch operations. Everything appeared in good order. Andy had contracted the calf crop to a buyer at an impressive advantage to the ranch. He was beginning to feel confident in his abilities.

During a conversation with Mr. Book, Andy discussed his forthcoming cow working, when he would wean and deliver his contracted calves. He told Mr. Book that he wanted to cull the herd pretty hard. Mr. Book stated that he felt the herd needed culling some and that he wished he could be there to work with Andy on this job. He joked that if cow prices were a little better, he would be more prone to agree. Mr. Book never told Andy not to do the culling, although he did express reservations about finding good replacements.

The following week, an old friend called Andy to see if Mucho Sacata would be interested in some young replacement cows he had. Andy knew the herd, and they were good cattle — just the right cross and quality he had in mind for the ranch. Andy went and looked at the cows and found them to be even better than he hoped. Andy's friend said he had other people interested in the cows and that if Andy wanted them, he had better decide right away. They haggled a little on price, but Andy felt that the price was well below the value of the cows.

Andy tried unsuccessfully to reach Mr. Bell that night. Mr. Book was enroute overseas and wouldn't be available for some time. Andy thought about his situation all the next day, then finally called his friend and told him he would take the cows. When he worked his herd, Andy did cull deeply to make room for the replacements.

Mr. Book called Andy shortly after his return to check on things and to discuss the cattle working. Andy told him what he had done about the culling and the replacements. Mr. Book was not pleased, even when Andy explained that he was sure next year's calf crop would be up considerably. Mr. Book told Andy, "Although I believe what you did will probably prove to be a wise move, Andy, I am just disappointed that you exceeded your authority. I must weigh this as a part of your performance evaluation that I don't like to see. We will now have to cancel our order for the feedlot due to the cash-flow constraint this will place on overall operations."

Just before the next deer/quail hunting season, Mr. Book hired a recent graduate as a wildlife biologist and told him to report to Andy with the objectives of improving the quality of the deer herd and increasing hunting revenues. Andy had kept good hunt records, which indicated that the buck–doe ratio was 1:6. The biologist's goal was 1:3. The problem was that the hunters would have to kill does and the current lessees refused to do so.

The biologist sent his report to Mr. Book. He, in turn, ordered Andy to get a new set of hunters who would kill does. Andy responded that current hunting revenues were above average, and if he contracted a new set of hunters, he would have to spend more of his personal time "cultivating and educating a new set of hunters." Mr. Book told him to do it anyway. This resulted in one of the ranch's highest calf and cow mortality that Mr. Bell could remember for the Mucho Sacata.

II. Pictorial Representation Summarizing the Situation

INSERT Fig A.3

III. Ranch Assets

- 1 office headquarters (1500 sq ft)
- 2 manager residences (2000 sq ft)
- 3 employee residences (1000 sq ft each)
- 5 pickups
- 2 cattle trailers
- 2 tractors with implements for row crops
- 1 grain combine
- 2 sets of corrals with complete working facilities
- 1 irrigation pump/well with 4¼-mile wheel lines
- 1 hay barn
- 1 tack barn and stable
- 30 miles of barbed-wire fence (15–30 years old)

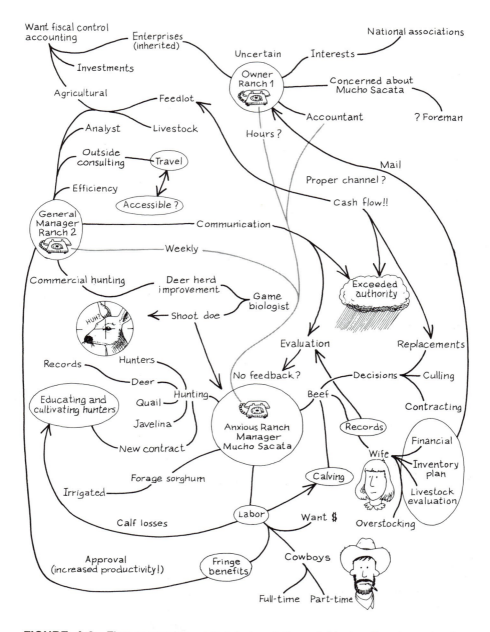

FIGURE A.3 First-generation pictorial description of Mucho Sacata and other ranches comprising owner's holdings.

IV. Resource Assessment

TABLE A.1 Resource Assessment: Mucho Sacata Ranch

Pasture	Acres	Recommended Stocking Rate (AC/AU)	Ecological Condition	AU
Pasture 1				
Sandy loam	800	20	Fair	40
Clay loam	1,200	30	Poor	40
	2,000			80
Pasture 2				
Sandy loam	2,000	30	Poor	67
Gravel ridge	3,000	40	Poor	75
Clay loam	1,000	25	Poor	40
Rumadero	500	25	Poor	20
	6,500			202
Pasture 3				
Sandy loam	1,000	20	Fair	50
Clay loam	1,100	20	Fair	55
Rumadero	300	18	Fair	16
	2,400			121
Pasture 4				
Sandy loam	500	25	Poor	20
Gravel ridge	2,000	35	Poor	59
	2,500			79
Grand Total	13,400		Grand Total	482
Pasture 5				
Irrigated Forage sorghum	400	—	—	—

V. Profit Centers of Mucho Sacata

1. Cow–Calf Operations

The cow herds are managed with a three-pasture, two-herd deferred rotation system in pastures 1, 3, and 4. Pasture 2 is so large it is managed in a continous grazing regime. Management wants to improve ecological conditions but cannot put together the cash flow to develop pasture 2 into a grazing system that would derive a positive ecological response. Management wants to maintain a saleable product that has a high degree of consistency from year to year. Their goal is to contract all cattle sight unseen to reduce the costs of gathering and sorting cattle for the contract bidders.

This ranch normally has 465 mature cows (hereford x brahman), 24 beefmaster bulls, and 70 breeding-age replacement heifers (contracted seven months prior to breeding season with a 60-day prebreeding season delivery). Calves are born from January through March (35 percent, 55 percent, 10 percent, respectively.) Breeding season is from April through June. Normal calf crop is 79 percent. Cow death loss is 1 percent, calf death loss is 3 percent from mature cows and 11 percent from replacement heifers. The heifers are delivered brucellosis-vaccinated, and bulls are tested for brucellosis 30 days prior to delivery. Calves are vaccinated with a four-way *Clostridium* vaccine, and cows are vaccinated for leptospirosis. All cattle are sprayed four times each year for ticks and flies, and mature cattle are wormed twice for internal parasites.

The calves are contracted when they are four months of age to a single-order buyer in California. Cull cattle are sold in the local auction. Bulls are purchased through registered bull sales year-round. Prices are hedged through the futures market. The ranch normally grosses $110,000 annually. Net profit was $30,990 last year.

Besides the three full-time cowboys, seven part-time cowboys are hired to assist with working of livestock during work periods. Records are kept on pregnancy status of the cows, calf weight, and relative ranking of calves' weight relative to other calves. All cattle are ear-tagged for identification.

Supplementation is carried out from January through March. Management does not have a good feel for proper amounts, but generally feeds one pound of cotton-seed meal pellets per day during this period (if no winter forbs develop) and 2.5 pounds per day if it is a dry winter. Stocking rates are maintained at a level where winter hay is not fed unless severe drought conditions develop. The supplement is contracted in late October or early December, depending on the nature of the cotton crop in Texas and availability of whole cotton seed at the local cotton gin.

2. Irrigated Forage Sorghum/Feedlot Operations

The 400 areas of irrigated cropland are used currently to produce forage for lease grazing by a local feedlot. Pasture is leased by the local livestock auction, which funnels cattle from the sale for temporary pasturage in the sorghum and then moves them into the feedlot. Logistics is critical, since the ranch manages the stockers for the auction, and close timing of water applications, standing crop levels, and communications with the auction must be maintained. One man is responsible for managing water applications. The cowboys assist in management of the stockers placed on the pasture, with the greatest activity concentrated on herd health problems associated with cattle that have been stressed due to shipment to the auction. This activity occurs from May through October, just prior to the hunting season. This operation netted the ranch $7,652 last year.

3. Commercial Hunt Program

Lease hunting is allowed from December 1 through January 31 and concentrates on white-tailed deer and quail. There is some bow hunting of javelina, which generates approximately $2000 per year, but it is very unstructured and could probably yield more money if management decided it was worth dealing with.

Quail hunting is a big moneymaker for the ranch, since $21,500 is attained from contract hunters on the 13,400 acres. Hunters are charged $50 per day and provided with a small bunkhouse. No hunts are guided. Only one set of rules is read to the hunters who hunt on the place. Dogs are allowed on the ranch for hunting purposes. Net profit was $11,000 last year.

Deer hunting requires more interaction with the hunters and grosses $84,000 dollars per year. However, the manager must essentially shut down his general cattle duties to cater to the hunters. He presents an evening orientation to each new group, which focuses on the ranch's population-management goals. The ranch is currently stocked at 12 acres per deer with a 1:6 buck–doe ratio. The goal of the ranch is to reduce the ratio to 1:3 in the next five years. This has led to a rigorously enforced rule requiring the hunting of does as well as bucks. Management has contracted all hunting with a nationwide retail chain, which provides hunts as a perk for outstanding blue-collar workers and high-performing executives. Total lease price is negotiated each year between top management of the retail firm and the ranch manager, who gets final approval of the contract from the general manager. A consulting game biologist is hired each year by the general manager to appraise herd status on all ranch holdings. These reports are made available to management of each ranch and executed by the on-site managers. The bunkhouse is available to parties of 15 hunters, four days each week, during the hunting season; i.e., 120 hunters come to the ranch to hunt each year. All animals harvested from the ranch are catalogued for antler measurement, weight, and age. The contract specifies a population goal that the hunters must seek to achieve. Hunters are escorted to and from their blinds each morning and evening. This requires the manager and two hands to get everyone out in a timely manner. The ranch provides cold storage of the harvested deer until the hunters leave the ranch. Net profit was $23,000 last year.

4. Overall Enterprise Net Profit for Last Year

Profit was $72,642 on net capital of $1.3 million.

VI. Ranch Land Utilization

(See facing page.)

VII. Regional Rainfall Profile

(See facing page.)

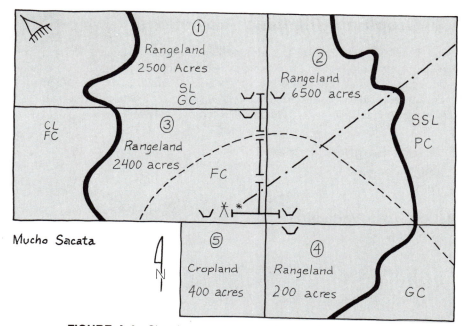

FIGURE A.4   Sketch map of Mucho Sacata Ranch land use.

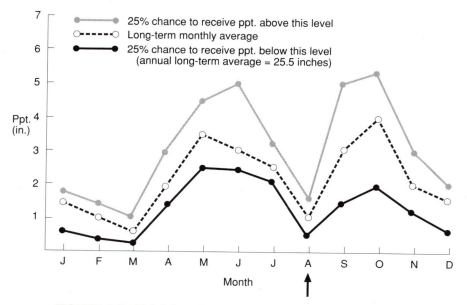

FIGURE A.5   Rainfall profile for Rio Grande Plains area, Texas.

VIII. Crop/Herd Planning Projections

Note: The Texas Agricultural Experiment Station makes available to farmers per-acre cost-return projections for common regional crop, crop-animal, and cultivation regimes. Based on prevailing or anticipated input costs and commodity prices, these worksheets are provided as a planning tool rather than a prediction regarding how any particular farm/ranch enterprise will perform.

1. Forage Sorghum, Irrigated

(See facing page.)

2. Cow–Calf Production

(See page 338.)

DATE	STAGE OF PRODUCTION	TYPE OF PROD.	PRODUCT NAME		NUMBER OF UNITS	WEIGHT PER HEAD	CASH NON- CASH	LANDLORD SHARE	BREAK EVEN PROD.
04/02/89		A	PASTURE*	SORGHUM	50.0000	.0000	C	.00	Y
05/02/89		A	PASTURE*	SORGHUM	100.0000	.0000	C	.00	Y
06/02/89		A	PASTURE*	SORGHUM	100.0000	.0000	C	.00	Y
07/02/89		A	PASTURE*	SORGHUM	100.0000	.0000	C	.00	Y
08/02/89		A	PASTURE*	SORGHUM	100.0000	.0000	C	.00	Y
09/02/89		A	PASTURE*	SORGHUM	100.0000	.0000	C	.00	Y
10/02/89		A	PASTURE*	SORGHUM	50.0000	.0000	C	.00	Y

DATE	STAGE OF PRODUCTION	TYPE OF INPUT	INPUT NAME		NUMBER OF UNITS	CASH NON- CASH	FIXED OR VARI.	LANDLORD SHARE
11/06/88		M	SHREDDING		1.0000			.00
11/11/88		M	CHISELING		1.0000			.00
11/16/88		M	DISC OFFSET	12 FT	1.0000			.00
11/21/88		M	PLANING	LAND	.2000			.00
11/26/88		M	DISC OFFSET	12 FT	.2000			.00
02/06/89		E	PHOSPHATE		60.0000	C	V	.00
02/06/89		M	APPLY.FERTILIZER		1.0000			.00
02/11/89		E	NITROGEN (ANHY)		60.0000	C	V	.00
02/11/89		M	ANHYDROUS APPL.		1.0000			.00
02/12/89		E	SEED	SORGFORG	40.0000	C	V	.00
02/12/89		M	DRILLING		1.0000			.00
03/16/89		O	IRRIGATION		4.0000			.00
05/01/89		M	PICKUP TRUCK	3/4 TON	21.0000			.00
05/16/89		E	NITROGEN (ANHY)		60.0000	C	V	.00
05/16/89		M	ANHYDROUS APPL.		1.0000			.00
06/16/89		O	IRRIGATION		4.0000			.00
07/01/89		E	MISC ADMIN O/H		.5000		F	.00
07/16/89		E	NITROGEN (ANHY)		60.0000	C	V	.00
07/16/89		M	ANHYDROUS APPL.		1.0000			.00
07/21/89		O	IRRIGATION		4.0000			.00
08/26/89		O	IRRIGATION		4.0000			.00
11/01/89		K	LAND - CASH RENT SORGHUMI		1.0000	C	F	.00

FORAGE SORGHUM FOR GRAZING, IRRIGATED
Southwest Texas District-13 (Wintergarden Region)
1989 Projected Costs and Returns per Acre

GROSS INCOME Description	Quantity	Unit	$ / Unit	Total	Your Estimate
PASTURE* SORGHUM	600.000	days	0.3000	180.00	_____
Total GROSS Income				180.00	_____

VARIABLE COST Description	Quantity	Unit	$ / Unit	Total	
PHOSPHATE	60.000	lb.	.200	12.00	_____
NITROGEN (ANHY)	60.000	lb.	.102	6.12	_____
SEED	40.000	lb.	.320	12.80	_____
NITROGEN (ANHY)	60.000	lb.	.102	6.12	_____
NITROGEN (ANHY)	60.000	lb.	.102	6.12	_____
Fuel & Lube - Machinery		Acre		10.96	_____
- Irrigation		Acre		44.49	_____
Repairs - Machinery		Acre		2.37	_____
- Irrigation		Acre		9.63	_____
Labor - Machinery	2.415	Hour	4.501	10.87	_____
- Irrigation	1.600	Hour	3.799	6.08	_____
Interest - OC Borrowed	25.465	Dol.	0.110	2.80	_____
Total VARIABLE COST				130.36	

Break-Even Price, Total Variable Cost $ 0.21 per days of PASTURE*

GROSS INCOME minus VARIABLE COST				49.64	

FIXED COST Description	Unit	Total	
MISC ADMIN O/H	acre	8.00	_____
Machinery and Equipment	Acre	31.48	_____
Irrigation	Acre	51.63	_____
Land	Acre	40.00	_____
Total FIXED Cost		131.11	

Break-Even Price, Total Cost $ 0.43 per days of PASTURE*

Total of ALL Cost	261.47	_____
NET PROJECTED RETURNS	-81.47	_____

Grazing is based on the number of Animal Unit Days.

FIGURE A.6 Cost-return projections, forage sorgum, irrigated: Texas Winter Garden region. (Courtesy of Texas Agricultural Experiment Station.)

Projections for Planning Purposes Only
Not to be Used without Updating after April 8, 1989.

B-1241(L13)

COW-CALF PRODUCTION, UNIMPROVED BRUSH COUNTRY
Southwest Texas District (13)
1989 Projected Costs and Returns per Head
==

PRODUCTION Description		Quantity	Unit	$ / Unit	Return	
CULL BULLS	BEEF	0.01Hd 12.000	cwt.	55.0000	4.42	
CULL COWS	BEEF	0.10Hd 9.500	cwt.	52.0000	49.40	
DEER LEASE		22.000	acre	2.5000	55.00	
HEIFER CALVES		0.26Hd 4.100	cwt.	87.0000	92.74	
STEER CALVES		0.39Hd 4.500	cwt.	96.0000	168.48	

Total GROSS Income   370.04
==

OPERATING INPUT or CUSTOM OPERATION					
Description	Input Use	Unit	$ / Unit	Cost	
COTTONSEED CAKE	180.000	lb.	0.140	25.20	
MISCELLANEOUS COW-CALF	1.000	head	5.000	5.00	
SALES COMMISSION	0.770	head	9.000	6.93	
SALT & MINERALS	45.670	lb.	0.280	12.79	
VET. MEDICINE	2.000	head	5.000	10.00	
WATER FACILITY REPAIR	1.000	head	2.000	2.00	
CUSTOM HAULING COW-CALF	0.750	head	8.000	6.00	
Fuel				2.48	
Lube				0.02	
Repair				0.86	

Total OPERATING INPUT and CUSTOM OPERATION Costs   71.28
==

Residual returns to capital, ownership
labor, land, management, and profit   298.77
==

CAPITAL INVESTMENT Description	Quantity Invested	Unit	Rate of Return	Cost	
Interest - IT Equity	1028.751	Dol.	0.080	82.30	
Interest - OC Borrowed	12.333	Dol.	0.120	1.48	

Total CAPITAL INVESTMENT Costs   83.78
==

Residual returns to ownership, labor,
land, management, and profit   214.99
==

OWNERSHIP COST Description (Depreciation, Taxes, and Insurance)	Cost	
Machinery and Equipment	11.77	
Livestock	16.59	

Total OWNERSHIP Costs   28.36
==

Residual returns to labor, land, management, and profit   186.63
==

LABOR COST Description	Input Use	Unit	Average Rate	Cost	
Machinery and Equipment	1.525	Hr.	4.500	6.86	
Other	7.470	Hr.	3.987	29.78	

Total LABOR Costs   36.65
==

Residual returns to land, management, and profit   149.98
==

LAND COST Description	Input Use	Unit	Rate of Return	Cost	
PASTURE NATIVE					
Annual Lease	22.000	Acre	3.200	70.40	

Total LAND Costs   70.40
==

Residual returns to management and profit   79.58
==

-WARNING- No Management Cost Specified

==
Residual returns to profit   79.58
==
Total Projected Cost of Production   290.46

Spring calving, 78% calf crop, 3% death loss on cows, 13% replacement rate,
10,000 acre ranch, 450 animal units.

FIGURE A.7   Cost-return projections, cow-calf production, Texas Winter Garden
region. (Courtesy of Texas Agricultural Experiment Station.)

Case 3. A Papua New Guinea Co-op

Prepared by Dr. George E. B. Morren, Jr., Rutgers University

Outline of Materials

I. Community leader interview
II. Historical and situation summary
III. List of principal crops
IV. Map of area
V. Research bibliography

I. COMMUNITY LEADER INTERVIEW

SCENARIO

Amusep is a village leader and Baptist circuit pastor who has been the principal actor in the development and implementation of this local modernization plan. At the time of this interview, he was in his mid-forties with a wife and two children. In addition to his community and religious activities, he maintained a life-style similar to that of any other community member. He and his family support themselves as subsistence farmers and pig herders; they also try out newly introduced crops and animals such as citrus, chilis, and poultry. In this interview, he provides a personal overview of development, but he starts with a capsule description of his group's contact with the outside world and what he and others have thought about it along the way. Most importantly, he speaks of the unfolding of a community plan.

We were living on the May River when we first encountered white skins. I was perhaps eight years old at the time. We called them *sebrip*, which means "different smell" or "smell of soap" because their skins or clothing had a distinctive odor. You could smell them at great distances, so people were able to run away and hide. We really did not know what this meant. We suspected that the black policemen might be our younger half-brothers born in the afterlife, but we were very frightened.

Later, some Miyanmin went to Atbalmin over there. The little girl who came on the plane with you is from the very village. The Miyanmin went and killed them. I too went to the fight. The *kiap* [administration officer] came and caught us and took us by force to Wewak. We were there for about six months. Then the present aid post orderly (APO), another young man (who died some time ago) who also became an APO, and two more who have since died, both carpenters, and I myself—five together—we all went to Madang. The older men stayed [in jail] in Wewak and the government sent the five of us to the Lutheran Mission in Madang . . . 1957 . . . the 17th of September.

I was around 20 years of age after three years at Madang—so I was 17 when this all happened. And after three years we came back to Telefomin [in 1961]. We went to the Baptist Mission to work, attend Bible school, and I was in charge of the store at the same time. You know the Pasuwe store there now; I worked at the old one and then it got very large.

Before we went to Wewak I underwent intitiation in the spirit house. . . . I'm not sure when it was. There were five distinct times [rituals], but I did not do the sixth and last. We were taken into the spirit house and told the lore of adult men—women couldn't hear these things. The reason for [initiation] is like this; if the adults don't take us into the spirit house then we could never grow up to be true adults ourselves—we would be of no account—that is what we thought.

One reason for this—there are some game animals—they aren't sacred now—we can eat them freely. But formerly some game was taboo, like the possum *kwiyam*, certain snakes, other animals we couldn't eat; one kind of wild fowl with red legs *sena*, the fish with whiskers *fini*, these things were for adult men. If you hadn't been initiated, you couldn't eat these things. So you had to be initiated.

But my son won't be initiated because we have heard of the new ways. School is somewhat a substitute. Although there are many customs we will keep, we will lose that one [of initiation].

I started to think about things in Madang. I thought about bringing the word of God to everyone, about peace, about not being afraid of the white man, and about the good ways of the white skins, about schools and what I had been taught. We couldn't fight anymore and we didn't have good things of our own. So we had to go and build an airstrip because we did not have good roads for cars—so the airstrip was the most important thing.

In Madang we saw everything of the white man—ships and cars and all the activities there. The mission took us to see everything in the town and all the livestock the *didiman* [agricultural extension workers] were raising. And we saw the hospital and its work. And we traveled by ship to the Rae Coast, Lae, the Markham River, and we came back. And this started me to thinking.

Nothing happened quickly; all has come slowly. The first thing was the big speech [announcement] and teaching some of the young men *tokpisin* [a new language used all over the country]—teaching about God and *tokpisin* were the first steps. Second was teaching young men to write. Then I thought about building the airstrip—around 1965. Then we thought about persuading authorities to start a primary school. Everything went very slowly.

We would wait until the *kiap* came to conduct the census and we would talk about it and it took us something like four years to do it. The school started in 1974—before [national] independence.

Since then we have been thinking about ways of supporting kids in school. That's next and we are "out of breath." . . . [We need] some kind of business in the village or some other way of getting the money. It won't come quickly; it's hard—six months, a year, two years. If it doesn't come quickly we'll wait years.

For my own kids, I want them to go to school to gain knowledge. We'll see what happens after that. You have to take things one step at a time. I admire a lot of things in the white man's way and I like a lot of our own good ways too, but some white skins' ways and some of our ways are not good.

Some of the people working for the government have done good things for us; they too believe that the community should be improved. I've seen business people, people from different government departments — *didiman* — and if they try to help us I admire them. There are good people and bad people with us too.

(Note: My wife, Janet Gardner, and I conducted this interview in 1981. It is reprinted with amendments from Morren, 1986:289–90.)

II. Historical and Situation Summary

Whites first entered the region of Papua New Guinea, in which the Miyanmin dwell, around 1914, but there is no evidence of direct contact until 1936. Then a party of gold prospecters made brief contact, long enough to trade some cloth and tools for foodstuffs and show off an amphibian airplane. In 1939, they had a violent brush with a government patrol, resulting in the loss of 16 Miyanmin and one member of the patrol. Beyond these early contacts, the Miyanmin began to gain in-depth experience of the outside world in 1959, when 25 men from several allied villages were arrested, tried, convicted, and sentenced to prison terms for acts they committed in the course of intertribal warfare. Their experiences were varied. All were flown from the interior of New Guinea to the coast, a terrifying but ultimately valuable experience. The older men enjoyed 'modern living' at Wewak, a town on the north coast of New Guinea, while they followed the prison routine: They were given clothing, strange foods, and medical treatment; they interacted with acculturated New Guineans, as well as some whites, and most learned *pidgin*, a trade language that has evolved to span differences of language and culture; they cleared brush and cut grass along the roads and beside the town airport.

Five younger boys who were sent to the mission vocational school had comparable experiences and, in addition, received formal instruction in reading, writing, arithmetic, and Bible study, as well as job training in health care or carpentry. One of them, named Amusep, was quickly identified as possessing unusual intellectual and leadership abilities and was encouraged to train for pastoral duties. The boys also traveled to other coastal towns and saw a richer version of civilization than their brothers in jail.

Meanwhile, things were not static in the home communities of the boys and men who had been detained. This marked the beginning of more than a decade of very high mortality due to newly introduced diseases, such as influenza, which still prevailed when research began with the Miyanmin. The deaths provoked high anxiety in the area, a general sense of dread and insecurity. And, since people believe that the deaths of adults are actually due to sorcery carried out by members of traditional enemy groups, their feelings of outrage, hostility, and frustration at not being allowed to take revenge by going to war were also high.

People were concerned about the fate of those who had been taken from them and took every opportunity to gain information from authorities. A local man named Beliyap, who had been appointed Headman by officials, would periodically take the three-day walk over the mountains to the district headquarters to plead for news. Finally, the officer in charge responded, placing Beliyap on a plane for Wewak. Beliyap visited the prisoners and exchanged news with them. He also "got loose" around town and collected some novel plants; seed, cuttings, suckers, tubers, and so on — to bring home. Papaya, tomato, several new taro and sweet potato varieties, pineapple, the commercial banana variety, and coconut were included. Only the last failed in the mountain habitat of the Miyanmin.

When the older men finally were released and sent home, some expected to find their world transformed with new facilities, such as hospitals and schools. Of course, they were disappointed in this and even more distressed to find that parents, siblings, or kids had died during their absence. The important thing to note is that these people, representing several villages, shared a vision of an improved future. One of the young men, who had spent his detention at the vocational school, returned to his home group to start a medical aid post. Amusep, belonging to a neighboring village-community, was to become the formulator, leader, and main proponent of the consensus that emerged around the overall modernization plan, including efforts to develop business opportunities (described in the text).

No outsider observed all of the foregoing events directly or the details of this community's early-1960s "debate about feasible and desirable change." Based on current knowledge of this society, something can be said about how this was likely to have advanced. Among these people, public airings of issues of importance to the community occur in several different settings. There are large public gatherings, connected with feasts and dances as well as with emergencies, such as funerals and impending hostilities between groups. Although individual oratory ("the big speech") is a prominent feature, a lot of discussion also occurs in smaller groups of men, who gather around a hearth to share food and talk. In a consensual democracy characterized by open and wide-ranging discussion, status nevertheless looms large. Some people, such as very prominent men and occasionally an elderly woman, have "louder voices" than others. I do not mean this literally, but in the sense of being accorded greater social recognition than others for their gender, age, accomplishments, and reputation. They are listened to and thus have a greater role in shaping a consensus. Traditionally, shamans served as facilitators by promulgating in a ritual context a decision that had emerged from an earlier open discussion. In modern times, similar things happen in the small churches, where people assemble on Sundays and on other occasions, and pastors have assumed facilitative roles.

Debates are also common in small groups, residential hamlets of a few families, or even within extended families, such as a group of brothers who form a typical unit of cooperation. Visitors from other groups may also provide occasions and topics for discussion, something that occurred frequently in the 1960s as members of this innovative group attempted to convince others of the soundness of their program. Discussions often dealt with specifics, such as giving up the system of food taboos. In one, a man from a more traditional Miyanmin group expressed fears of the supernatural sanctions he would face if he violated a particular taboo. A very robust middle-aged man from a more "progressive" group responded by expanding his chest and flexing his biceps. "I ate meat from a pig raised by my own daughter, and look at me." A person violating such a rule is supposed to waste away.

In addition, Amusep traveled to other villages to preach about God and the new ways and also to recruit young men to attend the bush school he ran. Later, these young men were to serve as leaders and facilitators of change in their own communities.

When research commenced among the Miyanmin in 1968, these things were all in place. In addition, the quest for business and money was quite visible, though only marginally successful. A large airstrip was nearing completion and was already in use; on average, it received one mission flight per month. Typically, this flight would bring a missionary, who would hold a service and operate a store, and a nurse, who conducted a clinic for women and children and selected patients for evacuation to central health-care facilities.

Around the airstrip, people had established large plantations, extensive plantings of pineapple and commercial grade banana in a pattern that was quite new to them. There were also smaller test plots of other introductions, mainly vegetables. The expectation was that buyers would beat a path to the community's door. Sales, however, were severely limited. The monthly mission aircraft might be able to take a few bunches of bananas and a sack of papaya and pineapple back to headquarters. The researcher bought what he could consume of the new as well as traditional vegetables. Other, more casual visitors, such as patrol officers, did the same for themselves and their police and cargo porters. It certainly did not amount to much.

Small numbers of men also were lured into contract labor. At this time, labor recruiters were ranging farther and farther afield in search of naive and willing men to serve a two-year period working on a coastal coconut plantation or a highland tea or coffee farm. The benefits included the opportunity to travel and gather experience, new cultivars, odds and ends of material goods, and a very modest amount of cash, rarely more than a hundred Australian dollars.

Access to health care remained a big issue into the 1970s. In the late 1960s, Amusep's village community was somewhat more favored than many other Miyanmin communities due to the monthly missionary flying clinic and the ability of the clinic to dispense medicines such as penicillin. The aid post started by another detainee, referred to earlier, was approximately a half-day's walk away from this village.

When the researcher prepared to leave the Miyanmin late in 1969, Beliyap appeared at the head of a delegation and said, "Morren, after you leave, we want to replace you with a doctor and a businessman!"

At this time, it was clear that the community was beginning to serve as a model and catalyst, first for other Miyanmin groups, then for certain former enemy tribes. At least one other group was starting their own airstrip project. The pace of change accelerated shortly thereafter.

National independence, which occurred in 1974, had real meaning for this most remote Papua New Guinea village. This meaning was reinforced by the spread throughout the 1970s of a pan-regional movement that sought to reduce the barriers between tribes to forge a common identity. This movement, called *Rabaibal* (or "revival" in English) is based on local interpretations of Christian beliefs and practices. One of the movement's original priorities was to wrest control of churches from the established Australian Baptist mission organization, which had, through the 1950s and 60s, successfully converted many groups in the area to Christianity. *Rabaibal* also succeeded in converting groups that had previously resisted mission efforts.

The Miyanmin were in many respects already well advanced when *Rabaibal* sprang up, but they identified their own aspirations with those of the movement and embraced regionalism as well as some other features that dove-tailed with their established initiatives. They also shared with other groups a concern about the possible benefits of mining projects advancing in the area.

During the 1970s the community had begun to transform itself into a small jungle town. Around 1971, the Baptist mission told people that they would be given responsibility for the management of the airstrip store, which up until then had been open intermittently during scheduled mission flying visits. At first people took this pronouncement quite literally and physically uprooted the *building* (which was technically a clinic) and carried it to the top of a low divide, much closer to where people actually lived. This was consistent with their established mobility. Subsequently, the store was formally chartered as a co-op and, later still, rebuilt once again near the airstrip with a community water tank installed as well.

They also persuaded the government to set up a primary school, a major undertaking, which involved flying in materials for the permanent headmaster's residence, as well as mobilizing the community to construct classrooms and other amenities from local materials. Around the same time, the mission built a permanent clinic building, though it was still only staffed intermittently. The formerly distant community that had an aid post was persuaded to relocate to a site near the airstrip. The airstrip itself was lengthened and generally improved. A rural development team directed the construction of a steel cable suspension bridge for pedestrian traffic. The result of all this was that an immense settlement sprang up around the airstrip. In 1981, it contained 58 family houses and 20 other buildings and could shelter more than 400 people. Included was a neighborhood of several families from a former enemy group, who were allowed to live there and grow food so that their kids could go to the community school.

Agriculture was also changing incrementally. The grand plans of the 1960s for commercial agriculture were pretty much in ruins. For example, the old pineapple plantations were overgrown, as were later chili trials, which had been suggested by agricultural extension workers. The new fruit and vegetable stock was still available, but in people's subsistence fields or kitchen gardens in smaller quantity. The production of sweet potato, formerly a marginal crop for these people, was expanding in response to some of the other changes. Due to over-exploitation, building materials were harder to find in the forest close to the large settlement and many game animals had been seriously reduced, either by overhunting or habitat disruption.

These changes have been costly in terms of natural and human resources, as well as in creating a need for money. Previously self-sufficient for virtually all their needs, people are now caught up in an expanding spiral of demand for goods and services from outside the community: clothing; food items, such as canned fish and rice; soap; salt; tools, such as axes and machetes; radios; batteries; health care; travel; and education. The last item bears the biggest ticket of all. Primary education costs a modest $3.00 per year, whereas academic high school tuition now exceeds $150 per year, exclusive of transportation and subsistence.

In 1980, despite the ambitions of the people, ways to earn money in the community were minimal. Church pastors, nurses, and aid post orderlies received small stipends. Casual employment might be obtained cutting grass on the airstrip, providing porterage to sporadic government patrols, or working for the visiting researcher. The alternative way for men to earn significant amounts of money was through contract labor on distant coastal or highlands plantations — perhaps $100 to $200 per year could be saved. But few wanted to absent themselves from home for more than a year or two, and a real hardship was imposed on the families of married men who were away. Hence the desire, even a sense of urgency, to improve local earnings.

III. Principal Crops

1. Traditional and Long-Established Cultigens

- *Colocasia esculenta*   taro (50 plus varieties)
- *Allocasia*   taro
- *Impomoea batatas*   sweet potato (present in New Guinea for 250 years)
- *Musa*   banana (10 traditional varieties)
- *Dioscorea alata*   winged yam
- *Dioscora spp.*   various cultivated and wild yams
- *Saccarum offincinarum*   sugar cane
- *Saccarum edule*   pitpit
- *Setaria palmifolia*   mountain pitpit
- *Rungia klossi*   indigenous spinach
- *Amaranthus tricolor*   amaranth spinach
- *Zingiber officinale*   ginger
- *Hibiscus manihot*   abika spinach
- *Pandanus conoideus*   oil pandanus
- *Metroxylon sago*   sago palm
- *Sechium edulae*   chalote/choko squash

- *Lagenaria siceraria* bottle gourd
- *Tricosanthes pulleana* climbing squash
- *Ficus damaropus* mountain breadfruit
- *Ficus wassa, F. spp.* various domesticated and wild figs
- *Psophocarpus tetragonolobus* winged bean
- *Lablab vulgaris* bean
- *Nicotiana tobacum* tobacco

2. Cultigens and Cultivars Introduced Since 1950

- *Manihot esculenta* cassava/manioc/tapioca
- *Xanthosoma sagittifolium* tannia
- *Curcurbita maxima* pumpkin
- *Passiflora edulis* passion fruit
- *Lycopersicon esculentum* tomato
- *Ananas comocus* pineapple
- *Carica papaya* papaya
- *Musa* Cavendish var.
- *Allium cepa* onion
- *Zea mays* corn
- *Daucus carota* carrot
- *Brassica olaracea* cabbage
- *Arachis hypogaea* peanut
- *Coffea sp.* coffee
- *Citrus limon* lemon
- *Citrus sinensis* orange

IV. Map of Area
(See facing page.)

V. Research Bibliography

Barrau, Jacques. *Subsistence Agriculture in Melanesia.* Honolulu: Bernice P. Bishop Bulletin 219, 1958.

Gardner, Janet P. "Papua New Guinea: A Model Nation with Problems," *The Plain Dealer Magazine,* July 5:6–7, 10–12, 1981.

Jackson, Richard, *Ok Tedi: The Pot of Gold,* Port Moresby: University of Papua New Guinea, 1981.

Jorgensen, Dan. "Life on the Fringe: History and Society in Telefomin," In *The Plight of Peripheral People in Papua New Guinea,* Vol. I, *The Inland Situation.* R. Gordon (ed.). Cambridge, MA: Cultural Survival, 1981.

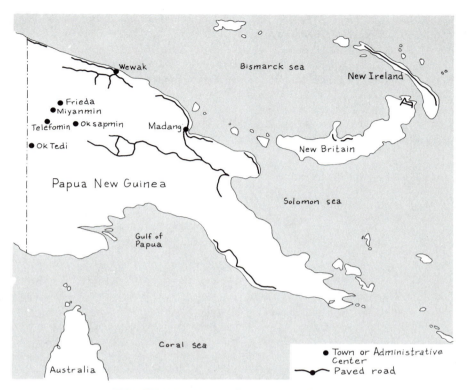

FIGURE A.8 Map of Papua, New Guinea.

Morren, George E. B., Jr. "From Hunting to Herding: Pigs and the Control of Energy in Montane New Guinea" In *Subsistence and Survival: Rural Ecology in the Pacific*. T. Bayliss-Smith and R. Feachem (eds). London: Academic Press, 1977.

Morren, George E. B., Jr. "Seasonality Among the Miyanmin" *Mankind* 12(1):1-12, 1979.

Morren, George E. B., Jr. "A Small Footnote to the 'Big Walk': Development and Change among the Miyanmin of Papua New Guinea" *Oceania* LII(1):39–65, 1981.

Morren, George E. B., Jr. *The Miyanmin: Human Ecology of a Papua New Guinea Society*. Ames, IO: Iowa Sate University Press, 1986.

Morren, George E. B., Jr., and David Hyndman. "The Taro Monoculture of Central New Guinea" *Human Ecology* 15(3):301–315, 1988.

Purseglove, J. W. *Tropical Crops: Monocotyledons*. Vols. I & II. London: Longman, 1972.

Steensberg, Axel. *New Guinea Gardens: A Study of Husbandry with Parallels in Prehistoric Europe*. London: Academic Press, 1980.

Stilltoe, Paul. *Roots of the Earth: Crops in the Highlands of Papua New Guinea*, Manchester: Manchester University Press, 1983.

Wang, Jaw-Kai. *Taro: A Review of* Colocasia esculenta *and its Potentials*. Honolulu: University of Hawaii Press, 1983.

Ward, R. G., and A. Proctor. *South Pacific Agriculture: Choices and Constraints — South Pacific Agricultural Survey, 1979*. Canberra: Australian National University Press; Manilla: Asian Development Bank, 1980.

Glossary

NOTE: Based on commonly occurring words or phrases in the text, this glossary has been compiled from technical sources such as those cited in the chapter bibliographies and standard dictionaries. Dictionary definitions are particularly useful to get behind the meanings of technical terms that embody ideas *and need not always be thought of rigidly. Boldface words and phrases in the definitions have their own entries, which should be consulted.*

abstract separated from matter, practice, or particulars; considered apart from any specific embodiment; ideal; opposite of **concrete.**

abstract conceptualization analysis of experience in order to understand it; to give meaning to something; one pole of the prehension axis of the **learning cycle** (the other pole is **concrete experience**) characterizing a range of ways in which people respond to experience; mentally mapping or modeling an experience.

accommodation the action of adaptation or adjustment; self-adaptation, obligingness; an arrangement of a dispute, a settlement, compromise; anything that supplies a need (see **accommodative learning**).

accommodative learning one of David Kolb's four **learning styles**, oriented toward assessing experience by immersing oneself fully in it and grasping and organizing meaning as seen by others; takes action by moving quickly to make specific changes; combines orientations toward **concrete experience** and **active experimentation** on the axes of the **learning cycle**.

active characterized by action, practical, spontaneous, energetic, brisk, diligent; opposite of passive, contemplative, speculative, theoretical (see **active experimentation** and **active learning**).

active experimentation explicitly testing ideas against reality in order to establish their validity, reliability, and usefulness; one pole of the transformation axis of the **learning cycle** (the other pole is **reflective observation**), characterizing a range of ways in which people respond to experience.

active learning the full involvement of the learner in the process of gaining experience, knowledge, and understanding in everyday life, as well as in formal educational contexts; the self-reliant seeking of experience and competence; opposite of passive, dependent learning.

actor(s) a person who acts or performs; one of six features identified during the development of a human activity system model definition; the *A* in the CATWOE mnemonic; the people and groups who will be responsible for carrying out the human activities envisioned in an improved state of affairs.

affect subjective mental dispostion toward something; pertaining to feelings.

analyst/facilitator see **facilitator**

applied science inquiry also called technology development; an inquiry methodology using **basic science** principles in order to create appropriate solutions to real-world problems; the methodology assumes that the whole is equal to or the same as the sum of its parts and proceeds through a series of stages based on the principles of **reduction**, **experimentation**, refutation, repeatability.

apprehension the perception or grasp of experience, basic facts, and impressions through our senses or from external sources of information.

assimilation the conversion of an external input such as experience, knowledge, or food into one's own being.

assimilative learning acquiring knowledge by imposing a mental framework on experience; producing knowledge of or explaining experience by formulating theories and building concepts; one of David Kolb's four learning styles; combines orientations toward **reflective observation** and **abstract conceptualization** in the **learning cycle**.

attitudinal change modification of the bases of judgment; alteration in **cognition**, **affect**, or **intention** used by people to assess their experience.

basic science inquiry a methodology for investigating the nature of a property or phenomenon; the approach assumes that the whole is equal to or the same as the sum of its parts and uses the basic principles of **reduction**, **experimentation**, refutation, and repeatability; the analyst is responsible for specifying the problem(s) to be investigated, the purpose(s) of the investigation, and the tentative explanation(s).

boundary that which serves to indicate, physically or by convention, the limits of anything; a defining property of a system separating internal from environmental features; in human activity systems, the boundary discussions include the description of the management functions for each subsystem and the system as a whole, the definition of what is to be controlled or regulated external to and within the system, and the interrelationships (e.g., limits on inputs, outputs, information sharing, services rendered, products shared, etc.); a central feature of a **human activity system** model in order for it to be identified as a formal systems model.

cartooning a technique developed by the Checkland group that may be used during stage 2 of **soft systems inquiry**; a way to visualize structure, process, and climate relationships and key features of a problematic situation.

change a difference that makes a difference; a substitution or succession of one element or activity in a situation in place of another, or any variation or alteration made in a situation.

climate the emotional response resulting from the quality of the relationship between structure and process in a situation; how well people feel things work.

closed boundary a low degree of system receptiveness to external inputs or to externally sharing system outputs.

cognition the action or faculty of knowing; the conversion of people's perceptions of the world into knowledge.

communication a system's ability to impart, convey, or exchange **information** between components in order to control internal and external functioning; primary types of communication include negative and positive **feedback** and **feedforward**.

communications the capability of an organization or person to extend knowledge to others by means of varous media.

comparison the act of comparing or likening one thing to another; stage 5 of **soft systems inquiry**; assessing the similarities and differences between features of a present **situation** and the features of a proposed future; the products developed in stages 1 and 2 are compared with the models resulting from stages 3 and 4.

component a part, an element of a system.

composite mind map see **mind map**

comprehending grasping or taking in with the senses and mind; conceiving or conceptually mapping experience as we care to understand it.

conceptualization the act of forming a mental conception, thought, notion, or idea (see **abstract conceptualization**).

concrete embodied in matter, actual practices, or a particular example; opposite of **abstract**.

concrete experience immersion in the real world of sensation, facts, impressions, or events involving things and people; the source of information about and understanding of a situation; a pole of the prehension axis of the **learning cycle** (the other pole is **abstract conceptualization**) characterizing a range of ways in which people respond to experience.

consensual decision making agreement by consensus in which people talk with each other and, through a sympathetic process of give and take, achieve genuine agreement on a topic of discussion, with all winners and no losers; contrasts with vote taking.

consensus general concord of different members of a group in effecting a given purpose.

control nearly synonymous with regulation, feedback, equilibration, adaptation, self-regulation; systems are conceived as having the capacity to maintain key internal components within an appropriate range of values in the face of external disturbance; there are (at least) two kinds of control, open- and closed-sequence; in open-sequence control, the order and magnitude of controlling action are fixed; in closed-sequence control, the order and magnitude of the controlling actions vary in relation to the nature of the changes in the state of the system; closed-sequence control is synonymous with negative feedback; see also **communication, feedback,** and **feedforward.**

convergence coming together or terminating at the same point or focus; arriving at an unambiguous decision.

convergent learning one of David Kolb's four learning styles; oriented to approaching experience by applying theoretical frameworks; action is taken by using the frameworks to identify and solve problems; combines orientations toward **abstract conceptualization** and **active experimentation** on the axes of the **learning cycle.**

cost (of an action) that which is given (or given up) in order to effect something; some costs are subtractive and readily quantifiable, such as monetary cost; others are fractionating and not easy to quantify because they are qualitative, representing faculties, options, or future oportunities used up or committed, and other effects that are not easy to reverse, e.g., opportunity cost, loss of flexibility. Still other costs, such as those commonly referred to as externalities, may be seemingly unrelated to the actions initiated by the people selected for study but are nevertheless "paid" by them.

currency units of measurement employed in an analysis to characterize quantitative changes in or flows between variables and/or relationships; e.g., money, energy, information.

customer(s) a person with whom one has regular dealings; one of six features identified during the development of a **human activity system** model definition; the *C* in the CATWOE mnemonic; the people who will be the anticipated beneficiaries and those who may be affected adversely if the human activity system were put into practice.

cycle a pattern of interaction characterized by a circular or mutual exchange of matter and the conservation of **information**; see also **flow.**

debate the act of discussing or examining a question from more than one perspective; stage 6 of soft systems inquiry; the raising of questions related to proposed changes (as found in the human activity systems models and accompanying narratives) and consideration of arguments from all points of view (**Weltanschauungen**—Ws) represented among participants.

debate arena the context in which a debate about an improved future state takes place.

debate participants key people in a situation who are involved in an examination of questions related to proposed changes in the way things are currently done and organized; the **actors**, **customers**, and **owners** of the present and proposed future state of affairs must be represented.

debate techniques procedures for carrying out an examination of questions and arguments regarding the **desirability** and **feasibility** of proposed changes embodied in the **human activity system** model(s) and accompanying narrative.

decision processes the procedures for making decisions in an improved human activity system, including the specification of who makes what decisions; a central feature of a HAS model in order for it to be identified formally as a systems model.

desirability positive judgments on the part of people involved in a situation that the changes suggested in the human activity system model(s) are an improvement over their current experience and that they are wanted; the desirability of anything is judged on the basis of a person and/or group's **Weltanschauungen**, or world views.

divergence the act of proceeding in different directions from each other or from a common point; differing from a standard or norm.

divergent learning one of David Kolb's four styles of learning; oriented toward immersing oneself in experience and grasping and organizing meaning as seen by others and from a variety of perspectives; action is taken by weighing the validity and relevancy of the meanings attributed to a situation; combines orientations toward **concrete experience** and **reflective observation** in the **learning cycle.**

emergent properties a major proposition of systems thinking that the whole is different from the sum of its parts, with the difference being the emergent property; emergent properties uniquely pertain to particular hierarchical levels of a given system.

energy the ability to perform work; a quantity having the dimensions mass times velocity squared (MV^2) or its thermal equivalency; strictly speaking, energy flows from a concentrated (negentropy) form to a dispersed (entropy) form while materials cycle; a **currency** employed in certain systems analyses.

environment one of six features identified during the development of a definition of a **human activity system** model; the E in the CATWOE mnemonic; the sum of the uncontrollable external positive and negative forces that the actors in the system take as givens and must accommodate in some way; resources and constraints external to a system.

environmental effects the environmental and other external consequences of an activity, including the operation of a **human activity system**, in an improved future state of affairs; a feature of a HAS model that must be specified in order for it to be identified as a formal systems model.

experiment the action of trying anything; a test; a trial; a procedure adopted in uncertainty whether it will answer the purpose; an operation undertaken in order to discover something unknown, to test a hypothesis, or to establish or illustrate some known truth; practical proof.

facilitation the act of providing help and guidance; according to the tenets of experimental learning and **soft systems inquiry**, facilitative intervention guides people through a process whereby they gain self-reliance in inquiry and action, learning to create their own knowledge to use for improving problematic situations; contrasts with authoritative intervention, involving the act of giving orders, directions, and instructions; see **intervention styles**.

facilitator he or she who uses facilitative, as opposed to authoritative, approaches to intervention, group organization, learning, and action; professionals engaged in assisting others to use **soft systems inquiry** into a problematic situation; the role implies helping and supporting the people involved in a situation to understand it in new ways and to act to foster beneficial change; a change agent or catalyst; referred to throughout the book as an analyst/facilitator to suggest the dual responsibility of guidance and engagement in an articulated assessment, development, and implementation process.

feasibility positive judgments on the part of the people involved in a situation that proposals for change suggested in a **human activity system** model are possible to put into practice; the feasibility of something is based on a person's or group's assessment of environmental and internal constraints to change; see **environment**.

feedback a type of **communciation** in a system; a compensatory response to an effected dispacement; negative feedback, also known as closed-sequence control or deviation compensation, attempts to accomplish equilibration; positive feedback, also called deviation amplification, brings about change; see **communication** and **control**.

feedforward a type of communication in a system; an anticipatory response to a possible future displacement; predicting and possibly correcting in advance anticipated disruptions of a system.

flexibility uncommitted potentiality for change; according to the economics of flexibility, costs are fractionating, not subtractive. The loss of flexibility is due to escalating responses or subsystems; see **cost**.

flow a pattern of interaction characterized by unidirectional exchanges of matter or **energy** and lost **information**.

hard systems inquiry an articulated inquiry methodology based on systems thinking; quantitative modeling of the present situation using optimization or maximization assumptions; "hard" means that the problems, goals, or end states addressed by inquiry are readily defined by the analyst, who then moves into stages of inquiry aimed at developing solutions; useful methodology for allocation decisions.

hierarchy a body of persons, things, or types ranked or graded according to their level of complexity or logical order; unique properties emerge at each hierarchical level that are not predictable from a study of the entities without (above) or within (below); communication exists between the entities or systems at different levels, such that an action at one level can influence other levels in the hierarchy.

holism the practice of viewing the world conceptually as consisting of structured wholes or systems that maintain their identity or integrity under a range of conditions and exhibit certain general properties emerging from their wholeness; the major propositions are that everything is connected with everything else and that the whole is different from the sum of its parts.

human activity system (HAS) a conceptualized purposive system that expresses some purposeful human activity; it is conceptual in the sense that it is a not a description of actual, current, real-world activity, but is an intellectual formulation used in a **debate** about possible changes that might be introduced into a real-world **problematic situation** leading to improved future human activity; its design and formulation are the tasks of stages 3 and 4 of **soft systems inquiry**.

iconic or **physical model** a representation that mimics a real object or phenomenon; purposes for developing physical models include to catch basic characteristics of something at a point in time, to test the characteristics of something under varying conditions, to change the scale of something; see **model**.

implementation the act of providing something necessary to make a thing complete; to put into practice; the name of stage 7 of **soft systems inquiry**.

implementation plan a detailed scheme of action to put into practice the **human activity system(s)** designed and tested in stages 4 through 6 of **soft system inquiry**.

information the action of telling or the fact of being told of something; that of which one is told; organization, predictability, concentration, or storage of matter and/or energy; negentropy; see **communication, energy,** and **flow**.

input a resource needed by a system or subsystem; a specific factor to which a system or subsystem must be accommodated; typical resource inputs include information, materials, money, human resources, and services expressed qualitatively and quantitatively where possible; a central feature of a HAS model in order for it to be identified as a formal systems model.

intention the action of directing the mind or attention to something; the faculty of understanding; the aim or intended consequences of an action.

interaction relationship between components of a system; mutual or reciprocal action, communication, or controlling influence.

intervention styles habitual ways in which outsiders purposefully interfere in the lives of other people and groups; may be invited or uninvited, **facilitative** or authoritative.

learning the process whereby individuals acquire and internalize experience, language, social behavior, responses, understanding, meaning, knowledge, actions, plans, and so on.

learning cycle a circular model of learning formulated by combining a **concrete experience** ↔ **abstract conceptualization** (or prehension) axis with a **reflective observation** ↔ **active experimentation** (or transformation) axis; according to the model, learning, problem solving, knowledge creation, and purposive action are dimensions of the same process.

learning styles inventory (LSI) a multiple-item, self-description questionnaire developed by David Kolb to test an individual's orientation toward finding out about and taking action on experience; the finding-out and taking-action dimensions are combined to produce four styles of learning, each connected with definite learning competencies; see **learning cycle**.

measure of performance a precise definition of the amount, rate, or quality of output expected of a system or subsystem; a central feature of a HAS model in order for it to be called a formal systems model.

mental map a learned abstract framework that people use to impose order and meaning on their experience to transform it into knowledge.

methodology an explicit, ordered, nonrandom way of carrying out a process of inquiry; intermediate in status between a philosophical (or theoretical) formulation and a **technique** (or specific procedure); has phases or stages that act as guides to the inquiry process; rests on sound, explicit philosophy (what, why, values) that gives the investigator a sense of the "rules of the game"; each stage within a methodology employs techniques chosen in light of the philosophy behind the methodology and stage of inquiry.

mind mapping also called spider diagramming or dendrogramming; a technique developed by Tony Buzan for visually displaying patterns of thought; a technique useful during stage 1 of **soft systems inquiry** for recording conversations so that patterns and associations of the respondent's ideas are displayed; a technique that may be used during stage 2 for developing a composite of the themes of concern heard from various people in a given situation (called a composite mind map).

model a simplification or reduction, representation, or copy of reality carried out according to a standard; conceptual abstractions that represent a simple or complex reality.

modeling the act of formulating a model; a mental construction or reconstruction process used to make sense out of phenomena that are made explicit to others through the use of one or more symbolic languages (e.g., narratives, pictures, mathematical equations, charts, diagrams, replicas, games); modeling is used in **soft systems inquiry** to design improved **human activity system** models and, in some cases, to design computer-based quantitative models of the kind that ask "If we changed this feature, then what are apt to be the changes in other features?"

mutual learning the relationship between **facilitator** and clients, teacher and students, characterized by joint participation in a process of inquiry leading to purposeful action; see **facilitation, facilitator, learning,** and **soft systems inquiry.**

objectify to make objective, to treat or deal with facts without distortion by personal feelings or prejudices; sometimes to dehumanize persons subjected to curiosity, investigation, or other scrutiny.

observation the act of taking notice; the action of observing scientifically a phenomenon in relation to its cause and effect, or phenomena in regard to their mutual relations, as they occur in nature; the opposite of **experiment.**

open boundary indicates a high degree of system receptiveness to external inputs and to sharing system outputs with entities in its environment; see **boundary, closed boundary, input,** and **output.**

opportunity a time, juncture, or condition of things favorable to an end or purpose; a newly emerged advantage; experience of a condition associated with concerns, as well as positive feelings; opposite of **problem.**

optimize the act or intention to achieve the amount or degree of something most favorable to an end.

optimization the assumption in hard systems analysis that the purpose of a system is to **optimize;** see **satisfycing.**

output any product of a system or subsystem; typical outputs include information, services, materials, money, products, and waste; typically expressed qualitatively and quantitatively where possible; a central feature of a HAS model in order for it to be identified as a formal systems model.

owner(s) one of six features identified during the development of a definition of a **human activity system** model; the O in the CATWOE mnemonic: the persons or groups to whom the designers wish to give the power to cause the system to cease to exist or to be altered in an improved state of affairs, or the persons or groups who already possess such power.

paradigm the explanatory and meaningful body of knowledge existing at a given time that scientists use to explain their observations; similar to **mental map** and **weltanschauung**.

primary task the mission of a group, enterprise, or organization in a given situation; summary of a group's reason for being, its basic nature, what it functions to do, its niche or role; typically expressed in operational terms (i.e., human activities and outputs); designates one kind of theme of concern voiced by people in a situation.

problem a change that threatens a person's well-being or survival and is perceived to; experiences associated with negative feelings; a difficult question; a concern; a thing that is difficult to understand; in inquiry generally, an analytical focus on a problem (rather than a problematic situation) means to limit or narrow attention to a few things (or one thing) defined as the primary difficulties(y).

problematic situation a set of circumstances comprised of people (as individuals and in groups); a historical context that bears on the present; key human activities; themes of concern and opportunity; decision-making structures and processes; physical and biological environmental factors; a political, economic, and social context; and relational climate(s); during stages 1 and 2 the focus of inquiry is on describing a problematic situation where the themes of concern (problems) are but one aspect of what is assessed.

procedural change alterations in the functional elements of situations such as communication flows, reporting relationships, decision making, resource utilization, product and service monitoring, and control functions; alteration in a a particular way of doing something.

process a series of activities and events with a definite output; a key feature of a situation described during stages 1 and 2; the way things are done by people within the constraints of structure; the relatively fast or constantly changing features of a situation.

property see **quality**.

quality a characteristic peculiar to a thing; qualities are not readily quantifiable; among the properties or qualities of systems are multiple determination, **flexibility**, resilience, stability, and sustainability.

reductionism a prevailing logic of inquiry that rests on the premise that the sum of the parts equals the whole; i.e., if each part of a problem can be explained or solved, then the parts can be put back together, and it will be an adequate explanation or solution of the whole; one of four fundamental premises of basic and applied science inquiry methodologies (the others are experimentation, refutation, and repeatability).

reflection the act of turning or casting thought upon some object; meditation; deep or serious consideration; the mode, operation, or faculty by which the mind has knowledge of itself and its operations, or by which it deals with the ideas received form sensation and perception.

reflective observation the action to evaluate the adequacy of different perspectives in assessing experience, identifying or solving a problem, explaining a phenomenon, or improving a situation; one pole of the transformation axis of the **learning cycle** (the opposite pole is **active experimentation**) characterizing a range of ways in which people respond to experience; see **observation** and **reflection**.

rich picture the comprehensive description of a current situation (for the elements to be described, see **problematical situation**); a rich picture synthesizes the complexity and interrelationships of elements in a situation without reducing the amount or kind of complexity experienced.

root definition see **system definition.**

satisfycing acting or intending to achieve the amount or degree of something that is satisfactory or adequate to an end.

situation see **problematic situation** and **rich picture**.

soft systems inquiry an inquiry methodology that uses systems thinking; "soft" means the problems and goals are messy and complex and treated as not easy to define because there is no general agreement on problems, goals, or purposes; the analyst is reponsible for facilitating an inquiry process whereby the articulation of the complexity of the problematic situation and of improved future states is developed by and with the people involved; a methodology useful for inquiry into complex, real-world situations and for strategic planning.

strategy the technique adopted for a hard systems solution; a set of practices having a discrete outcome or objective; e.g., a resource management strategy; also, systems analysts' methodological decisions regarding such matters as (1) level at which to begin an analysis, (2) unit(s) of analysis, (3) currency, and the like, which in some sense determine the kinds of questions the analyst is able to ask.

structural change alterations in the framework of institutions and organizations; see **structure**.

structure a key feature of a problematic situation; the slow-to-change features of a situation; relatively durable physical, biological, and social patterns and institutions associated with a situation in a particular place and time; includes aspects of the natural or modified environment, the designed environment of buildings, roads, and other infrastructures, social patterns of political, economic, and social institutions, and corporate and community organizations.

subsystem a designated part of a system; a set of key human activites grouped around one of the functions comprising a **human activity system** model; a further development and more detailed description of the transformation as developed in stage 3 of **soft systems inquiry**: the answer to the question "What activities are needed in order for the transformations specified in the HAS model to occur?"; each subsystem has all the properties of a system (i.e., actors, owners, customers, environment, world view, transformation, inputs, ouputs, boundaries, measures of performance); a central feature of a HAS model in order for it to be identified as a formal systems model.

symbolic model a conceptualization of something that is hard to observe or cannot be readily observed, such as a relationship or property; a formulation of how things might function using a graphical standard.

system a set of parts that behave in a way that an observer has chosen to view as coordinated to accomplish one or more purposes.

system definition the basic design for a conceptual model of a HAS, consisting of a transformation statement and the definitions of other elements of the CATWOE: synonymous with Checkland's "root definition."

systems thinking conceptualizing the complexity and dynamism of the world in terms of holism, means of measurement and control, emergent properties, hierarchical structure, and communication processes.

technique (as a procedure in an inquiry methodology) a step-by-step operation producing a similar result irrespective of the person using it; usually several techniques are applied in each stage of an investigation consistent with the overall philosophy of the particular inquiry **methodology**.

technology development inquiry see **applied science inquiry**.

transformation a process of change; in systems thinking, systems tranform themselves continuously; thinking systemically means conceptualizing the transformation processes at work in a given situation, as well as the transformation processes needed to improve a situation; the focus of the principal task of stage 3 of **soft systems inquiry**; a summary of improved human activity in the future.

unit of analysis a logical type of thing in a hierarchy of things such as the following series: individual, population, community; genotype, isolate, variety, subspecies, species, genus; household, neighborhood, community, municipality, county, state, nation.

variable a dimension of a component or interaction in a system that has more than one value.

Weltanschauung(en) a German word only partially translated into English as world view(s) or paradigm(s); consists of our experience, feelings, emotions, attitudes, values, beliefs, morals, tastes, intelligence, and store of knowledge; expressed, among other ways, in the meanings given to situations and what improvements are preferred; one of the key features of a situation to be described during stages 1 and 2 of **soft systems inquiry**; one of six key features identified during the development of a definition of a human activity system model in stage 3; the *W* in the CATWOE mnemonic; the criteria of judgment used by model designers to assess the **desirability** of a given **human activity system** as an improvement to the present situation.